GENETIC ENGINEERING OF CROP PLANTS

Proceedings of Previous Easter Schools in Agricultural Science, published by Butterworths, London

*SOIL ZOOLOGY   Edited by D. K. McL. Kevan (1955)
*THE GROWTH OF LEAVES   Edited by F. L. Milthorpe (1956)
*CONTROL OF PLANT ENVIRONMENT   Edited by J. P. Hudson (1957)
*NUTRITION OF THE LEGUMES   Edited by E. G. Hallsworth (1958)
*THE MEASUREMENT OF GRASSLAND PRODUCTIVITY   Edited by J. D. Ivins (1959)
*DIGESTIVE PHYSIOLOGY AND NUTRITION OF THE RUMINANT   Edited by D. Lewis (1960)
*NUTRITION OF PIGS AND POULTRY   Edited by J. T. Morgan and D. Lewis (1961)
*ANTIBIOTICS IN AGRICULTURE   Edited by M. Woodbine (1962)
*THE GROWTH OF THE POTATO   Edited by J. D. Ivins and F. L. Milthorpe (1963)
*EXPERIMENTAL PEDOLOGY   Edited by E. G. Hallsworth and D. V. Crawford (1964)
*THE GROWTH OF CEREALS AND GRASSES   Edited by F. L. Milthorpe and J. D. Ivins (1965)
*REPRODUCTION IN THE FEMALE ANIMAL   Edited by G. E. Lamming and E. C. Amoroso (1967)
*GROWTH AND DEVELOPMENT OF MAMMALS   Edited by G. A. Lodge and G. E. Lamming (1968)
*ROOT GROWTH   Edited by W. J. Whittington (1968)
*PROTEINS AS HUMAN FOOD   Edited by R. A. Lawrie (1970)
*LACTATION   Edited by I. R. Falconer (1971)
*PIG PRODUCTION   Edited by D. J. A. Cole (1972)
*SEED ECOLOGY   Edited by W. Heydecker (1973)
 HEAT LOSS FROM ANIMALS AND MAN: ASSESSMENT AND CONTROL   Edited by J. L. Monteith and
    L. E. Mount (1974)
*MEAT   Edited by D. J. A. Cole and R. A. Lawrie (1975)
*PRINCIPLES OF CATTLE PRODUCTION   Edited by Henry Swan and W. H. Broster (1976)
*LIGHT AND PLANT DEVELOPMENT   Edited by H. Smith (1976)
 PLANT PROTEINS   Edited by G. Norton (1977)
 ANTIBIOTICS AND ANTIBIOSIS IN AGRICULTURE   Edited by M. Woodbine (1977)
 CONTROL OF OVULATION   Edited by D. B. Crighton, N. B. Haynes, G. R. Foxcroft and G. E. Lamming
    (1978)
 POLYSACCHARIDES IN FOOD   Edited by J. M. V. Blanshard and J. R. Mitchell (1979)
 SEED PRODUCTION   Edited by P. D. Hebblethwaite (1980)
 PROTEIN DEPOSITION IN ANIMALS   Edited by P. J. Buttery and D. B. Lindsay (1981)
 PHYSIOLOGICAL PROCESSES LIMITING PLANT PRODUCTIVITY   Edited by C. Johnson (1981)
 ENVIRONMENTAL ASPECTS OF HOUSING FOR ANIMAL PRODUCTION   Edited by J. A. Clark (1981)
 EFFECTS OF GASEOUS AIR POLLUTION IN AGRICULTURE AND HORTICULTURE
    Edited by M. H. Unsworth and D. P. Ormrod (1982)
 CHEMICAL MANIPULATION OF CROP GROWTH AND DEVELOPMENT   Edited by J. S. McLaren (1982)
 CONTROL OF PIG REPRODUCTION   Edited by D. J. A. Cole and G. R. Foxcroft (1982)
 SHEEP PRODUCTION   Edited by W. Haresign (1983)
 UPGRADING WASTE FOR FEEDS AND FOOD   Edited by D. A. Ledward, A. J. Taylor and R. A. Lawrie
    (1983)
 FATS IN ANIMAL NUTRITION   Edited by J. Wiseman (1984)
 IMMUNOLOGICAL ASPECTS OF REPRODUCTION IN MAMMALS   Edited by D. B. Crighton (1984)
 ETHYLENE AND PLANT DEVELOPMENT   Edited by J. A. Roberts and G. A. Tucker (1985)
 THE PEA CROP   Edited by P. D. Hebblethwaite, M. C. Heath and T. C. K. Dawkins (1985)
 PLANT TISSUE CULTURE AND ITS AGRICULTURAL APPLICATIONS   Edited by Lindsey A. Withers and
    P. G. Alderson (1986)
 CONTROL AND MANIPULATION OF ANIMAL GROWTH   Edited by P. J. Buttery, N. B. Haynes and
    D. B. Lindsay (1986)
 COMPUTER APPLICATIONS IN AGRICULTURAL ENVIRONMENTS   Edited by J. A. Clark, K. Gregson
    and R. A. Saffell (1986)
 MANIPULATION OF FLOWERING   Edited by J. G. Atherton (1987)
 NUTRITION AND LACTATION IN THE DAIRY COW   Edited by P. C. Garnsworthy (1988)
 MANIPULATION OF FRUITING   Edited by C. J. Wright 1989

These titles are now out of print but are available in microfiche editions

Genetic Engineering of Crop Plants

G. W. LYCETT
Lecturer in Plant Molecular Biology, University of Nottingham School of Agriculture, Sutton Bonington

D. GRIERSON
Head, Department of Physiology and Environmental Science, University of Nottingham School of Agriculture, Sutton Bonington

BUTTERWORTHS
London Boston Singapore Sydney Toronto Wellington

 PART OF REED INTERNATIONAL P.L.C.

First published 1990

© Contributors 1990

British Library Cataloguing in Publication Data

Lycett, G. W. (Grantley)
Genetic engineering in crop plants.
1. Crops. Genetic engineering
I. Title   II. Grierson, D. (Donald)

ISBN 0-408-04779-8

Library of Congress Cataloging in Publication Data

Applied for
ISBN 0-408-04779-8

Composition by Genesis Typesetting, Borough Green, Sevenoaks, Kent
Printed and bound in Great Britain by Courier International Ltd, Tiptree, Essex

PREFACE

The 49th Nottingham Easter School in Agricultural Science, the first on the topic of Genetic Engineering of Crop Plants, was held at Sutton Bonington from April 17–21 1989. This conference, which was attended by over 200 participants from 18 countries, provided a timely opportunity to review and discuss many of the exciting recent developments in this rapidly expanding field.

Many of the results reported at the meeting are recorded in the contribution to this book and the editors would like to thank all the authors for their splendid cooperation in producing it. The participants heard of the progress and successes in generating transgenic plants resistant to herbicides, viruses and insects, and the impressive extension of genetic engineering techniques to a wider range of crop species. Success in understanding the problems of developmental gene regulation were reported, including studies on regulator genes themselves. In addition the opportunities and uses of the newer antisense technology were also discussed. A notable feature of the conference was the clear message, not yet fully appreciated by those working outside the field, that research on transgenic plants allows knowledge from cell and molecular biology, physiology, biochemistry and genetics, to be combined in a concerted new approach to plants. Foremost in the minds of the contributors was the objective of increasing our fundamental understanding of how plants function, which will provide the knowledge base for future exploitation.

A highlight of the meeting was an evening discussion session led by Chuck Gasser and Lothar Willmitzer, both of whom were outstanding in the degree of audience participation that they engendered. Topics discussed included means of maximizing gene expression and in particular the role of introns, and also the nature and importance of position effects in the expression of foreign gene constructs. The possible role of the scaffold associated regions in this raised considerable interest. The effect on normal academic research of the possible unavailability of clones produced in commercial laboratories or as part of industrially funded projects was a concern raised. Whilst this is a topic that will continue to be discussed, academic participants were generally reassured.

The social aspects of the meeting also went well. A break in the middle of the meeting for a visit to the world-famous Chatsworth House in Derbyshire was greatly enjoyed by those who attended. The compact setting of the campus, with sessions, bar dining room and accommodation within 100 yards was appreciated by many participants and no doubt contributed to the lively and friendly discussions in the bar following the evening sessions and the conference dinner.

The success of the meeting was in no small part due to the participants and those who gave financial support, and the organizers wish to thank them both. Thanks are due to David Baulcombe, Roger Beachy, Don Boulter, Ted Cocking, Simon Covey, Gunter Feix, Chuck Gasser, John Gray, Jan Leemans, Danny Llewellyn, Fritz Schöffl, Wolfgang Schuch and Lothar Willmitzer for chairing sessions. We would also like to acknowledge the willing and valuable help of Elizabeth Horwood, Mandy Newman and many other members of the Department who contributed so much to the organization of the meeting. Financial support in the form of grants to offset the costs or bursaries for younger scientists was provided by: Roussell-Uclaf, International Society for Plant Molecular Biology, ICI Seeds, CIBA-Geigy Corporation, Monsanto Company, Bayer A.G., The Genetical Society, The Royal Society, Shell, Unilever Research and Engineering Division, Agricultural Genetics Co. Ltd., Dalgety PLC Group Research Laboratory and BP Nutrition Ltd.

Grantley W. Lycett and Don Grierson
Sutton Bonington

CONTENTS

1

ALL SORTS OF PLANT GENETIC MANIPULATION

EDWARD C. COCKING
Plant Genetic Manipulation Group, University of Nottingham, Nottingham, UK

Introduction

It is only about four years since the 41st Easter School was planned around the theme of the plant tissue culture revolution (Withers and Alderson, 1986). At that meeting it was highlighted that this revolution was stemming largely from the fact that plant cell and tissue culture was providing the foundation for the exploitation of genetic engineering. Since then the subject has rapidly developed as forecast, and this 49th Easter School is concerned with the actual genetic engineering of crop plants.

Plant genetic engineering is defined by some as the isolation, introduction and expression of foreign DNA in the plant. This narrow definition focuses on recombinant DNA technology. However, in considering plant genetic manipulation a broader definition is preferred which focuses on the cellular level of organization and involves the interfacing of all aspects of cell biology, molecular biology and gene-transfer procedures (Sharp, Evans and Ammirato, 1984). This definition reflects the enormous scope for the many research activities that have been intertwined to apply such basic information to plant genetic manipulations. The genetic engineering tools of tissue culture, somaclonal and gametoclonal variation, cellular selection procedures and recombinant DNA are either indirectly or directly concerned with the enhanced expression and transfer of genes.

An essential problem in using interspecific hybridization in plant breeding programmes is the low probability of obtaining in one individual the desired combination of genes from the parental species. The process of speciation leads to the development of reproductive isolation barriers that maintain the integrity of species by restricting the flow of genes from one to another (Hadley and Openshaw, 1980). Experience has shown that before attempting to utilize more sophisticated technology to enable such gene flow it may be advantageous to pursue efforts to obtain sexual hybrids to their limit. For instance, a major effort at the sexual hybridization of African violet (*Saintpaulia ionantha*) with the alpine wild species of African violet (*Saintpaulia shumensis*) produced, from an extensive crossing programme, two seeds that germinated to produce hybrid plants that exhibited lower temperature requirements for flowering than *Saintpaulia ionantha* whose optimum temperature for flowering is 21°C. This led to the production of the Endurance (cold preference) lines of African violet which preferred 5°C cooler cultural temperatures and which were marketed in the early 1980s by T. C. Rochford in collaboration with the Plant Genetic Manipulation Group here at Nottingham (Bilkey, 1981). It should also not be forgotten that the 'miracle' rices such as IR8 and IR36 were produced by a sophisticated sexual hybridization programme (Cocking, 1989).

Somaclonal variation

Somaclonal variation, which arises as a consequence of tissue culture has been found in essentially all plant species that have been regenerated from tissue culture. As recently pointed out by Scowcroft and Larkin (1988) the question is frequently put: 'Among the array of somaclonal variants derived from tissue culture, are the mutants the result of changes at pre-existing loci or do entirely new mutants occur?' Certainly the former is true, but the evidence for entirely new mutants is, at least, only circumstantial. They also point out that some genetic events, which occur at a very low rate spontaneously, occur far more frequently during tissue culture and for this reason give the impression that somaclonal variation 'creates' new genes.

One of the major potential benefits of somaclonal variation in plant genetic manipulations was seen to be the opportunity to create additional genetic variability in co-adapted, agronomically useful cultures, without the need to resort to hybridization or the production of transgenic plants. An additional option, namely to introgress alien genes from wild relatives into crop species emerged from the observation that chromosome rearrangements occur in plants regenerated from tissue culture (Scowcroft and Larkin, 1988).

Recent studies on somaclonal (protoclonal) variation in the seed progeny of plants regenerated from rice protoplasts evaluated a wide range of phenotypic characters (Figure 1.1). A unidirectional shift was revealed in the means of data recorded for the protoclones, for a range of characters studied, away from the direction of previous selection history of the variety, just like quantitative mutations. However, some individuals showing positive variability could be preferentially selected from the population (Abdullah *et al.*, 1989). These detailed assessments of protoclonal variation in rice were greatly facilitated by the availability of a single cell, protoplast system capable of efficient regeneration into plants by somatic embryogenesis. The development of finely-divided fast-growing suspension lines was found to be essential for efficient protoplast isolation and division and for subsequent plant regeneration (Abdullah, Cocking and Thompson, 1986). A recent review has documented the fact that from protoplasts of 212 higher plant species, representing 96 genera of 31 families, cell colonies and calli have been developed, which regenerate into embryo-like structures, embryoids or shoots. Sometimes when cultured on appropriate media, these structures are able to develop into plants (Roest and Gilissen, 1989).

Transgenic plants by protoplast fusion

Somatic cell hybridization by protoplast fusion overcomes sexual incompatibility barriers and creates a novel cytoplasmic mix as organelles of both fusion partners come together in a common milieu after protoplast fusion. Somatic hybrids provide unique opportunities to investigate hybridization through cytoplasmic transfer by

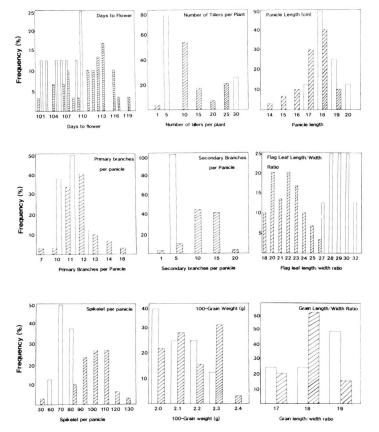

Figure 1.1 Protoclonal variation in a range of phenotypic characters in the seed progeny of plants regenerated from rice protoplasts. Control plants were from seed of the same variety. ☐ Control; ▨ Protoclones

protoplast fusion. This method enables the production of transgenic plants with new combinations of cytoplasmic genes between species in one step, thereby circumventing a lengthy sexual backcross programme. It also provides novel combinations in the case of both sexually compatible and sexually incompatible crosses (Kumar and Cocking, 1987).

Fusions may be carried out, efficiently using millions of protoplasts as required, either electrically (Watts and King, 1984) or using polyethylene glycol possessing a low carbonyl content (Chand *et al.*, 1988). Either procedure can result in a high fusion frequency with good resultant heterokaryon viability and a low level of clumping. Direct isolation of fusion products using fluorescence-activated cell sorting is now available, enabling the rapid isolation of large numbers of hybrid cells (Alexander *et al.*, 1985). Direct observation of the electrofusion of pairs of protoplasts is also possible using microfusion procedures in droplets of medium using an inverted microscope fitted with suitable electrodes (Schweiger, Dirk and Koop, 1987). This procedure is more technically demanding and difficult to scale up, but it avoids the need to select fusion products.

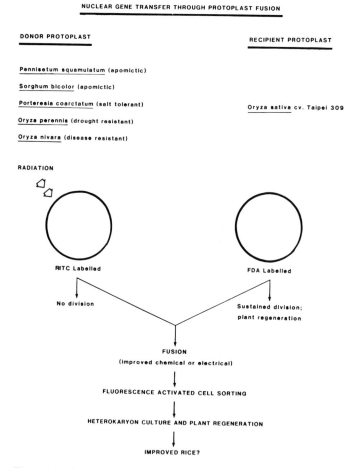

NUCLEAR GENE TRANSFER THROUGH PROTOPLAST FUSION

DONOR PROTOPLAST

RECIPIENT PROTOPLAST

Pennisetum squamulatum (apomictic)

Sorghum bicolor (apomictic)

Porteresia coarctatum (salt tolerant)

Oryza perennis (drought resistant)

Oryza sativa cv. Taipei 309

Oryza nivara (disease resistant)

RADIATION

RITC Labelled

FDA Labelled

No division

Sustained division; plant regeneration

FUSION
(improved chemical or electrical)

FLUORESCENCE ACTIVATED CELL SORTING

HETEROKARYON CULTURE AND PLANT REGENERATION

IMPROVED RICE?

Figure 1.2   Opportunities for nuclear gene transfer into cultivated rice from alien species by protoplast fusion. FDA: fluorescein diacetate; RITC: rhodamine isothiocyanate

Protoplast fusion has been used for genetic manipulation in various species, particularly in the Solanaceae and the Cruciferae. However, until very recently, little progress has been made with the cereals due to the lack of procedures available for efficient plant regeneration from protoplasts. In the case of rice, now that an efficient, reproducible system is available for the regeneration of fertile plants from protoplasts (Finch *et al.*, 1989), numerous exciting possibilities are arising both in connection with nuclear gene transfer (Figure 1.2) and cytoplasmic gene transfer, mainly in relation to cytoplasmic male sterility (cms) (Figure 1.3). These developments have been greatly helped by the use of iodoacetamide to inhibit the division of one of the parental rice protoplast systems in the case of nuclear gene transfer (Terada *et al.*, 1987; Hayashi, Kyozuka and Shimamoto, 1988), with γ-irradiation of the other protoplast system in the case of cytoplasmic hybrid rice production (Akagi *et al.*, 1989; Yang *et al.*, 1989). There is not, however, a ready availability of reliable markers for cms in rice. Recently it has

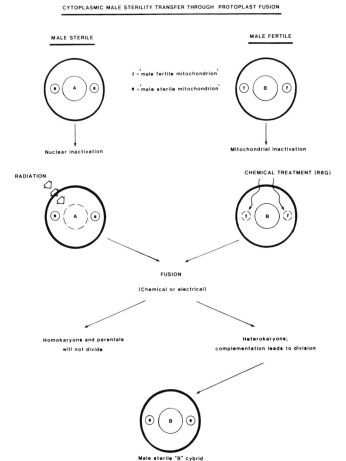

Figure 1.3 Opportunities for cytoplasmic male sterility transfer by protoplast fusion. R6G: Rhodamine 6G

been shown that small mitochondrial DNA molecules of wild abortive cytoplasm in rice are not necessarily associated with cms (Saleh *et al.*, 1989).

As recently critically discussed (Hinnisdaels *et al.*, 1988), interest in the application of fusion technology has tended to move from the creation of novel hybrid plants to chromosome transfer and gene introgression. However, in potato there continues to be considerable interest in somatic hybridization as an additional method of improving potato cultivars, highlighted by the difficulty of applying conventional techniques to potato breeding because of their high level of sterility. An assessment of work so far on somatic hybridization is that in most instances insufficient numbers of fusion products have been selected and regenerated into whole plants for an adequate evaluation of hybrid fertility. For instance, from 21 experiments only three yielded somatic hybrid plants *Lycopersicon esculentum* (+) *Lycopersicon peruvianum* ($2n = 6x = 72$). These were hexaploid, fertile and set seed after self-pollination (Kinsara *et al.*, 1986). This is an example where protoplast fusion has resulted in the production of fertile somatic hybrid plants, when sexual crosses (albeit produced with difficulty) in this important crop usually do not produce fertile plants. Analysis of an aneuploid somatic hybrid *Petunia parodii* ($2n = 2x = 14$) and *P. parviflora* ($2n = 2x = 18$) with 31 chromosomes has shown that the hybrid behaves as an allotetraploid. This somatic hybrid between these two sexually incompatible species was sterile. A detailed chromosomal analysis showed that there had been an interchange of segments between one of the large satellited *P. parodii* chromosomes and a *P. parviflora* chromosome of medium size, followed by loss of one of the interchange products (White and Rees, 1985).

The possibility of the transfer of only part of the genome in somatic hybridization and thereby the facilitation of the maintenance of fertility has been extensively examined. Irradiation studies have shown that irradiation directs the process of chromosome elimination, but that the amount of donor DNA undergoing elimination after fusion remains variable and random. More encouragingly the fusion of somatic protoplasts with gametic (pollen tetrad) protoplasts has resulted in the production of triploid plants with good self- and cross-fertility (Pirrie and Power, 1986). O'Mara (1940) first proposed the concept of combining one set of chromosomes (n) of an alien parent A (genome designated as AA) with two full sets of chromosomes ($2n$) of a crop parent B (genome designated as BB) for synthesizing $3n$ hybrids (ABB). Pental and Cocking (1985) proposed that triploid plants (ABB) could be produced by fusing protoplasts isolated from microspores at the tetrad stage (n) of species A with protoplasts isolated from somatic cells ($2n$) of species B. Synthesis of triploids through gameto-somatic fusions has an advantage over the conventional sexual crosses in bypassing pre- and post-zygotic sexual incompatibility barriers. It has been suggested that the ability to produce such interspecific triploids possessing the haploid genome of the alien species may facilitate the transfer of only part of the genome (perhaps one or a few chromosomes) from an alien species into the cultivated species (Pental and Cocking, 1985). It has also been suggested that an alternative may be actual chromosome-mediated transformation (Grisbach, 1987). *Petunia hybrida* proto-plasts were genetically transformed via microinjection of chromosomes isolated from *Petunia alpicola*. This resulted in changes in the relative activity of various enzymes in the flavonoid biosynthetic pathway in the plants regenerated and these changes were inherited in a Mendelian manner.

Transgenic plants by transformation

The general topic area of foreign genes in plants, their transfer, structure, expression and applications has recently been comprehensively reviewed (Weising, Schell and Kahl, 1988), and also specifically in relation to gene transfer in cereals (Cocking and Davey, 1987). At this Easter School it was useful to present some perspective in relation to gene transfer into plants, and also to assess the present limitations to the production of transgenic plants by transformation.

It has not as yet been possible to utilize *Agrobacterium* to mediate gene transfer into cereals, either by direct interaction of bacteria with the wounded plant or with explants, or by co-cultivation with protoplast-derived cells and suspension cultures. The current failure to achieve *Agrobacterium*-induced transformation in cereals, and also in some legumes and other crops, has led to increased interest in assessing other transformation procedures for the production of transgenic plants including the use of DNA-coated microprojectiles, microinjection of DNA (using special single cell, protoplast injection procedures, Schweiger, Dirk and Koop, 1987) and direct uptake of DNA into protoplasts stimulated either electrically or chemically. Each approach has its own specific limitations. In general what is becoming increasingly evident is that the actual delivery of DNA into cells either by microprojectiles, microinjection or direct uptake through an exposed plasma membrane is not a remaining problem. Rather, the remaining problem is the actual competence of the cell receiving the DNA to express that DNA; and for the development of the plant from transformed cells to enable the DNA to become adequately incorporated into the plant germ line and transmitted to subsequent generations.

Immature soyabean (*Glycine max*) embryos from commercially important cultivars were targets for rapidly accelerated, DNA-coated gold particles. Cell transformation rates in the embryos of the order of 1 in 10^5 were observed (Christou, McCabe and Swain, 1988). It was suggested that plants derived from treatment of intact tissues were likely to be chimaeras of both transformed and non-transformed cells which will complicate the identification or selection of rare transformation events; recovery of transformants among self-pollinated progeny of treated plants may be effective, however, if transformation of the germ line can be achieved at practical rates. McCabe *et al.*, 1988 also used particle acceleration by electric discharge to introduce DNA-coated free particles into meristems of immature soyabean seeds. Approximately 2% of shoots derived from these meristems were chimaeric for expression of the gene, and only one plant was produced from the progeny of chimaeric plants. Similar problems appear to be arising in the application to cereals of the somewhat analogous DNA microinjection approach successfully employed for the microinjection of selectable marker genes with microspore-derived proembryos of oilseed rape that produced transgenic chimaeras (Neuhaus *et al.*, 1987). Marker genes have been microinjected into microspores of wheat, rice, maize and barley and the sexual offspring of regenerated plants analysed for their transgenic nature with negative results (Potrykus *et al.*, 1989).

It is clear that at present the only reliable transformation in cereals yielding transgenic plants is based on direct gene transfer using protoplast-based protocols. Experience at Nottingham has been that the availability of a rice protoplast system from cell suspension cultures capable of efficient plant regeneration through somatic embryogenesis has enabled fertile transgenic rice plants to be produced by electroporation-mediated plasmid uptake into protoplasts (Zhang *et al.*, 1988). Even when using this single cell protoplast system only about 1 in 10^5 protoplasts plated became transformed, yet because efficient fertile plant regeneration from protoplasts was possible transgenic plants were readily recovered. Similar results have been reported in a range of japonica rice varieties using protoplasts isolated from cell suspension cultures (Toriyama *et al.*, 1988; Zhang and Wu, 1988; Shimamoto *et al.*, 1989). These studies are now enabling the inheritance of foreign genes introduced into transgenic rice to be evaluated. Protoplasts are also particularly useful for visualizing mRNA expression. Video image analysis of tobacco protoplasts electroporated with luciferase mRNA demonstrated a wide range in the level of expression of this marker (Gallie, Lucas and Walbot, 1989).

These studies on the production of transgenic crop plants are currently being combined with the construction of restriction fragment length polymorphism (RFLP) linkage maps for most major crop plants (Tanksley *et al.*, 1989).

Integration of RFLP techniques into plant breeding, including procedures leading to the production of transgenic plants, promises among other things to allow the transfer of novel genes from related wild species perhaps making it possible to clone genes of complex traits including even those whose products are unknown, such as genes for disease resistance or stress tolerance.

Genetic manipulation of complex traits

The ability of legumes to establish a symbiotic association with rhizobia resulting in the formation of effective, nitrogen fixing root nodules is another example of a complex plant genetic trait. There is also genetic control by the rhizobia of nodule formation in this complex interaction process. As recently cogently argued by Sprent (1989), on current evidence there appears to be no clear reason why non-legumes should not nodulate with rhizobia, given the appropriate environment and genetic background.

Recently, in an effort to try to identify the key steps controlling legume nodulation by rhizobia, it was observed that treatment of root hairs of clover seedlings with a cell wall degrading enzyme removed a barrier to *Rhizobium*-host specificity (Al-Mallah, Davey and Cocking, 1987). This result indicated that the cell wall at the tip of roots hairs was playing a key role in the control of rhizobial specificity, and that its removal enabled this control to be bypassed. This observation led to experiments on non-legumes in which the cell walls at the tips of root hairs of both rice and wheat were similarly removed by enzymatic treatment utilizing a cell wall degrading mixture of cellulase of pectolyase (Cocking, 1985). Nodular structures were induced on rice roots when the roots of 2-day-old seedlings were treated with this cell wall degrading enzyme mixture, followed by inoculation with rhizobia in the presence of polyethylene glycol (Al-Mallah, Davey and Cocking, 1989a). Nodular structures were also produced on the roots of wheat seedlings when they were similarly treated (Figure 1.4) (Al-Mallah, Davey and Cocking, 1989b). Nodules, similar to those induced on rice and wheat by *Rhizobium* developed on oilseed rape under comparable conditions. Electron microscopy confirmed the presence of *Rhizobium* both within and between cells of the nodules. Such oilseed rape nodules exhibited very low nitrogen fixation as assessed by acetylene reduction. These results demonstrated that enzymatic treatment of the root system of non-legumes was providing a 'port of entry' for rhizobia, probably the exposed plasma membrane at the tip of root hairs. These results have identified the cell wall at the tip of root hairs as a target site for genetic manipulation of the complex plant trait controlling nodule formation by rhizobia in both legumes and non-legumes.

This new approach to unravelling the key steps in the genetic control of the complex traits of bacterial interactions has also been shown to enable *Agrobacterium rhizogenes* to induce tumour-like outgrowths on rice roots when enzyme-PEG treated rice seedlings are inoculated with supervirulent *A. rhizogenes* R1601. When thin sections of these tumour-like outgrowths were examined in the electron microscope, agrobacteria were observed between, but not within, some of the cells of the outgrowths (Al-Mallah, Davey and Cocking, 1989b). While molecular biological and biochemical evidence for genetic transformation is required, this result suggests a possible new approach to cereal transformation and the production of transgenic plants. This is particularly likely in the light of the

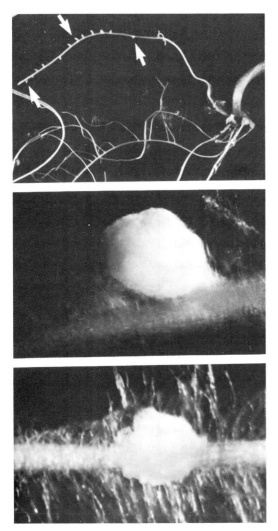

Figure 1.4 Nodular structures (arrowed) on the roots of a wheat seedling following enzyme treatment and inoculation of the seedling root with a mixture of rhizobia in the presence of polyethylene glycol ($\times 1.0$). The other two are illustrations of these wheat nodules at higher magnification ($\times 13.5$)

observation that a factor inducing *Agrobacterium tumefaciens vir* gene expression is present in monocotyledonous plants; this observation suggests that T-DNA processing, and possibly its transfer, should take place when *Agrobacterium* invades suitable tissues of monocotyledonous plants (Usami *et al.*, 1988).

At this Easter School, many facets of the genetic engineering of crop plants were considered, and many of the objectives centred on the genetic manipulation of complex traits. What is increasingly clear is that significant advances will only come from a better understanding of plant physiology and developmental biology coupled with a 'broad spectrum' approach to plant genetic manipulations.

Acknowledgements

Original research described in this review was supported in part by grants from the Rockefeller Foundation and the Overseas Development Administration.

References

Abdullah, R., Cocking, E. C. and Thompson, J. A. (1986) Efficient plant regeneration from rice protoplasts through somatic embryogenesis. *Bio/Technology*, **4**, 1087–1090

Abdullah, R., Thompson, J. A., Khush, G. S., Kaushik, R. P. and Cocking, E. C. (1989) Protoclonal variation in the seed progeny of plants regenerated from rice plants. *Plant Science*, (in press)

Akagi, H., Sakamoto, M., Negishi, T. and Fujimura, T. (1989) Construction of rice cybrid plants. *Molecular and General Genetics*, **215**, 501–506

Alexander, R. G., Cocking, E. C., Jackson, P. J. and Jett, J. H. (1985) The characterization and isolation of heterokaryons by flow cytometry. *Protoplasma*, **128**, 52–58

Al-Mallah, M. K., Davey, M. R. and Cocking, E. C. (1987) Enzymatic treatment of clover roots hairs removes a barrier to *Rhizobium*-host specificity. *Bio/Technology*, **5**, 1319–1322

Al-Mallah, M. K., Davey, M. R. and Cocking, E. C. (1989a) Formation of nodular structures on rice seedlings by rhizobia. *Journal of Experimental Botany*, **40** 473–478

Al-Mallah, M. K., Davey, M. R. and Cocking, E. C. (1989b) A new approach to the nodulation of non-legumes by rhizobia and the transformation of cereals by agrobacteria using enzymatic treatment of root hairs. *International Journal of Genetic Manipulation in Plants*, **5**, 1–7

Bilkey, P. (1981) An assessment of the suitability of *Saintpaulia* for plant genetic manipulation. *PhD Thesis*, University of Nottingham

Chand, P. K., Davey, M. R., Power, J. B. and Cocking, E. C. (1988) An improved procedure for protoplast fusion using polyethylene glycol. *Journal of Plant Physiology*, **133**, 480–485

Christou, P., McCabe, D. E. and Swain, W. F. (1988) Stable transformation of soybean callus by DNA-coated gold particles. *Plant Physiology*, **87**, 671–674

Cocking, E. C. (1985) Protoplasts from root hairs of crop plants. *Bio/Technology*, **3**, 1104–1106

Cocking, E. C. (1989) Rice: the Chinese connection. *Proceedings of the Royal Institution*, **61** (in press)

Cocking, E. C. and Davey, M. R. (1987) Gene transfer in cereals. *Science*, **236**, 1259–1262

Finch, R. P., Lynch, P. T., Jotham, J. P. and Cocking, E. C. (1989) Isolation, culture and genetic manipulation of rice protoplasts. In *Biotechnology in Agriculture and Forestry*, vol. 14, (ed. Y. P. S. Bajaj), Springer Verlag, Berlin

Gallie, D. R., Lucas, W. J. and Walbot, V. (1989) Visualizing mRNA expression in plant protoplasts: factors influencing efficient mRNA uptake and translation. *The Plant Cell*, **1**, 301–311

Griesbach, R. J. (1987) Chromosome-mediated transformation via microinjection. *Plant Science*, **50**, 69–77

Hadley, H. H. and Openshaw, S. J. (1980) Interspecific and intergeneric hybridization. In *Hybridization of Crop Plants*, (eds W. R. Fehr and H. H. Hadley), American Society of Agronomy and Crop Science Society of America, Madison, Wisconsin, pp. 133–159

Hayashi, Y., Kyozuka, J. and Shimamoto, K. (1988) Hybrids of rice (*Oryza sativa* L.) and wild *Oryza* species obtained by cell fusion. *Molecular and General Genetics*, **214**, 6–10

Hinnisdaels, S., Negrutui, I., Jacobs, M. and Sidorov, V. (1988) Plant somatic hybridization: evaluations and perspectives. *Newsletter, International Association for Plant Tissue Culture*, **55**, 2–20

Kinsara, A., Patnaik, S. N., Cocking, E. C. and Power, J. B. (1986) Somatic hybrid plants of *Lycopersicon esculentum* Mill. and *Lycopersicon peruvianum* Mill. *Journal of Plant Physiology*, **125**, 225–234

Kumar, A. and Cocking, E. C. (1987) Protoplast fusion: a novel approach to organelle genetics in higher plants. *American Journal of Botany*, **74**, 1289–1303

McCabe, D. E., Swain, W. F., Martinell, B. J. and Christov, P. (1988) Stable transformation of soybean (*Glycine max*) by particle acceleration. *Bio/ Technology*, **6**, 923–926

Neuhaus, G., Spangenberg, G., Mittelsten-Scheid, O. and Schweiger, H. G. (1987) Transgenic rapeseed plants obtained by the microinjection of DNA into microspore-derived embryoids. *Theoretical and Applied Genetics*, **74**, 30–36

O'Mara, J. G. (1940) Cytogenetic studies on Triticale I. A method for determining the effects of individual *Secale* chromosomes on *Triticum*. *Genetics*, **25**, 401–408

Pental, D. and Cocking, E. C. (1985) Some theoretical and practical possibilities of plant genetic manipulation using protoplasts. *Hereditas* (suppl.), **3**, 83–93

Pirrie, A. and Power, J. B. (1986) The production of fertile, triploid somatic hybrid plants (*Nicotiana glutinosa* (n) + *N. tabacum* (2n) via gametic-somatic protoplast fusion. *Theoretical and Applied Genetics*, **72**, 48–52

Potrykus, I., Datta, S. K., Neuhaus, G. and Spangenberg, G. (1989) Approach to cereal transformation via microinjection into microspore-derived and into zygotic proembryos. *Journal of Cellular Biochemistry*, suppl. 13D (UCLA Symposia on Molecular and Cellular Biology, M.001, p. 228)

Roest, S. and Gilissen, L. J. W. (1989). Plant regeneration from protoplasts: a literature review. *Acta Botanica Neerlandica*, **38**, 1–23

Saleh, N. M., Mulligan, B. J., Cocking, E. C. and Gupta, H. S. (1989) Small mitochondrial DNA molecules of wild abortive cytoplasm in rice are not necessarily associated with cms. *Theoretical and Applied Genetics*, **77**, 617–619

Schweiger, H. G., Dirk, J. and Koop, H. U. (1987) Individual selection, culture and manipulation of higher plant cells. *Theoretical and Applied Genetics*, **73**, 769–783

Scowcroft, W. R. and Larkin, P. J. (1988) Somaclonal variation. In *Applications of Plant Cell and Tissue Culture*: CIBA Foundation Symposium 137. John Wiley and Sons, Chichester, pp. 21–35

Sharp, W. R., Evans, D. A. and Ammirato, P. V. (1984) Plant genetic engineering: designing crops to meet food industry specifications. *Food Technology*, (Feb.) 112–119

Shimamoto, K., Terada, R., Izawa, T. and Fujimoto, H. (1989) Fertile transgenic rice plants regenerated from transformed protoplasts. *Nature*, **338**, 274–276

Sprent, J. I. (1989) Tansley review no. 15. Which steps are essential for the formation of functional legume nodules? *New Phytologist*, **111**, 129–153

Tanksley, S. D., Young, N. D., Paterson, A. H. and Bonierbale, M. W. (1989) RFLP mapping in plant breeding: new tools for an old science. *Bio/Technology,* **7**, 259–266

Terada, R., Kyozuka, J., Nishibayashi, S., Shimamoto, K. (1987) Plantlet regeneration from somatic hybrids of rice (*Oryza sativa* L.) and barnyard grass (*Echinochloa oryzicola* Vasing). *Molecular and General Genetics,* **210**, 39–43

Toriyama, K., Arimoto, Y., Uchimiya, H. and Hinata, K. (1988) Transgenic rice plants after direct gene transfer into protoplasts. *Bio/Technology,* **6**, 1072–1074

Usami, S., Okamoto, S., Takebe, I. and Machida, Y. (1988) Factor inducing *Agrobacterium tumefaciens vir* gene expression is present in monocotyledonous plants. *Proceedings of the National Academy of Sciences, USA,* **85**, 3748–3752

Watts, J. W. and King, J. M. (1984) A simple method for large-scale electrofusion and culture of plant protoplasts. *Bioscience Reports,* **4**, 335–342

Weising, K., Schell, J. and Kahl, G. (1988) Foreign genes in plants: transfer, structure, expression and applications. *Annual Review of Genetics,* **22**, 421–477

White, J. and Rees, H. (1985) The chromosome cytology of a somatic hybrid petunia. *Heredity,* **55**, 53–59

Withers, L. A. and Alderson, P. G. (1986) *The Tissue Culture Revolution* Butterworths, London

Yang, Z. Q., Shikanai, T., Mori, K. and Yamada, Y. (1989) Plant regeneration from cytoplasmic hybrids of rice (*Oryza sativa* L.). *Theoretical and Applied Genetics,* **77**, 305–310

Zhang, W. and Wu, R. (1988) Efficient regeneration of transgenic plants from rice protoplasts and correctly regulated expressions of the foreign gene in the plants. *Theoretical and Applied Genetics,* **76**, 835–840

Zhang, H. M., Yang, H., Rech, E. L., Golds, T. J., Davis, A. S. and Mulligan, B. J. (1988) Transgenic rice plants produced by electroporation-mediated plasmid uptake into protoplasts. *Plant Cell Reports,* **7**, 379–384

2

Prevention of viruses

COAT PROTEIN-MEDIATED PROTECTION AGAINST VIRUS INFECTION*

RICHARD S. NELSON
Noble Foundation, Plant Biology Division, Ardmore, Oklahoma, USA

PATRICIA A. POWELL† and ROGER N. BEACHY
Washington University, Plant Biology Program, Department of Biology, St Louis, Missouri, USA

Coat protein-mediated protection and cross-protection

The discovery that expression of viral coat protein (CP) in transgenic plants protects these plants against virus infection is, in some ways, a natural extension of work by previous researchers on cross-protection. Reported in 1929 by McKinney, cross-protection is defined, in its simplest form, as the protection of a plant against pathogen invasion due to prior inoculation of the plant with another pathogen. Although the phenomenon was first observed for related plant viruses (*see* reviews by Fulton, 1986; Ponz and Bruening, 1986; Urban *et al.*, 1990) similar observations have been made for viroids (Fernow, 1967; Niblett *et al.*, 1978). The inclusion of the protection between viroids under the umbrella of cross-protection presents a dilemma since viroids do not code for the synthesis of a capsid protein and there is evidence, to be discussed below, which implicates the involvement of CP in cross-protection between viruses (Sherwood and Fulton, 1982). It is clear, therefore, that cross-protection is defined more by the final result than by the mechanism of protection.

With regard to plant viruses and cross-protection, much research has been directed toward determining whether the presence of the CP, the RNA, or both from the protecting virus provides protection. Zaitlin (1976) and Sarkar and Smitamana (1981), using protecting tobacco mosaic virus (TMV) strains which produced a defective CP or no CP at all, respectively, reported that protection occurred when plants harbouring these strains were challenged with a related virus. These results appear to be in conflict with those of Sherwood and Fulton (1982) who showed that CP of the protecting virus was important for protection. Sherwood (1987) resolved much of the controversy by showing that the CP-minus TMV strain used by Sarkar and Smitamana did indeed provide protection against superinfection by a TMV strain, but that greater protection was observed when a CP-plus strain was used as the protecting virus. The protection afforded by the

*This contribution has been published in the Proceedings of the NATO advanced Research Workshop on *Recognition and Response in Plant–Virus Interactions*, ed. R. S. S. Fraser, Springer Verlag, Heidelberg.

†Present address: Salk Institute for Biological Studies, Plant Biology Laboratory, La Jolla, California, USA

13

CP-minus strain was non-specific since infection by turnip mosaic virus (TuMV), a potyvirus, was also inhibited. The protection provided by the CP-expressing strain was specific for a related virus since there was an approximately five-fold greater level of protection against a second TMV strain compared with that observed against TuMV. This and other work (Zinnen and Fulton, 1986) suggest, therefore, that cross-protection may be composed of two or more mechanisms.

One problem with interpreting the results from cross-protection experiments is that replication of a viral genome is involved. Viral replication and its regulation are complicated processes which are poorly understood. Researchers are left to identify mutants which lack a function associated with a specific gene product and then to correlate this loss with a loss in protection. Mutants which have lost the ability to replicate cannot, of course, be analysed for cross-protection. In contrast, the tools of molecular biology and genetic transformation allow the researcher to express fragments of the viral genome throughout transgenic plants or in specific tissues and then to evaluate the ability of these fragments to provide protection in the absence of replication of the protecting virus.

Coat protein-mediated protection

FIRST PROPOSALS FOR GENETICALLY ENGINEERED PROTECTION

Hamilton (1980) was one of the earliest researchers to propose that expressing viral sequences in transgenic plants could be a method to protect plants against virus infections. Others followed with more refined ideas as to which viral sequences could potentiate protection (Sequeira, 1984; Beachy *et al.*, 1985; Sanford and Johnston, 1985). Based on earlier models of cross-protection proposed by de Zoeten and Fulton (1975) and Sherwood and Fulton (1982), Sequeira (1984) suggested that insertion and expression of the CP gene in plants might result in protection. Beachy *et al.* (1985), in addition to discussing the use of a CP gene for protection, discussed the potential for using genes that encode antisense viral sequences to provide protection. Sanford and Johnston (1985) suggested that the replicase of the protecting virus might bind to 'a replicase attachment site' but not replicate the virus thereby preventing the challenge virus replicase from binding. All of these ideas were based on possible explanations for cross-protection put forth previously by plant pathologists.

TRANSGENIC PLANTS AND COAT PROTEIN-MEDIATED PROTECTION

By the early to mid-1980s researchers in several laboratories were cloning plant viruses and inserting portions of the viral genome into plants under the control of heterologous promoters to test whether expression of the inserted sequences could provide protection against virus infection. Bevan, Mason and Goelet (1985) produced transgenic tobacco plants that expressed the CP of TMV, but expression levels were low (approximately 0.17 μg/g fresh weight) and no results were presented indicating protection of these plants after virus challenge. Beachy *et al.* (1985) reported expression of TMV-CP in transgenic tobacco plants at levels greater than ten times those reported by Bevan, Mason and Goelet (1985) and Powell Abel *et al.* (1986) showed that these plants were protected after challenge with TMV. Many reports quickly followed showing the usefulness of CP gene expression in providing protection in a number of other plant:virus combinations (Loesch-Fries *et al.,* 1987; Tumer *et al.,* 1987; van Dun, Bol. and van Vloten-Doting, 1987; Cuozzo *et al.,* 1988; Hemenway *et al.,* 1988; Stark and Beachy, 1989). Thus, of the 28 plant virus families consisting of ssRNA without envelopes (for latest presentation of virus families *see* Wilson, 1989), CPs from viruses within seven of the families have now been shown to provide protection against virus infection in transgenic plants. Practical application of this technology is promising since transgenic tomato plants that expressed the TMV-CP gene were protected in field experiments under conditions of high infection pressure. Protection was reflected in the fruit yields which were 0.25–3 times greater than those from inoculated plants not expressing the CP gene (Nelson *et al.,* 1988; Gaser and Fraley, 1989). Recently, results have been presented showing that expression of CP from potato virus Y in transgenic potato plants is effective in providing protection against aphid transmission of that virus (Lawson *et al.,* 1989).

Sites and potential mechanisms of CP-mediated protection

While it is clear that CP-mediated protection is moving towards application in a commercial setting, it is less clear where and how the protection is effected in transgenic plants. Information is, however, beginning to appear which addresses these issues (Table 2.1). Many of the studies cited in Table 2.1 were conducted to identify the extent of the protection and its applicability for practical use and not necessarily to determine the mechanism(s) of protection *per se.* However, even these results often give clues exposing the underlying mechanism(s).

SITE OF PROTECTION

Protection by CP expression appears to exert its effect at several sites. There is a decrease in lesion numbers on inoculated leaves of CP(+) plants for every system tested to date. Thus, one site of protection is at the primary infection site (*see* Table 2.1 for references). For TMV, there is a decrease in necrotic or chlorotic lesions in transgenic CP(+) plants that are local lesion (*Nicotiana tabacum* cv. Xanthi 'nc') or systemic (cv. Xanthi) hosts (Nelson, Powell-Abel and Beachy, 1987). Although there is a trend indicating that the decrease in lesion numbers is greater in the local lesion host compared with the systemic host, too few experiments have been conducted under rigorously controlled conditions to give much validity to the observation. Thus, any suggestion that the local lesion locus may interact with the CP gene product to enhance protection requires further proof. To determine whether expression of the TMV CP gene stimulates synthesis of potentially protective pathogenesis-related (PR) proteins, TMV CP(+) local lesion and systemic hosts have been analysed (Carr, Beachy and Klessig, 1989). Expression of PR proteins is known to be induced by pathogens and other antagonists (*see* Fritig, 1990) and, although debated, the expression of the PR1 group has been shown to be correlated with protection against spread of TMV in tobacco (*see* van Loon, 1985 and references within). PR1 protein accumulation was not enhanced by expression of the TMV CP gene in the systemic host (Carr, Beachy and Klessig, 1989). It was enhanced in the transgenic local lesion host compared with controls

Table 2.1 COMPILATION OF RESULTS AND CONCLUSIONS IN THE LITERATURE CONCERNING CP-MEDIATED PROTECTION IN PLANTS

Issue addressed	System studied[a]	Phenotype of CP-mediated protection	Conclusions drawn	References
Site of protection	T/T/Tb A/A/Tb TS/TS/Tb P/P/Tb	Decrease in lesion numbers (chlorotic, necrotic, starch) on inoculated leaves after challenge with virus from which the CP was isolated	Protection manifested at the primary infection site	Loesch-Fries et al. (1987); Nelson, Powell-Abel and Beachy (1987); Tumer et al. (1987); van Dun, Bol and van Vloten-Doting (1987); Hemenway et al. (1988) van Dun et al. (1988)
	T/T/Tb	Decrease in virus accumulation in systemically infected leaves in spite of virus accumulation in inoculated leaves equal to that found in controls	Protection manifested after primary infection and inhibiting systemic spread of the infection	Wisniewski et al. (1989) Nelson, Powell and Beachy (this Chapter)
Mechanism of protection	T/T/Tb A/A/Tb A/A/Al	Decrease in viral CP and (+) and (−) strand viral RNA accumulation in challenged leaves and protoplasts.	Primary infection blocked early in infection; for TMV, blockage is at or before virus replication.	Register and Beachy (1988); Halk et al. (1989)
	T/T/Tb A/A/Tb P/P/Tb	Less protection in inoculated leaves and protoplasts challenged with viral RNA or pH 8.0 treated virions except for studies involving PVX	Interaction between the challenge CP and its viral RNA is required for maximum protection. Indirectly indicates that RNA:RNA interactions between the transgenic CP transcript and the challenge virus RNA are not providing much protection against primary infection. Results for PVX indicate that mechanisms other than those working in TMV and AlMV systems may provide significant protection against PVX-RNA	Loesch-Fries et al. (1987); Nelson, Powell-Abel and Beachy (1987); van Dun, Bol and van Vloten-Doting (1987); Hemenway et al. (1988); Register and Beachy (1988) van Dun et al. (1988)
	A/A/Tb T/T/Tb	Protection is correlated with CP levels and not CP-RNA levels; a threshold level of CP expression is apparently required for protection	Protection requires interaction between the CP expressed by the transgenic plants and the challenge virus. More than a low level of biologically active CP is required for protection	Loesch-Fries et al. (1987); van Dun et al. (1988); Powell et al. (1989a)

17

T/T/Tb	CP expressed in transgenic plants forms aggregates; aggregates of CP provided greater protection in protoplasts than non-aggregated CP	Aggregation state of CP (i.e. its quaternary structure) plays an important role in primary protection	Register and Beachy (1989); Wilson (1989)
A/-/Tb T/-/Tb T/-/Tm S/-/Tb C/-/Tb	Protection generally declines as the amino acid sequence homology of the challenge virus CP with the virus which supplied the CP for transformation declines. For TMV- and SMV-CP expressing plants, protection is observed against viruses with CPs having relatively low amino acid sequence homologies with the protecting CPs	Protection relies on an undetermined degree of amino acid sequence homology and it is clear that at least some of the protection is related to tertiary and/or quaternary structures	Loesch-Fries et al. (1987); Nelson et al. (1988); van Dun et al. (1988); Anderson et al. (1989); Quemada, Slightom, and Gonsalves (1989); Register and Beachy (1989); Stark et al. (1989); Stark and Beachy (1989)
T/T/Tb T/T/Tm	Continuous high temperatures result in a decline in CP accumulation but no CP-mRNA. A decline in protection in tobacco but not tomato plants expressing TMV-CP was observed under these conditions	Implicates the role of the CP and not the mRNA in protection. The difference between responses of tobacco and tomato plants indicates potential differences in host response	Nejidat and Beachy (1989)
T/T/Tb	Grafted stem and leaf sections of CP(+) plants can block systemic spread of virus in CP(−) tissue after challenge; primary infection not inhibited in non-expressing tissue	Implies that protection against systemic spread is effected by expression in tissue through which the challenge virus must move	Wisniewski et al. (1989)
T/T/Tb	Expression of the entire viral genome in plants provides greater protection than CP expression alone; resistance against challenge with viral RNA	Expression of the entire genome may utilize antisense inhibition or other mechanism(s) suggested to produce a greater level of cross-protection than that seen in plants expressing the CP gene only	Yamada et al. (1988)

[a] Virus from which CP gene was derived/challenge virus/host plant.

T/TMV = tobacco mosaic virus; A/AlMV = alfalfa mosaic virus; TS = tobacco streak virus; P/PVX = potato virus X; C = cucumber mosaic virus; S = soyabean mosaic virus; Tb = tobacco; Al = alfalfa; Tm = tomato; - = various viruses.

but the level was at least 25-fold lower than in an untransformed local lesion host infected with TMV. Whether this increase is sufficient to increase protection in the CP(+) local lesion host above that of the systemic host requires not only proof that the hypersentive host is more resistant to challenge than the systemic host, but that PR1 expression in PR1-expressing transgenic plants increases protection against virus infection. The effect of CP gene expression on the expression of other PR proteins has not been analysed.

Prevention of systemic spread by CP-mediated protection has been reported in several examples of CP-mediated resistance but has been studied in detail only for the TMV/tobacco system (Wisniewski *et al.*, 1989). In this study, CP(+) or CP(−) systemic host plants were inoculated with TMV or TMV-RNA and infection was followed in the inoculated and systemically infected leaves. Previously, Nelson, Powell-Abel, and Beachy (1987) showed that inoculation with TMV-RNA effectively overcame protection at the primary infection site in terms of lesion numbers. A follow-up experiment was conducted to determine whether virus accumulation within established primary infection sites differed between leaves from CP(+) and CP(−) tobacco plants (Figure 2.1). Virus inoculum concentrations varied in this experiment to ensure both similar lesion numbers on CP(+) and CP(−) leaves and a range of lesion numbers for each plant line for linear correlation analysis. Results from this work (Figure 2.1) and from Wisniewski *et al.* (1989) show that virus accumulation within primary lesions increases to essentially the same level in both CP(+) and CP(−) plants (note V/L values in Figure 2.1). In spite of similar lesion numbers and amounts of virus per lesion in inoculated leaves, spread was decreased in CP(+) plants compared with CP(−) plants as determined by immunoassay of tissue sections and extracts from systemic leaves and stems (Wisniewski *et al.*, 1989). These results clearly indicate that a second site exists in CP(+) plants providing protection against virus infection beyond the primary site.

MECHANISM(S) OF PROTECTION

Intensive research efforts during the last 2 years are gradually revealing the mechanism(s) of CP-mediated protection. Much of the data gathered to date must be evaluated cautiously however, because (1) many of the published results address primary or general protection rather than systemic protection; (2) the general conclusions drawn from Table 2.1 often depend upon observations from very few reports; and (3) reports of plants expressing TMV or alfalfa mosaic virus (AlMV)-CP predominate in the literature and they may or may not reflect the mechanism(s) of protection in other systems (e.g. plants expressing potato virus X(PVX)-CP and inoculated with viral RNA are still protected, *see* Table 2.1 for reference). With these precautions in mind, the information presented in Table 2.1 for plants that express the TMV-CP can be summarized as follows.

First, there is a strong indication that there is a blockage of the infection process prior to the production of the (−) strand viral RNA (Register and Beachy, 1988). Second, maximum protection is much greater against TMV than against TMV-RNA and requires the production of the CP and not the CP transcript *per se* in the transgenic plant (Nelson, Powell-Abel and Beachy, 1987; Register and Beachy, 1988; Powell *et al.*, 1989a; Nejidat and Beachy, 1989). This also applies to the protection against AlMV observed for AlMV-CP expressing plants (Loesch-Fries *et al.*, 1987; van Dun, Bol, and van Vloten-Doting, 1987; van Dun *et al.*,

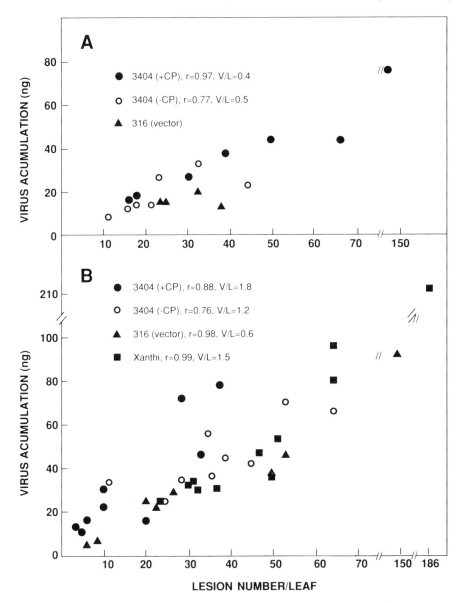

Figure 2.1 TMV accumulation versus visible chlorotic lesion numbers in leaves from CP(+) and CP(−) tobacco plants (*Nicotiana tabacum* cv. Xanthi) after inoculation with TMV and TMV-RNA (strain PV230). Plants were greenhouse grown and inoculation took place on fully expanded leaves 51 days after planting. Inoculum concentrations ranged from 0.1 to 10 μg/ml for TMV and 1 to 10 μg/ml for TMV-RNA. Inoculations were done on leaves as described by Nelson, Powell-Abel and Beachy (1987). Leaf tissue was harvested 5 days post-inoculation and extracts analysed by enzyme-linked immunosorbent assay (ELISA) using antibody against TMV. (A): plants inoculated with TMV. (B): plants inoculated with TMV-RNA.3404(+CP), CP-expressors, R₁ generation; 3404(-CP), non-expressors, R₁ generation; 316, vector control (i.e. plants transformed with vector not containing the TMV-CP coding sequence); Xanthi, untransformed parental line; r, linear correlation coefficient; V/L, virus accumulation/lesion. Each point represents a value from a single plant. No V/L was given for the 316 (vector) leaf samples shown in A due to the small sample size (4 plants) and lack of slope

1988). Third, Register and Beachy (1989) have shown that not only is CP expression important, but that the state of aggregation of the CP could affect the extent of protection in a protoplast transient assay system. The hypothesis that the quaternary structure of the protecting CP was important in protection was supported by the finding that plants that expressed the CP gene of TMV were resistant to challenge by tobacco mild green mosaic virus which has only 67% amino acid sequence homology compared with the protecting CP (Stark *et al.*, 1990; Nejidat and Beachy, unpublished results). Similar results were obtained in CP-mediated resistance against potyviruses (Stark and Beachy, 1989). Fourth, Wisniewski *et al.* (1989) demonstrated that by grafting TMV CP(+) stem and leaf tissue on to CP(−) root and shoot stocks that CP expression prevented systemic spread but not primary infection of the challenge virus. This systemic protection appears to involve the CP and not the CP transcript since Powell *et al.* (1989a) observed no protection either at the primary infection site or systemically when only TMV CP-transcript was synthesized.

Gene constructs that express RNA complementary to CP mRNA (i.e. antisense RNA) might be expected to provide greater protection than (+) sense CP transcripts since they could, in theory, hybridize to both (+) sense genomic and subgenomic viral RNAs. However, antisense gene constructs gave only minimal protection in the PVX, CMV, and TMV systems (Cuozzo *et al.*, 1988; Hemenway *et al.*, 1988; Powell *et al.*, 1989b). Powell *et al.* (1989b) were able to show that only sequences complementary to the 3' untranslated region of TMV RNA, rather than the CP-coding sequence, were required for protection. Although protection via antisense RNA is readily overcome, antisense protection may have some ability to limit secondary infections under field situations. Other researchers have investigated the use of antisense sequences against the 5' or 3' end of viral genomes in order to provide protection (Beachy *et al.*, 1985; Crum *et al.*, 1988; Valle *et al.*, 1989) and this work is continuing.

Conclusions

The research results summarized here greatly extend our understanding of how CP-mediated protection is implemented. They may also shed light on the mechanism of cross-protection, as it is suggested that CP-mediated protection is part of cross-protection. It is of interest to note that expression of the entire TMV viral genome in transgenic tobacco plants, which would closely represent cross-protection, results in a highly effective protection against TMV infection (Yamada *et al.*, 1988). Since these plants are nearly completely protected against challenge by TMV-RNA it is likely that its effectiveness is due to more than simply higher levels of CP in the leaves prior to challenge inoculation. Thus, for these plants and for plants that are cross-protected by classical methods, antisense inhibition and/or other mechanism(s) may be involved to a greater extent than in plants protected by CP gene expression. To understand better the implications of the results of Yamada *et al.* (1988), further work involving expression of portions of the TMV genome other than the CP and 3'-untranslated region should be pursued.

Future work on CP-mediated protection should address how the CP expressed in the transgenic plant prevents or retards virus replication. Site-directed mutations of the CP and further studies on virus strains known to have high structural homology but low sequence homology with the transgenic CP could further define the protein conformation necessary for maximum protection. Namba, Pattanayek and Stubbs (1989) have recently determined the structure of intact TMV by X-ray fibre diffraction methods. They, as well as others (Register *et al.*, 1989), postulate that co-translational disassembly (Shaw, Plaskitt and Wilson, 1986) may be initiated by a change in Ca^{2+} concentrations and pH values within the cellular milieu. In the Namba report at least two potential Ca^{2+} binding sites were identified in subunit interfaces. They postulate that upon virus entry low Ca^{2+} concentration and high pH within the cell removes calcium and protons from the virion leading to a general destabilization and, potentially, to ribosome binding and co-translational disassembly. Experiments with CP-expressing protoplasts in the presence of varying Ca^{2+} concentrations may provide evidence linking co-translational disassembly and CP-mediated protection.

In addition to defining the mechanism of protection at the primary infection site, future work should address both the mechanism and specific site of protection against systemic spread of virus. Analysis of cells in the vascular region of CP(+) tissue prior to inoculation might show whether CP accumulates at high levels in this area and thus leads to greater protection. Analysis of cells around the vascular system of the inoculated leaves from both CP(+) and CP(−) plants would provide evidence for or against the hypothesis that transgenic CP interferes in some way with virus entry into the phloem of protected plants. These are but a few of the many areas which should be addressed to provide further understanding of CP-mediated protection in plants.

Acknowledgements

The authors wish to thank Allyson Wilkins for preparation of the manuscript and Susan Martin for graphics on Figure 2.1. Roger N. Beachy acknowledges support by a grant from the Monsanto Company.

References

Anderson, E. J., Stark, D. M., Nelson, R. S., Powell, P. A., Tumer, N. E. and Beachy, R. N. (1989) Transgenic plants that express the coat protein genes of TMV and AlMV interfere with disease development of some non-related viruses. *Phytopathology*, (in press)

Bevan, M. W., Mason, S. E. and Goelet, P. (1985) Expression of tobacco mosaic virus coat protein by a cauliflower mosaic virus promoter in plants transformed by *Agrobacterium*. *EMBO Journal*, **4**, 1921–1926

Beachy, R. N., Abel, P., Oliver, M. J., De, B., Fraley, R. T., Rogers, S. G. *et al.* (1985) Potential for applying genetic transformation to studies of viral pathogenesis and cross-protection. In *Biotechnology in Plant Science: Relevance to Agriculture in the Nineteen Eighties*, (eds M. Zaitlin, P. Day and A. Hollander), Academic Press, Inc., New York, pp. 265–275

Carr, J. P., Beachy, R. N. and Klessig, D. F. (1989) Are the PR1 proteins of tobacco involved in genetically engineered resistance to TMV? *Virology*, **169**, 470–473

Crum, C., Johnson, J. D., Nelson, A. and Roth, D. (1988) Complementary oligodeoxynucleotide mediated inhibition of tobacco mosaic virus RNA translation *in vitro*. *Nucleic Acids Research*, **16**, 4569–4581

Cuozzo, M., O'Connell, K. M., Kaniewski, W., Fang, R., Chua, N. and Tumer, N. E. (1988) Viral protection in transgenic tobacco plants expressing the cucumber mosaic virus coat protein or its antisense RNA. *Bio/Technology*, **6**, 549–557

de Zoeten, G. A. and Fulton, R. W. (1975) Understanding generates possibilities. *Phytopathology*, **65**, 221–222

Fernow, K. H. (1967) Tomato as a test plant for detecting mild strains of potato spindle tuber virus. *Phytopathology*, **57**, 1347–1352

Fritig, B. (1990) In *Recognition and Response in Plant-Virus Interactions*, (ed. R. S. S. Fraser). Springer-Verlag, Heidelberg, (in press)

Fulton, R. W. (1986) Practices and precautions in the use of cross protection for plant virus disease control. *Annual Review of Phytopathology*, **24**, 67–81

Gasser, C. S. and Fraley, R. T. (1989) Genetically engineering plants for crop improvement. *Science*, **244**, 1293–1299

Halk, E. L., Merlo, D. J., Liao, L. W., Jarvis, N. P., Nelson, S. E., Krahn, K. J. *et al.* (1989) Resistance to alfalfa mosaic virus in transgenic tobacco and alfalfa. In *Molecular Biology of Plant-Pathogen Interactions*, (eds B. J. Staskawicz, P. Ahlquist and O. Yoder), Alan R. Liss, Inc., New York, pp. 283–296

Hemenway, C., Fang, R., Kaniewski, W. K., Chua, N. and Tumer, N. E. (1988) Analysis of the mechanism of protection in transgenic plants expressing the potato virus X coat protein or its antisense RNA. *EMBO Journal*, **7**, 1273–1280

Lawson, C., Kaniewski, W., Haley, L., Rozman, R., Newell, C. and Tumer, N. (1989) Expression analysis and field performance of russet burbank potato genetically engineered for resistance to potato virus X and potato virus Y. *Horticultural Biotechnology Symposium Abstract*, p. 42

Loesch-Fries, L. S., Merlo, D., Zinnen, T., Burhop, L., Hill, K., Krahn, K. *et al.* (1987) Expression of alfalfa mosaic virus RNA 4 in transgenic plants confers virus resistance. *EMBO Journal*, **6**, 1845–1851

McKinney, H. H. (1929) Mosaic diseases in the Canary Islands, West Africa, and Gibraltar. *Journal of Agricultural Research*, **39**, 557–578

Namba, K., Pattanayek, R. and Stubbs, G. (1989) Visualization of protein-nucleic acid interactions in a virus: refined structure of intact tobacco mosaic virus at 2.9 Å resolution by X-ray fiber diffraction. *Journal of Molecular Biology*, **208**, 307–325

Nejidat, A. and Beachy, R. N. (1989) Decreased levels of TMV coat protein in transgenic tobacco plants at elevated temperatures reduces resistance to TMV infection. *Virology*, (in press)

Nelson, R. S., McCormick, S. M., Delannay, X., Dubé, P., Layton, J., Anderson, E. J. *et al.* (1988) Virus tolerance, plant growth, and field performance of transgenic tomato plants expressing coat protein from tobacco mosaic virus. *Bio/Technology*, **6**, 403–409

Nelson, R. S., Powell-Abel, P. and Beachy, R. N. (1987) Lesions and virus accumulation in inoculated transgenic tobacco plants expressing the coat protein gene of tobacco mosaic virus. *Virology*, **158**, 126–132

Niblett, C. L., Dickson, E. Fernow, K. H., Horst, R. K. and Zaitlin, M. (1978). Cross protection among four viroids. *Virology*, **91**, 198–203

Ponz, F. and Bruening, G. (1986) Mechanism of resistance to plant viruses. *Annual Review of Phytopathology*, **24**, 355–381

Powell-Abel, P., Nelson, R. S., De, B., Hoffman, N., Rogers, S. G., Fraley, R. T. *et al.* (1986) Delay of disease development in transgenic plants that express the tobacco mosaic virus coat protein gene. *Science*, **232**, 738–743

Powell, P., Sanders, P. R., Tumer, N., Fraley, R. T. and Beachy, R. N. (1989a) Protection against tobacco mosaic virus infection in transgenic plants requires accumulation of coat protein rather than coat protein RNA sequences. *Virology*, (in press)

Powell, P., Stark, D. M., Sanders, P. R. and Beachy, R. N. (1989b) Protection against tobacco mosaic virus in transgenic plants that express TMV antisense RNA. *Proceedings of the National Academy of Sciences, USA* **86**, 6949–6952

Quemada, H., Slightom, J. L. and Gonsalves, D. (1989) The ability of CMV-C coat protein to protect against other strains of CMV in transgenic plants. *Journal of Cellular Biochemistry*, suppl. 13D, 341

Register, J. C., III and Beachy, R. N. (1988) Resistance to TMV in transgenic plants results from interference with an early event in infection. *Virology*, **166**, 524–532

Register, J. C., III and Beachy, R. N. (1989) A transient protoplast assay for capsid protein-mediated protection: effect of capsid protein aggregation state of protection against tobacco mosaic virus. *Virology*, (in press)

Register, J. C., III, Powell, P., Nelson, R. S. and Beachy, R. N. (1989) Genetically engineered cross protection against TMV interferes with initial infection and long distance spread of the virus. In *Molecular Biology of Plant-Pathogen Interactions*, (eds B. J. Staskawicz, P. Ahlquist and O. Yoder), Alan R. Liss, Inc., New York, pp. 269–281

Sanford, J. C. and Johnston, S. A. (1985) The concept of parasite-derived resistance: deriving resistance genes from the parasite's own genome. *Journal of Theoretical Biology*, **113**, 395–405

Sarkar, S. and Smitamana, P. (1981) A proteinless mutant of tobacco mosaic virus: evidence against the role of a viral coat protein for interference. *Molecular and General Genetics*, **184**, 158–159

Sequeira, L. (1984) Cross protection and induced resistance: their potential for plant disease control. *Trends in Biotechnology*, **2**, 25–29

Shaw, J. G., Plaskitt, K. A. and Wilson, T. M. A. (1986) Evidence that tobacco mosaic virus particles dissassemble cotranslationally *in vivo*. *Virology*, **148**, 326–336

Sherwood, J. L. (1987) Demonstration of the specific involvement of coat protein in tobacco mosaic virus (TMV) cross protection using a TMV coat protein mutant. *Journal of Phytopathology*, **118**, 358–362

Sherwood, J. L. and Fulton, R. W. (1982) The specific involvement of coat protein in tobacco mosaic virus cross protection. *Virology*, **119**, 150–158

Stark, D. M. and Beachy, R. N. (1989) Protection against potyvirus infection in transgenic plants: evidence for broad spectrum resistance. *Bio/Technology*, (in press)

Stark, D. M., Register, J. C., III, Nejidat, A. and Beachy, R. N. (1990) Toward a better understanding of coat protein mediated protection. In *Plant Gene Transfer*, (eds C. J. Lamb and R. N. Beachy), Alan R. Liss, Inc., New York, (in press)

Tumer, N. E., O'Connell, K. M., Nelson, R. S., Sanders, P. R., Beachy, R. N., Fraley, R. T. *et al.* (1987) Expression of alfalfa mosaic virus coat protein gene confers cross-protection in transgenic tobacco and tomato plants. *EMBO Journal*, **6**, 1181–1188

Urban, L. A., Sherwood, J. L., Rezende, J. A. M. and Melcher, U. (1990) Examination of mechanisms of cross protection with non-transgenic plants. In

Recognition and Response in Plant–Virus Interactions, (ed. R. S. S. Fraser), Springer-Verlag, Heidelberg, (in press)

Valle, R. P. C., Lambert, M., Joshi, R. L., Morch, M. D. and Haenni, A. L. (1989) Strategies for interfering *in vivo* with the replication of turnip yellow mosaic virus. *Journal of Cellular Biochemistry*, suppl. 13D, 345

van Dun, C. M. P., Bol, J. F. and van Vloten-Doting, L. (1987) Expression of alfalfa mosaic virus and tobacco rattle virus coat protein genes in transgenic tobacco plants. *Virology,* **159**, 299–305

van Dun, C. M. P., Overduin, B., van Vloten-Doting, L. and Bol, J. F. (1988) Transgenic tobacco expressing tobacco streak virus or mutated alfalfa mosaic virus coat protein does not cross-protect against alfalfa mosaic virus infection. *Virology,* **164**, 383–389

van Loon, L. C. (1985) Pathogenesis-related proteins. *Plant Molecular Biology*, **4**, 111–116

Wilson, T. M. A. (1989) Plant viruses: a tool-box for genetic engineering and crop protection. *BioEssays,* **10**, 179–186

Wisniewski, L. A., Powell, P., Nelson, R. S. and Beachy, R. N. (1989) Local and systemic movement of tobacco mosaic virus (TMV) in tobacco plants that express the TMV coat protein gene. *Plant Cell*, (submitted)

Yamada, J., Yoshioka, M., Meshi, T., Okada, Y. and Ohno, T. (1988) Cross protection in transgenic tobacco plants expressing a mild strain of tobacco mosaic virus. *Molecular and General Genetics*, **215**, 173–175

Zaitlin, M. (1976) Viral cross protection: more understanding is needed. *Phytopathology*, **66**, 382–383

Zinnen, T. M. and Fulton, R. W. (1986) Cross-protection between sunn-hemp mosaic and tobacco mosaic viruses. *Journal of General Virology*, **67**, 1679–1687

3

THE MOLECULAR BIOLOGY OF SATELLITE RNA FROM CUCUMBER MOSAIC VIRUS*

DAVID BAULCOMBE, MARTINE DEVIC
The Sainsbury Laboratory, Norwich

and MARTINE JAEGLE
Department of Molecular Biology, Institute of Plant Science Research, Trumpington, Cambridge, UK

Introduction

Many viruses are associated with small, extra-genomic RNA molecules which may be considered as molecular parasites. These RNA species, known as satellite RNA (sat RNA), are parasites in the sense that they are not required by the virus but are themselves dependent on the virus for all functions necessary for propagation through the infected plant and transmission from plant to plant (Francki, 1985). A previous review suggested that an important feature of sat RNA is a lack of extensive homology with the helper virus (Murant and Mayo, 1982). However, since sat RNAs are recognized by replicase and other functions of the helper virus, it is likely that, even if there is no homology in the primary structure, features of the secondary or tertiary structure are shared with the helper virus.

Like other parasites, sat RNA may modify its host. In the instance of sat RNA associated with cucumber mosaic virus (CMV), this modification is observed as an effect on the symptoms produced by the infected plant. Depending on the strain of sat RNA, the genotype of the host plant and the strain of helper virus, the modification is either an amelioration of the viral symptoms or the induction of severe symptoms. The severe symptoms have sometimes been described as exacerbated viral symptoms. However, as they are quite distinct in nature from the viral symptoms it is more useful to consider the sat RNA-induced symptoms as different from the symptoms produced in a plant that is infected with virus alone. This point is illustrated by the effects of CMV infection on tobacco or tomato when the sat RNA of CMV strain Y (Takanami, 1981) is added to the inoculum. In the absence of sat RNA the virus produces mild mosaic symptoms on both plants and a fern leaf effect on tomato. The addition of sat RNA Y to the inoculum results in a bright yellow mosaic symptom on tobacco and a systemic necrosis on tomato. Sat RNA is referred to as virulent if it induces severe symptoms or benign if it ameliorates symptoms.

It has been demonstrated previously in transgenic plants transformed with a DNA copy of sat RNA from CMV coupled to a suitable promoter, that a polyadenylated form of the sat RNA was produced (Baulcombe *et al.*, 1986). This transcript had very little biological activity in its transcribed form as the sat RNA

---

*This contribution has been published in the Proceedings of the NATO Advanced Research Workshop on *Recognition Response in Plant–Virus Interactions*, ed. R. S. S. Fraser, Springer Verlag, Heidelberg

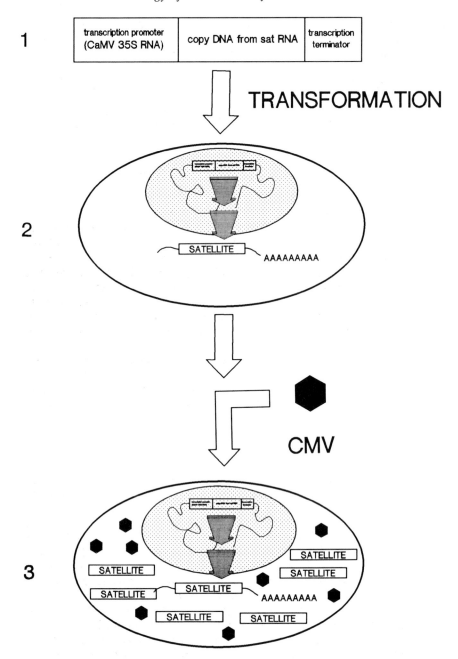

Figure 3.1   Plant transformation to express biologically active satellite RNA. The figure illustrates steps described in detail by Baulcombe *et al.* (1986). In step 1 a DNA copy of CMV sat RNA was coupled to a promoter and transcriptional terminator. Step 2 was the transfer, by *Agrobacterium*-mediated transformation, of the DNA into the nuclear genome of tobacco plants. The biological activity of the transcripts produced from the integrated DNA was demonstrated when the transformed plants were infected with CMV (step 3). The sat RNA replicated to a high level and was also encapsidated in virus particles

sequence was surrounded by non-viral sequence at the 5′ and 3′ ends (Figure 3.1) (unpublished data). However, when the transgenic plants were infected with CMV, the transcribed sat RNA was replicated and accumulated to high levels. This replicated sat RNA had the full biological activity of the natural sat RNA, including the ability to attenuate symptoms of CMV and also a related virus, tomato aspermy virus (TAV) (Harrison, Mayo and Baulcombe, 1987). Gerlach, Llewellyn and Haseloff (1986) have shown similar resistance against tobacco ringspot (nepo) virus resulting from expression of its sat RNA in transgenic plants. Therefore, this use of satellite RNA is a second example, together with the use of viral coat protein genes (Powell Abel *et al.*, 1986), in which expression of viral RNA in transgenic plants produces resistance to the effects of viral infection. The properties of the sat RNA effect are complementary to those of coat protein-mediated resistance, so that these two types of viral sequence may be a particularly effective combination when expressed in crop plants. For example, the sat RNA effect differs from coat protein-mediated resistance in that no new protein is produced and in its lack of sensitivity to both the level of sat RNA expression in the plant and the concentration of inoculum (Harrison, Mayo and Baulcombe, 1987). However, as the virulent species of sat RNA have a nucleotide sequence which is very similar to the benign sat RNAs (Palukaitis, 1988), it is not yet possible to use this genetically engineered resistance in the field. It is necessary to identify functional domains within the sat RNA so that a modified form can be expressed in plants without the risk of mutation or transmission from the protected plants on to nearby non-transformed plants.

A modified sat RNA will be useful also as a component of a viral 'vaccine' that can be applied to non-transformed plants. Tien Po *et al.* (1987) have shown already that a natural isolate of benign sat RNA of CMV is effective as the active component of a 'vaccine' used to protect field grown pepper plants against the effects of CMV.

Symptom amelioration

A model for the mechanism of satellite-mediated suppression of viral symptom production proposes a competition between the sat RNA and the helper virus for limiting amounts of replicase enzyme. Evidence for this model is based on analysis of the kinetics of helper virus RNA synthesis in cultured cells inoculated in the presence or the absence of sat RNA (Piazolla, Tousignant and Kaper, 1982). It was apparent that viral RNA synthesis was inhibited in the presence of sat RNA. There is also evidence from the analysis of steady state levels of both virus and viral RNA showing that sat RNA inhibits viral accumulation (Harrison, Mayo and Baulcombe, 1987). However, it is unlikely that this is the simple explanation of the amelioration effect. There are several examples, including TAV, which can be attenuated by sat RNA of CMV, in which symptom severity is not directly related to the amount of virus. Transgenic plants expressing sat RNA were capable of effective inhibition of TAV symptoms despite there being no reduction of the amount of viral RNA compared with control plants which were not producing sat RNA (Harrison, Mayo and Baulcombe, 1987). Similarly there was no effect on production of capsid protein (Harrison, Mayo and Baulcombe, 1987) (Figure 3.2). It is likely, therefore, that the satellite RNA has the capability to interfere either with the process of symptom induction by the virus, or of symptom production by the plant.

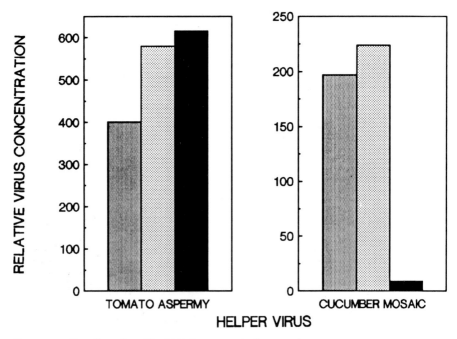

Figure 3.2   The effect of satellite RNA in transgenic plants on virus accumulation. Transgenic plants either expressing satellite RNA (satellite plant ■) or not expressing satellite RNA (control plants A ▨ and B ▨) were inoculated with satellite free helper virus. Virus concentration in systemically infected leaves was measured by inoculation of extracts on to a local lesion host (*Chenopodium quinoa*)

The sat RNA of tobacco ringspot virus also has the capability to attenuate the effects of viral infection without necessarily inhibiting the helper virus replication (Ponz *et al.*, 1987). In this instance the effect is even more striking than with the sat RNA of CMV and TAV, as the attenuated virus (cherry leafroll nepovirus) cannot replicate the sat RNA of tobacco ringspot virus.

Benign sat RNA of CMV can also protect against the effects of a virulent sat RNA, under certain conditions. This was demonstrated by inoculating plants first with a viral inoculum containing a benign sat RNA and, after this infection became established, with a viral inoculum supplemented with a virulent sat RNA. An experiment designed to test whether the transformed plants expressing satellite RNA could resist the virulent satellite in the same way is described in Figure 3.3. The virulent sat RNA was from the Y strain of CMV (Takanami, 1981) and its effects could be diagnosed by the production of yellow mosaic symptoms on the systemically infected leaves. The two strains of helper virus used in the experiment were either the Y strain of CMV, which is the natural helper of the Y sat RNA, or CMV KIN, which was not associated with a sat RNA when it was isolated (Harrison, Mayo and Baulcombe, 1987). The results summarized in Figure 3.3 showed a different result depending on the type of helper strain. When the CMV KIN was used as helper there was inhibition of Y sat, both at the level of symptoms and sat RNA accumulation. This suggests that the transcribed sat RNA is at an advantage over the sat RNA in the challenge inoculum, possibly because it is present in cells before the arrival of the virus. However, this advantage was

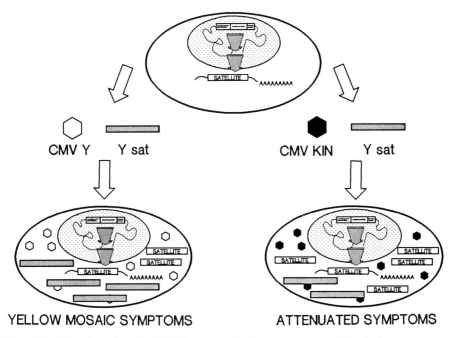

Figure 3.3 The effect of satellite RNA in transgenic plants on infections with a virulent sat RNA. The diagram illustrates schematically the accumulation of sat RNA and symptom production when the virulent Y sat RNA was introduced with different helper virus strains

overridden when the helper virus was the Y strain of CMV (Figure 3.3). A likely explanation for this second result proposes that an affinity between the CMV Y strain and the Y sat RNA gives the Y sat RNA an advantage in the competition with the sat RNA produced in transgenic plants. At the moment there is no obvious suggestion for the nature of this affinity. Perhaps an antisense interaction, which has been described by Rezaian and Symons (1986) (Figure 3.4), is stronger in the homologous combination of helper and sat RNA so that the Y sat RNA is encapsidated and transported from cell to cell more efficiently than with CMV KIN. Alternatively, the viral replicase may operate more efficiently in the homologous combination.

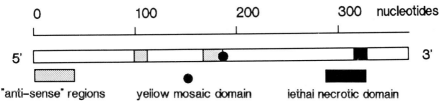

Figure 3.4 The domain structure of CMV Y sat RNA. The diagram illustrates the location of 'antisense', yellow mosaic inducing and lethal necrotic domains in the nucleotide sequence of CMV Y sat RNA. The antisense domain was defined by Rezaian and Symons (1986) and the symptom domains by Devic, Jaegle and Baulcombe (1989) and M. Devic, M. Jaegle and D. Baulcombe (unpublished data)

Symptom induction

It is clear that the formation of symptoms in response to the presence of sat RNA in an inoculum of CMV involves an interaction, either direct or indirect, between the sat RNA and components of the host plant. As different species of sat RNA induce different responses in the host plant it is likely that this interaction involves variable parts of the sat RNA molecule. These domains have been located in the sat RNA by creating hybrid molecules and correlating the phenotype with the presence of domains from a virulent sat RNA. The virulent sat RNA used in this study was the Y sat RNA described by Takanami (1981) as producing a lethal necrotic disease on tomato and a bright yellow mosaic on tobacco.

A series of hybrid molecules was produced and from the phenotype of these it was concluded that the domain responsible for induction of the yellow mosaic disease is located centrally in the molecule. The lethal necrotic domain was separate and towards the 3' end (Devic, Jaegle and Baulcombe, 1989).

It was not too surprising that the yellow mosaic domain was located within the central part of the molecule. The Y sat RNA has the unique property among known sat RNA species of inducing the yellow mosaic disease and has a similarly unique nucleotide sequence in the central region of the molecule. A better resolution of the yellow mosaic domain was obtained by a mutational analysis which identified a site where changing two nucleotides destroyed the yellow mosaic-inducing property of the molecule.

The localization of the domain responsible for the induction of the lethal necrotic disease was less obvious, as there was no particular area in which a sequence motif was unique to the virulent form of the sat RNA. However, there was a region in which all of the necrosis-inducing forms of sat RNA were identical, and where there was sequence variation in the benign forms. However, there was not a simple correlation between the presence of this sequence and the symptom-inducing phenotype, as each benign form varied from the virulent form at different nucleotide positions. The complexity of the situation was confirmed when various defined mutations were created in that region of the molecule. The simplest interpretation of the data is that the necrosis domain extends over several nucleotides and variation at several positions in that region has the potential to destroy the necrosis-inducing capability of the sat RNA.

Conclusions

The domain structure of sat RNA is summarized in Figure 3.4. This information will be used in the development of sat RNA for use in crop protection and for more fundamental investigation of the mechanism of symptom induction by sat RNA.

In the applied context, the new information suggests regions of the molecule which could be modified to produce a sat RNA that is not capable of inducing symptoms. Ideally, the modified sat RNA would differ from the virulent forms at several positions, so that back mutation would not readily occur. In the course of this work several mutant forms of the satellite have been created, illustrating that modified forms of the sat RNA are replicable, can move systemically in a plant and be transmitted by aphids from plant to plant. One mutation (M. Jaegle, unpublished data) involved a large deletion within the central part of the molecule. Surprisingly, a smaller deletion from within the region of the larger deletion did not accumulate when inoculated on to plants together with helper virus. It appears therefore, that any modifications must leave secondary or higher order structure features of the sat RNA molecule intact. Unfortunately, understanding of higher order structures involving the sat RNA molecule is not very sophisticated. Current models of sat RNA structure do not explain why the smaller deletion described above does not accumulate on plants and the possibility remains that interactions of sat RNA with other viral or host plant factors determine the structure of sat RNA *in vivo*.

In principle the sat RNA could induce symptoms via the action of small sat RNA-encoded peptides or directly as an RNA molecule. Currently it appears that the latter is the more likely alternative, as various mutations have been created which interrupt the small open-reading frames and which have no effect on the ability of the sat RNA to induce symptoms. How RNA molecules might induce symptoms is not immediately obvious. A feasible model proposes that sat RNA interferes in RNA-mediated processes of the healthy cell. It is now clear that there are several potential targets in this way, as RNA is known as a component in ribonucleases and protein transport complexes as well as in the protein synthetic apparatus.

In future work we shall investigate these RNA-mediated processes and look for proteins and other components that bind to sat RNA of CMV. The outcome will be a more profound understanding of the molecular biology of sat RNA so that it can be used safely in crop protection. It is also anticipated that the detailed knowledge of the pathogenic process will provide information about the healthy cell and suggest new methods of engineering disease resistance in plants.

References

Baulcombe, D. C., Saunders, G. R., Bevan, M. W., Mayo, M. A. and Harrison, B. D. (1986). Expression of biologically active viral satellite RNA from the nuclear genome of transformed plants. *Nature,* **321**, 446–449

Devic, M., Jaegle, M. and Baulcombe, D. C. (1989) Symptom production on tobacco and tomato is determined by two distinct domains of the satellite RNA of cucumber mosaic virus (strain Y). *Journal of General Virology*, **70**, 2765–2774

Francki, R. I. B. (1985) Plant virus satellites. *Annual Review of Microbiology*, **39**, 151–174

Gerlach, W. L., Llewellyn, D. and Haseloff, J. (1987) Construction of a plant disease resistance gene from the satellite RNA of tobacco ringspot virus. *Nature,* **328**, 802–805

Harrison, B. D., Mayo, M. A. and Baulcombe, D. C. (1987) Virus resistance in transgenic plants that express cucumber mosaic virus satellite RNA. *Nature,* **328**, 799–802

Murant, A. F. and Mayo, M. A. (1982) Satellites of plant viruses. *Annual Review of Phytopathology*, **20**, 49–70

Palukaitis, P. (1988) Pathogenicity regulation by satellite RNAs of cucumber mosaic virus: minor nucleotide sequence changes alter host responses. *Molecular Plant–Microbe Interactions,* **1**, 175–181

Piazzolla, P., Tousignant, M. E. and Kaper, J. M. (1982) Cucumber mosaic virus-associated RNA 5 – the overtaking of viral RNA synthesis by CARNA 5 and dsCARNA 5 in tobacco. *Virology,* **122**, 147–157

Ponz, F., Rowhani, A., Mircetich, S. M. and Bruening, G. (1987) Cherry leafroll virus infections are affected by a satellite RNA that the virus does not support. *Virology,* **160,** 183–190

Powell Abel, P., Nelson, R. S., De, B., Hoffmann, N., Rogers, S. G., Fraley, R. T. and Beachy, R. N. (1986) Delay of disease development in transgenic plants that express the tobacco mosaic virus coat protein gene. *Science,* **232,** 738–743

Rezaian, M. A. and Symons, R. H. (1986) Anti-sense regions in satellite RNA of cucumber mosaic virus form stable complexes with the viral coat protein gene. *Nucleic Acids Research,* **14,** 3229–3239

Takanami, Y. (1981) A striking change in symptoms on cucumber mosaic virus-infected tobacco plants induced by a satellite RNA. *Virology,* **109,** 120–126

Tien, P., Zhang, X., Qiu, B., Qin, B. and Wu, G. (1987) Satellite RNA for the control of plant diseases caused by cucumber mosaic virus. *Annals of Applied Biology,* **111,** 143–152

4

MOLECULAR ASPECTS OF CAULIFLOWER MOSAIC VIRUS PATHOGENESIS

SIMON N. COVEY, REBECCA STRATFORD, KEITH SAUNDERS,
DAVID S. TURNER and ANDREW P. LUCY
Department of Virus Research, Institute of Plant Science Research, John Innes Institute, Norwich, UK

Introduction

Recent advances in genetic manipulation have now reached the stage where agronomically-useful traits are being incorporated into recipient transgenic plants. Plant viruses represent a group of pathogenic genetic entities that cause significant crop losses worldwide and which are potential targets for the plant genetic engineer. Indeed, some successes in protecting plants from virus diseases have been reported and the future prospects for manipulating non-conventional plant resistance to viruses look promising. At present, the strategy adopted has relied on mimicking cross-protection against viruses by constructing transgenic plants expressing virus components. For example, protection has been conferred against virus infection in transgenic tobacco plants expressing the coat protein gene of tobacco mosaic virus (TMV) (Powell Abel *et al.*, 1986; Beachy *et al.*, Chapter 2), or satellite RNAs which interfere with virus multiplication (Harrison, Mayo and Baulcombe, 1987; Baulcombe, Devic and Jaegle, Chapter 3). A further potential novel strategy of virus control would be to exploit those systems where systemic virus infection is restricted as a result of a hypersensitive response mediated by specific host genes although knowledge of the nature and function of such genes is not yet sufficient for exploitation by the genetic engineer. An alternative strategy for the development of novel virus control measures, is to understand the molecular processes in successful or permissive virus infections rather than in those where establishment of a systemic virus infection is prevented by cross-protection or a hypersensitive response.

Since both host and virus genes are involved in development of a productive virus infection, understanding how their products interact at the molecular level to produce symptomatic effects at the whole plant level might lead to the identification of specific targets for manipulation. Symptom development is a complex series of phenomena arising both from specific host/virus interactions and from secondary effects of virus multiplication upon plant metabolism. It is the former of these that is likely to be most accessible and offer potential for manipulation.

Our approach to investigating virus infection with a view to characterizing host and virus genes involved in determining viral pathogenesis has been to exploit cauliflower mosaic virus (CaMV) infection of brassica plants. CaMV is one of the best characterized plant viruses at the molecular level and it has a number of properties that make it particularly suitable for molecular studies of viral pathogenesis.

Cauliflower mosaic virus

CaMV is the type member of the caulimoviruses, a group of plant viruses that are characterized by possessing a circular double-stranded DNA genome of about 8000 base pairs (for reviews *see* Covey, 1985; Gronenborn, 1987). An unusual feature of the DNA isolated from caulimovirus virions is the presence of site-specific discontinuities which are short regions of strand displacement produced during replication. The host range of CaMV is generally restricted to the family Cruciferae in nature, although experimental infection of some solanaceous plants by certain strains of the virus has been achieved. Infection of plants by CaMV can be readily achieved by mechanical inoculation, although in nature the virus is transmitted by aphids. Some aphid non-transmissible mutants of CaMV have been isolated or generated by *in vitro* mutagenesis of wild-type strains. A large number of CaMV variants, defined by restriction enzyme site polymorphisms and some differing in symptom characteristics, have been described (*see* Covey, 1985).

Several of these strains or isolates of CaMV have been sequenced and all show a very similar organization of open-reading frames. Of the eight open-reading frames capable of encoding polypeptides of 10 K molecular weight or greater, six have been shown to be expressed into protein *in vivo* and are described as viral genes (Figure 4.1). The product of gene I bears some amino acid similarity to the 30 K product of TMV which is involved in mediating virus movement within plants. Moreover, the CaMV gene I product has also been shown to accumulate close to plasmodesmata (Linstead *et al.*, 1988) suggesting that it might serve a similar role in cell-to-cell spread. Gene II specifies an 18 K polypeptide required for aphid acquisition from, and transmission to, host plants. The gene III product is found in association with virions and is probably a structural protein as is the polypeptide derived from gene IV, the major coat protein of CaMV. The viral polymerase, a reverse transcriptase with an associated protease activity, is encoded by gene V. Gene VI produces the virus inclusion body protein which is a major constituent of the amorphous structures which accumulate within CaMV-infected cells. The role of the inclusion bodies has not been fully characterized although they seem to be associated with virus replication and assembly; they might also be sites of virion sequestration. The genetic organization of the CaMV genome shares significant similarities with that of animal retroviruses with genes III/IV, V, and VI equivalent to retrovirus *gag, pol,* and *env* genes.

The discontinuous CaMV DNA molecule encapsidated in virions is not the template for virus gene expression. This important process, which plays a central role in regulating the virus multiplication cycle, occurs from a supercoiled (SC) form of the virus genome in the nucleus of infected plants (Olszewski, Hagen and Guilfoyle, 1982). The CaMV SC DNA is a major component of the virus minichromosome together with host histones and probably other host proteins. Expression of the CaMV minichromosome is effected by host RNA polymerase II and mediated by two virus transcription promoters that have features typical of eukaryotic promoters in general. The 35S RNA promoter in particular, but also the 19S RNA promoter, isolated from the CaMV genome, can be used to control

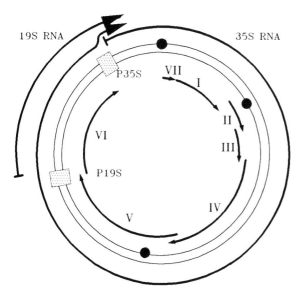

Figure 4.1 Functional map of the CaMV genome. The inner arrows represent viral open-reading frames of which six have been shown to encode polypeptides with the following demonstrated or putative functions: gene I, putative transport protein; gene II, aphid acquisition polypeptide; gene III, virion structural protein; gene IV virion coat protein; gene V, reverse transcriptase and protease; gene VI, inclusion body protein. The filled circles on the double-stranded virion DNA are the sites of discontinuities and the half-tone boxes represent the 35S and 19S transcription promoters which regulate expression of the two major CaMV transcripts (outer arrows)

expression of heterologous genes genetically manipulated into non-CaMV host transgenic plants (*see* Benfey and Chua, 1989). These promoters have been widely used to drive high levels of expression in a manner that is generally considered to be constitutive and so occurs in most tissues. During CaMV infection, the 19S RNA promoter controls transcription of the mRNA which is translated to produce the gene VI inclusion body protein. The 35S RNA promoter controls transcription of a slightly greater than genome length molecule that appears to serve two special roles in the CaMV multiplication cycle: the first of these is as a polycistronic messenger RNA, while the second is as the template for viral replication by reverse transcription (*see* Gronenborn, 1987).

The CaMV multiplication cycle

Examination of the CaMV multiplication cycle might suggest where specific points of interaction between virus and host might occur and towards which attention can be drawn to identify possible targets for genetic manipulation. The current view of the CaMV multiplication cycle in a single cell is summarized diagrammatically in Figure 4.2, although there are additional key phases at the whole plant level. At the initial inoculation site, virus particles (or viral DNA if mechanically-inoculated) enter a cell and uncoating occurs, presumably at a very early stage, as a prelude to replication. Nothing is yet known about virion uncoating of CaMV but this might specifically involve the host in an active process since CaMV virions are structurally extremely stable. After an initial phase of virus replication, movement of progeny virus to cells around the inoculation site can give rise to local lesions; this type of short distance movement presumably involves cell-to-cell spread via the plasmodesmata. At some stage in the initially-inoculated leaf, virus particles or DNA must enter the vascular tissue in which rapid systemic movement occurs. Following long distance movement of virus in the vascular tissue, young leaves emerging from the stem apex become systemically infected. Dissemination of virus

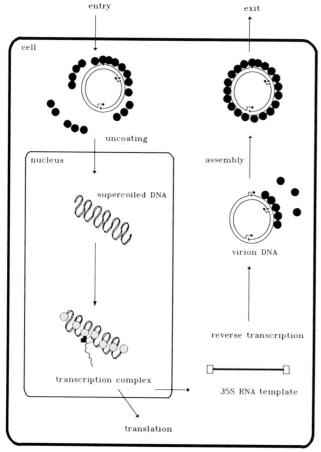

Figure 4.2   CaMV multiplication cycle in a single cell. Virus particles enter the cell during systemic infections probably through the plasmodesmata and the virion DNA is liberated as particles become uncoated. The DNA moves to the nucleus where it is converted to a supercoiled form and associates with host histones and probably other proteins to result in an episomal transcription complex. Viral polyadenylated RNA transcripts synthesized in the nucleus move to the cytoplasm for translation into viral polypeptides, and the 35S RNA also serves as a template for replication by reverse transcription. This process is thought to occur in cytoplasmic replication complexes associated with the inclusion bodies. Replication and assembly of progeny virions are also believed to be linked. Virions leave the cell either through the plasmodesmata or following acquisition by feeding aphids

in young developing leaves could occur either by cell-to-cell spread or by division of cells containing virus received from the vascular tissue.

After uncoating in a single cell (Figure 4.2), the viral DNA moves to the nucleus where the discontinuities are repaired and SC DNA is generated presumably by host nuclease and topoisomerase enzymes. This SC DNA then associates with histones to produce an episomal minichromosome which is transcribed into viral RNA under the direction of the host transcription machinery. Viral polyadenylated transcripts are transported to the cytoplasm where messenger RNAs are translated into viral proteins. The 35S RNA also moves to the cytoplasm to become a template for replication by reverse transcription. This process, which is believed to occur in replication complexes adjacent to virus inclusion bodies (*see* Hull, Covey and Maule, 1987), is primed by a host methionyl initiator tRNA and replication is effected by the virus-encoded reverse transcriptase. Nascent virion DNA probably associates with coat protein subunits during replication; completed progeny virions are usually sequestered within the inclusion bodies.

This complex multiplication cycle has several key processes at which specific molecular interactions with the host are likely to occur (Table 4.1). It is these molecular processes that eventually result in symptomatic effects expressed at the whole plant level. To understand this relationship, we have adopted two molecular

Table 4.1 POTENTIAL KEY SITES OF INTERACTION BETWEEN CaMV AND HOST PLANTS

Site	Process
Plasmodesmata	Movement between adjacent cells
Cytoplasm	Virus uncoating
Vascular tissue	Long distance movment of virus
Cytoplasm/nucleus	Organelle targeting of viral DNA
Nucleus	Generation of supercoiled DNA by host topoisomerase
Nucleus	Generation of viral minichromosome
Nucleus	Transcription of viral minichromosome
Nucleus/cytoplasm	Processing and transport of viral transcripts
Cytoplasm	Translation and reverse transcription of viral RNA
Cytoplasm	Virion packaging and sequestration in inclusion bodies

genetic approaches which have exploited genetic variation in the virus and in the host to identify possible points of contact. Genetic variability in the CaMV genome in respect of the symptom characters defined by infection of one particularly susceptible host plant species (*Brassica campestris* – turnip) was used to locate specific symptomatic determinants on the viral genome (Stratford and Covey, 1988; Stratford *et al.*, 1988). The second approach exploited genetic variability in the host to identify differences in response to CaMV infection and to attempt to associate this with specific molecular events in the CaMV multiplication cycle.

CaMV symptom genetic determinants

Our strategy for characterizing the symptom genetic determinants of CaMV was to genetically-engineer hybrid virus genomes from parental strains expressing contrasting symptom characters in turnip plants, and to follow segregation of these

Figure 4.3   Effects of CaMV infection on turnip leaves. Leaf A is from a healthy plant. B and C are leaves from plants infected with the severe CaMV strain Cabb B-JI showing early vein clearing and later leaf chlorosis, respectively. Leaf D was taken from a plant in the later stages of infection with the mild strain Bari 1

characters in chimaeras (Stratford and Covey, 1988). The two strains chosen for this study were Cabb B-JI, a typical severe CaMV strain, and Bari 1, a mild strain which has been described as the least typical CaMV strain in terms of restriction enzyme site similarities. In Cabb B-JI infections, systemic vein clearing symptoms appear rapidly (10–14 days) after inoculation and these symptoms spread to younger leaves over the subsequent 1–2 days. Leaves eventually develop severe chlorosis, accompanied by a reduction in leaf chlorophyll (Figure 4.3), and plants become severely stunted. In contrast, plants infected by the mild strain Bari 1 develop systemic vein clearing symptoms more slowly (20–25 days) and the symptoms take longer (5–6 days) to spread to younger leaves than in Cabb B-JI infections. Leaves of Bari 1-infected plants do not become chlorotic but develop dark green islands of tissue (Figure 4.3) which can be associated with a slight increase in chlorophyll content compared with non-infected plants. Plants infected with the Bari 1 CaMV strain exhibit only mild stunting (Table 4.2).

As is the case with TMV and some other plant viruses, the mild CaMV strain Bari 1 can confer cross-protection on turnip plants against superinfection by the severe Cabb B-JI strain. However, the difference in timing of symptom development between the two strains can be exploited to study how they interact in mixed infections. When the two strains were co-inoculated on to turnip plants, systemic symptom development in leaves was typical of Cabb B-JI infections suggesting dominance of this strain over the mild strain. Presumably, this was because Cabb B-JI infections develop more rapidly than those of Bari 1. When plants were first inoculated with Bari 1 and then with Cabb B-JI after a delay of 1–2 weeks, unusual symptoms were observed with characteristics of both strains in that some regions of the leaves had dark green islands of tissue typical of Bari 1 infections while others showed chlorosis typical of Cabb B-JI. This suggests that different groups of cells in the leaves contained either one strain or the other but not both. Moreover, from this phenomenon it can be inferred that leaf coloration

Table 4.2 SYMPTOM CHARACTERS PRODUCED BY WILD-TYPE AND HYBRID CaMV GENOMES

Virus genome	Vein clearing (days post inoculation)	Systemic spread (days)	Leaf colour	Plant stunting	Leaf chlorophyll†
(Healthy plants	–	–	green	–	1.0)
Cabb B-JI	10–14	1–2	yellow	severe	0.41
Bari 1	20–25	5–6	dark green islands	mild	1.12
pBaji 1	10–14	1–2	yellow	severe	0.39
pBaji 6	20–25	5–6	dark green islands	mild	1.07
pBaji 2	10–14	1–2	pale green	severe	0.58*
pBaji 3	10–14	1–2	pale green	severe	0.61*
pBaji 21	10–14	5–6	yellow	hypersevere	0.43
pBaji 20	10–14	1–2	yellow	hypersevere	0.42
pHD 5	10–14	5–6	yellow	severe	0.46
pCa K1	10–14	1–2	yellow	severe	ND
pBaji 10	30–40 20–25**	5–6	dark green islands	mild	1.09
pHD 1	20–25 10–14**	1–2	yellow	mild severe	0.45
pHD 3	20–25	5–6	dark green islands	mild	1.08

† Values expressed as a fraction of healthy leaves (1.47 ± 0.18 mg chlorophyll/g fresh wt).
* Significantly different from Cabb B-JI at 95% confidence level;
** subsequent passaging; ND, not determined.

symptomatic effects originate at the cellular level and not by physiological changes at the whole plant level. With prior inoculation by Bari 1 and a delay 4 weeks before subsequent inoculation with Cabb B-JI, symptoms produced were typical of Bari 1 infections demonstrating induction of cross-protection.

To map specific symptom genetic determinants which defined differences in the infection characteristics between the severe and mild CaMV strains, hybrid genomes were cosntructed either by recombinational rescue *in vivo* or by restriction fragment exchange *in vitro*. The former approach was necessary because of the relatively few common restriction enzyme sites between the two strains which would allow comprehensive exchanges by the latter method. Recombination *in vivo* can be induced by co-inoculation of plants with complementary fragments of the CaMV genome that individually are not infectious. Selection for recombination between the complementary fragments is on the basis of restoration of infectivity. A total of 10 hybrids between Bari 1 and Cabb B-JI were constructed to study the symptom genetic determinants and one Cabb B-JI mutant (Figure 4.4). The symptom characters produced upon infection of turnip plants by the hybrids are summarized in Table 4.2. Symptom characters resulting from infections by the hybrids fell into one of three categories. These were first, hybrids producing characters indistinguishable from either of the parental CaMV strains; second, those in which one or more symptom character segregated with a particular genome domain; third, chimaeras which produced novel symptom characters unlike those of either parental strain.

In hybrids pBaji 1 and pBaji 6, which were nearly reciprocal constructs and where exchanged fragments were located in the gene V/VI region, parental wild-type symptom characters, including leaf chlorosis, were observed (Table 4.2).

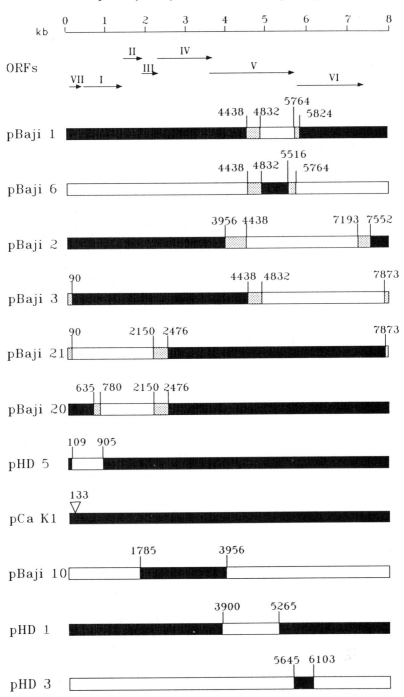

Figure 4.4 Hybrid genomes genetically-engineered between severe (Cabb B-JI) and mild (Bari 1) strains of CaMV. Contribution of the parental strains Cabb B-JI (▇) and Bari 1 (▭) are shown and in those hybrids generated by recombinational rescue *in vivo* the regions of uncertainty where crossover occurred is shown in halftone (▭)

This indicated that the exchanged fragments in each case did not specifically contribute to a particular symptom character. However, in hybrids pBaji 2 and pBaji 3 where large segments were exchanged encompassing gene VI and the two intergenic regions, infections exhibiting novel leaf coloration effects, which were intermediate between the two parental symptoms, were observed. This showed specific involvement of a particular genome domain in determining the leaf coloration symptom character. In other hybrids, where various exchanges had been achieved between nucleotides (nt) 90 and 6103, no effect on leaf coloration was observed. Comparison of all the hybrids together suggested that the determinant affecting leaf symptoms was located between nt 6103 and 90 containing part of gene VI and the large intergenic region including the 35S RNA promoter.

The location of genetic determinants influencing the timing of initial symptom appearance could be deduced from infections by CaMV hybrids pBaji 2, pBaji 3 and pBaji 21. This character mapped to between nt 2150 and 4438 since this region was common to each chimaera, plus a 217 bp fragment from nt 7873 to 90. Plant stunting determinants must be located between nt 7873 and 4438 since hybrids pBaji 2 and 3 produced the wild-type Cabb B-JI character (*see* Table 4.2). Involvement of sequences in determining stunting can be discounted between nt 109 and 905, and nt 1785 and 3956 because hybrids pHD 5 and pBaji 10 produced stunting similar to Cabb B-JI and Bari 1, respectively. This suggests potential involvement of sequences from nt 7873 to 90, nt 905 to 1785, and nt 3956 to 4438. In infections with hybrids pBaji 20 and pBaji 21 which contain Bari 1 DNA covering the domain between nt 905 and 1785, a novel stunting effect, which was more severe than that induced by either parental strain, was observed indicating a specific role, at least, for this region.

The symptoms produced in infections with pBaji 20 and pBaji 21 suggested that the character defining rate of systemic symptom spread is located between nt 7873 and 780. To test this, we attempted to convert the Cabb B-JI rate of symptom spread character to that of Bari 1 following exchange of a 797 bp *Spe*I-*Acc*I fragment between nt 109 and 905 to generate hybrid pHD 5 (*see* Figure 4.4). Infection of turnip plants with pHD 5 produced the Bari 1 slow rate of systemic symptom spread but all the other symptom characteristics were typical of wild-type Cabb B-JI infections. In fact, the genetic determinant influencing rate of symptom spread must be located between nt 109 and 780 since pBaji 20 produced the wild-type Cabb B-JI rate of spread. To determine possible involvement of gene VII sequences (between nt 109 and 302), plants were infected with a clone (pCa K1 kindly donated by Dr C. J. Woolston) derived from the Cabb B-JI genome but which contained a premature stop codon in gene VII at nt 133. However, the rate of symptom spread in plants infected with pCa K1 was the same as that of wild-type Cabb B-JI suggesting that expression of gene VII was not required to determine this character.

A map summarizing the location of symptom genetic determinants that differ between the two CaMV strains is shown in Figure 4.5. These regions do not define all potential sequences that might contribute to a specific symptom character, but those that are responsible for defining differences between strains Cabb B-JI and Bari 1. However, the ability to delimit these virus genome regions might suggest how specific virus gene products might function at the molecular level to determine symptomatic effects at the whole plant level. For instance, the difference in rate of spread of systemic vein clearing symptoms between the two strains mapped to part of gene I which has been implicated in mediating cell-to-cell spread of CaMV.

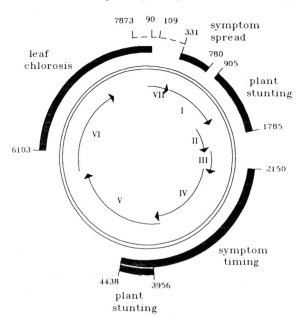

Figure 4.5   Symptom genetic determinants of CaMV. The regions contributing to specific symptom characters that differ in infections between CaMV strains Cabb B-JI and Bari 1 are shown. The characters were mapped according to the segregation of symptom characters in infections of turnip plants by the CaMV hybrids. The broken lines represent regions of uncertainty which might contribute to symptom characters but this has not yet been determined

Comparison of the nucleotide sequences of the two strains in this part of gene I has shown only eight amino acid differences of which six are conservative (Stratford and Covey, unpublished data). This suggests a highly specific function of the gene I product in its interaction with the host. One speculative interpretation of this conclusion is that the CaMV gene I polypeptide interacts with a host receptor, close to or within plasmodesmata, to mediate virus movement between cells. Clearly, this represents a likely target for genetic manipulation to control virus spread within a plant by blocking or changing the specificity of the putative plant receptor for virus factors.

Plant stunting mapped to multiple locations, one of which also included part of gene I although this was the C-terminal rather than N-terminal portion responsible for influencing virus movement as described above. Plant stunting might also be consistent with an effect due to the activity of a transport protein in controlling virus distribution within the plant. Perhaps different domains of this polypeptide influence either short distance movement from cell-to-cell or long distance movement in vascular tissue. A second domain affecting stunting was mapped to part of gene V, the reverse transcriptase gene, which was also involved in initial timing of symptom appearance. This suggests that rate of virus replication is an important factor in determining certain symptom characters, as might be expected.

Our experiments mapped the genome domain controlling chlorosis (leaf coloration and chlorophyll content) to part of gene VI and the 35S RNA promoter region. Since hybrid crossover points in the latter domain produced novel leaf coloration effects, we must conclude that there is a specific role of the intergenic region in this character. However, our mapping was not sufficient to confirm specific involvement of gene VI sequences *per se*. It has been previously suggested that CaMV gene VI determines key aspects of symptom character (*see* Daubert, 1989), although no clear distinction was drawn between gene VI and the intergenic region. Moreover, the CaMV gene VI protein has been shown to induce a symptomatic effect when expressed in transgenic tobacco plants (Baughman, Jacobs and Howell, 1988; Takahashi, Shimomoto and Ehara, 1989) although its toxicity and relationship to leaf chlorosis in host plants remains to be established. Even so, it might represent a further point of interaction between host and virus polypeptides.

Host regulation of CaMV

A second approach to investigating host/virus interactions was undertaken to determine to what extent different tissues and host plant species were capable of supporting specific phases of the CaMV multiplication cycle. This approach was suggested from observations of the activity of CaMV in cultured callus cells derived from infected turnip leaf tissue (Rollo and Covey, 1985) in which the virus replication cycle appeared to be interrupted at a specific phase. We have adapted and developed a high resolution two-dimensional gel electrophoresis technique coupled with Southern blotting to monitor the CaMV multiplication cycle by analysing the virus nucleic acid replication intermediates (Covey and Turner, 1986; Turner and Covey, 1988). This technique, together with use of Northern blotting to determine the levels of steady-state CaMV RNA species, provides a 'snapshot' of key viral processes and is illustrated in Figure 4.6.

Cellular nucleic acids, including CaMV replication cycle intermediates, are first isolated by phenol extraction. This method does not disrupt CaMV virions which require protease treatment to liberate the virion DNA. Nucleic acid is then fractionated in an agarose flat-bed slab gel first under non-denaturing conditions, and then, at 90° orientation, in a denaturing alkali medium. A variety of molecules with different topological conformations are resolved. In CaMV-infected turnip leaves actively producing virus, the virus-specific DNA forms characteristically contain a range of nascent DNA molecules generated by reverse transcription including single-stranded DNAs, molecules partially single-, and partially double-stranded, and a complex series of hairpin forms. The SC DNA component of the viral minichromosome is present in relatively small amounts but is highly active in generating RNA (*see* Figure 4.7).

In contrast, the DNA forms isolated from roots of infected turnip plants and resolved by 2-D gel electrophoresis show a very different pattern (Figure 4.6). There are very few of the intermediates indicative of active reverse transcription but a much-enhanced level of SC DNA. The absence of replication intermediates suggests that roots are not capable of supporting this phase of the multiplication cycle but can still generate the SC DNA transcription template. To determine more precisely the stage at which the multiplication had been interrupted, we analysed CaMV RNA transcripts by Northern blotting (Figure 4.7). In leaves, two major viral RNA species are typically observed during active replication: the 35S RNA reverse transcription template and the 19S mRNA for gene VI. However, these RNAs were virtually undetectable in roots even though considerably more of the

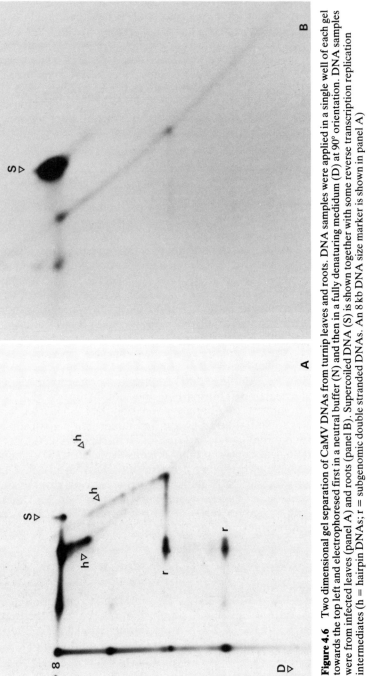

Figure 4.6   Two dimensional gel separation of CaMV DNAs from turnip leaves and roots. DNA samples were applied in a single well of each gel towards the top left and electrophoresed first in a neutral buffer (N) and then in a fully denaturing medidum (D) at 90° orientation. DNA samples were from infected leaves (panel A) and roots (panel B). Supercoiled DNA (S) is shown together with some reverse transcription replication intermediates (h = hairpin DNAs; r = subgenomic double stranded DNAs. An 8 kb DNA size marker is shown in panel A)

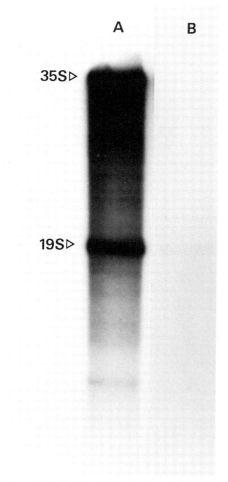

Figure 4.7 Northern hybridization of CaMV RNAs isolated from CaMV infected leaves (A) and roots (B)

transcription template DNA was present than in leaves. This indicates that expression of the CaMV minichromosome is regulated in a organ-specific manner in host plants.

To determine whether similar regulation occurred in other cell types from infected turnip plants, we also analysed the DNA forms from stem and callus tissue (generated by culture of infected turnip leaf discs for one month *in vitro*) by 2-D gel electrophoresis. Both tissues were found to contain less of the reverse transcription intermediates and elevated levels of SC DNA compared with leaves. The amount of viral RNA in these tissues was also approximately inversely proportional to the level of SC DNA such that roots contained very much less RNA but more SC DNA than leaves (Table 4.3). Thus, intact plants regulate the expression of the CaMV minichromosome in a tissue- or organ-specific manner. Moreover, the reduction in the level of viral transcripts, presumably a reflection of minichromosome

Table 4.3 RESPONSES OF *BRASSICA* HOSTS AND REGULATION OF THE CaMV MINICHROMOSOME

Tissue species	Symptoms	Minichromosome level	RNA level	Virus titre
Brassica campstris (aa)	Severe			
leaves		+	+++++	+++++
stems		+++	++	++
roots		+++++	<+	<+
callus		+++++	<+	<+
Brassica nigra (bb)	Severe	++	++++	++++
Brassica oleracea (cc)	Very mild	+++++	+	+
Brassica juncea (aabb)	Severe	++	++++	++++
Brassica napus (aacc)	Mild	+++	++	++
Brassica carinata (bbcc)	Mild	+++	++	++

transcription, is also associated with accumulation of SC DNA. This might be because the minichromosome is rapidly turned over when active but not so when it is not actively transcribed.

We also undertook a survey of a wide range of brassica hosts of CaMV to determine whether differing susceptibilities to infection were correlated with an ability to support specific phases of the virus multiplication cycle. Initially, we chose three species groups: *Brassica campestris* (genome descriptor aa) which includes turnip, the most frequently used experimental CaMV host, *Brassica oleracea* (genome descriptor cc) including all of the cole crops, and *Brassica napus* (genome descriptor aacc) including rape which is an allotetraploid species originally synthesized from *Brassica campestris* and *Brassica oleracea*. Of the three species, *Brassica campestris* plants are the most susceptible to the severe CaMV strain *Cabb B-JI*, *Brassica oleracea* the least susceptible in that they exhibit mild or no symptoms at the whole plant level; *Brassica napus* plants are intermediate in their response between the other two species.

On analysis by 2-D gel electrophoresis of the DNA forms present in leaves of the various brassica species, we found that cauliflower plants (*Brassica oleracea*), with low susceptibility to CaMV, exhibited a considerable accumulation of SC DNA (Figure 4.8) but contained very little viral RNA (*see* Table 4.3). This situation was very similar to that observed in less susceptible tissues of the *Brassica campestris* plants. *Brassica napus* exhibited an intermediate level of SC DNA and transcripts. From a broader survey of *Brassica* species (*see* Table 4.3) including *Brassica nigra* (genome descriptor bb) and the allotetraploid species *Brassica juncea* (genome descriptor aabb) and *Brassica carinata* (genome descriptor bbcc), we have found a close correlation between host susceptibility and the amount (inverse) and level of expression of the CaMV minichromosome. Thus, host genetic variability in expression of whole plant symptomatic effects can be linked with regulation of a specific stage of the virus multiplication cycle.

The tissue- and host-specific regulation of the CaMV minichromosome has more general implications for plant gene expression studies and functioning of the 19S and 35S RNA promoters in genetic manipulation constructs. These promoters are generally considered to be constitutively expressed when removed from the CaMV genome and expressed in non-host plants. However, there are some reports that the 35S RNA promoter linked to a GUS reporter gene is not expressed in all tissues

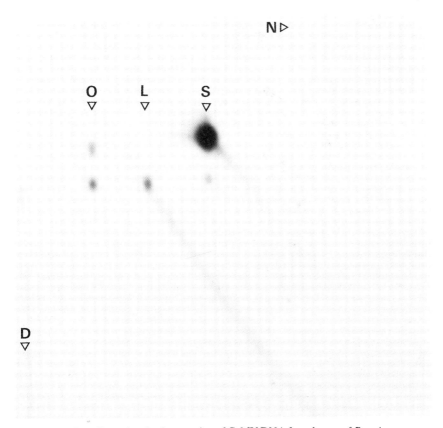

Figure 4.8 Two dimensional gel separation of CaMV DNA from leaves of *Brassica oleracea* (cauliflower). *See* Figure 4.6 for explanation. The predominant CaMV molecular form is the supersoiled DNA (S). There are also lesser amounts of genome-length linear (L) and open circular (O) DNA forms but no detectable reverse transcription replication intermediates

(Jefferson, Kavanagh and Bevan, 1987). Recently, it has been proposed (Benfey, Ren and Chua, 1989; *see* Benfey and Chua, 1989) that the enhancer domain of the 35S RNA promoter is composed of two distinct elements: one which effects root-specific expression, and a second controlling shoot-specific expression. The concerted functioning of both elements provides the constitutive expression frequently reported for this promoter. However, our findings suggest that regulation of these promoters, in the context of the whole CaMV genome, is more complex and that leaf-specific expression is dominant even when the 35S promoter is intact. From this we speculate that a further organ-specific enhancer-like element is located in some distal part of the CaMV genome and one which is not normally incorporated into genetic manipulation constructs containing the 35S RNA promoter.

Moreover, our observation of infections of host species that exhibit lower susceptibility to CaMV than *Brassica campestris* suggest that this distal 'host enhancer' is subject to regulation in leaves. One possible conclusion of our findings is that an organ-specific host regulatory *trans*-acting factor is involved in CaMV minichromosome expression. The fact that genetic variability of symptomatic response in different host plant species is reflected in differential activity of this putative regulatory protein might provide a genetic route to its isolation and characterization.

Conclusions

We have exploited genetic variability in CaMV and in host *Brassica* plants to understand how virus and host interactions at the molecular level produce specific symptomatic effects at the whole plant level. In the longer term, potential points of interaction, such as those suggested between the putative CaMV transport protein product of gene I and its plant receptor, might become targets for genetic manipulation of a wide range of virus diseases. Additionally, studies of such interactions should also lead to a greater understanding of processes like intracellular communication and the control of gene expression which have broader implications for crop modification by molecular means.

Acknowledgements

We are greateful to our colleagues for discussions, and RS gratefully acknowledges receipt of a Gatsby Foundation Research Studentship.

References

Baughman, G., Jacobs, J. D. and Howell, S. H. (1988) Cauliflower mosaic virus gene VI produces a symptomatic phenotype in transgenic tobacco plants. *Proceedings of the National Academy of Sciences USA*, **85**, 733–737

Benfey, P. N. and Chua, N-H. (1989) Regulated genes in transgenic plants. *Science*, **244**, 174–181

Benfey, P. N., Ren, L. and Chua, N-H. (1989) The CaMV 35S enhancer contains at least two domains which can confer different developmental and tissue-specific expression patterns. *EMBO Journal*, (in press)

Covey, S. N. (1985) Organization and expression of cauliflower mosaic virus. In *Molecular Plant Virology*, vol. 2 (ed. J. W. Davies), CRC Press, Boca Raton, Florida, pp. 121–159

Covey, S. N. and Turner, D. S. (1986) Hairpin DNAs of cauliflower mosaic virus generated by reverse transcription *in vivo*. *EMBO Journal*, **5**, 2763–2768

Daubert, S. D. (1989) Sequence determinants of symptoms in the genomes of plant viruses, viroids and satellites. *Molecular Plant–Microbe Interactions*, **1**, 317–325

Gronenborn, B. (1987) The molecular biology of cauliflower mosaic virus and its application as plant gene vector. In *Plant DNA Infectious Agents* (eds T. Hohn and J. Schell), Springer-Verlag, Vienna, pp. 1–29

Harrison, B. D., Mayo, M. A. and Baulcombe, D. C. (1987) Virus resistance in transgenic plants that express cucumber mosaic virus satellite RNA. *Nature*, **328**, 799–802

Hull, R., Covey, S. N. and Maule, A. J. (1987) Structure and replication of caulimovirus genomes. *Journal of Cell Science*, suppl. 7, 213–229

Jefferson, R. A., Kavanagh, T. A. and Bevan, M. W. (1987) GUS fusions: β-glucuronidase as a sensitive and versatile gene fusion marker in higher plants. *EMBO Journal,* **6**, 3901–3907

Linstead, P. J., Hills, G. J., Plaskitt, K. A., Wilson, I. G., Harker, C. L. and Maule, A. J. (1988) The subcellular location of the gene I product of cauliflower mosaic virus is consistent with a function associated with virus spread. *Journal of General Virology,* **69**, 1809–1818

Olszewski, N., Hagen, G. and Guilfoyle, T. J. (1982) A transcriptionally-active, covalently-closed minichromosome of cauliflower mosaic virus DNA isolated from infected turnip leaves. *Cell,* **29**, 395–402

Powell Abel, P., Nelson, R. S., De, B., Hoffmann, N., Rogers, S. G., Fraley, R. T. *et al.* (1986) Delay of disease development in transgenic plants that express the tobacco mosaic virus coat protein gene. *Science,* **232**, 738–743

Rollo, F. and Covey, S. N. (1985) Cauliflower mosaic virus DNA persists as supercoiled forms in cultured turnip cells. *Journal of General Virology,* **66**, 603–608

Stratford, R. and Covey, S. N. (1988) Changes in turnip leaf messenger RNA populations during systemic infection by severe and mild strains of cauliflower mosaic virus. *Molecular Plant–Microbe Interactions,* **1**, 243–249

Stratford, R., Plaskitt, K. A., Turner, D. S., Markham, P. G. and Covey, S. N. (1988) Molecular properties of Bari 1, a mild strain of cauliflower mosaic virus. *Journal of General Virology,* **69**, 2375–2386

Takahashi, H., Shimomoto, K. and Ehara, Y. (1989) Cauliflower mosaic virus gene VI causes growth suppression, development of necrotic spots and expression of defence-related genes in transgenic tobacco plants. *Molecular General Genetics,* **216**, 188–194

Turner, D. S. and Covey, S. N. (1988) Discontinuous hairpin DNAs synthesised *in vivo* following specific and non-specific priming of cauliflower mosaic virus DNA (+) strands. *Virus Research,* **9**, 49–62

5

GENETIC ENGINEERING OF CROPS FOR INSECT RESISTANCE USING GENES OF PLANT ORIGIN

VAUGHAN A. HILDER, ANGHARAD M. R. GATEHOUSE and DONALD BOULTER

Department of Biological Sciences, University of Durham, Durham, UK

Insects and crop losses

Crop plants are susceptible to attack by a wide range of herbivorous insects and, despite intervention of the farmers, it is estimated that around 13% of the total, world-wide crop production is lost to insect pests each year. In Third World countries there is often little alternative to accepting devastating losses in yield of food and cash crops to insect pests (Figure 5.1). Economic cultivation of many of the major crops in the 'developed' countries is dependent on massive inputs of chemical insecticides. It has been estimated that some 74 million acres were sprayed with insecticides to control moths alone (1984 figures). The desirability of introducing such quantities of some of these chemicals into the ecosphere is being increasingly questioned on environmental grounds. Some indication of the financial cost of so doing is given by the estimated approximately 4 billion dollars spent on chemical defence against insect pests of the three major crops grown worldwide (Table 5.1). These three crops account for about one-half of the world expenditure on insecticides (Barfoot and Connett, 1989).

Figure 5.1   Young maize plants attacked by *Spodoptera* larvae, Tanzania, 1981. (Photograph courtesy of Dr J. S. Kershaw)

Table 5.1 ESTIMATED WORLD-WIDE INSECTICIDE EXPENDITURE FOR PROTECTION OF COTTON, MAIZE AND RICE. VALUES ARE $ × 10^{-6}, 1987 EXPENDITURE.

	Cotton	*Rice*	*Maize*
Insecticide expenditure	1540	1040	490
Application expenditure	386	247	100
Total	1926	1287	590
Major pest	Lepidoptera	Homoptera	Coleoptera

Advantages of genetically engineered insect resistance

The potential of genetic engineering to address this problem, by incorporating insect resistance into the crop plants themselves, has been widely recognized (Meeusen, 1986; Gatehouse and Hilder, 1988; Barfoot and Connett, 1989). Some of the perceived advantages of this approach include:

1. Protection would be provided continuously and, therefore, *when* required to obtain maximal control of pests, without the need to predict accurately when this occurred.
2. Protection would be provided *where* required, including those plant tissues, such as roots, undersides of leaves and insides of pods, which are difficult to treat with chemical insecticides.
3. Insecticidal activity would be *restricted* to those insects which actually attacked the plant. Beneficial and harmless insects should not be directly affected.
4. The protectant would be *confined* within the plant, obviating, or at least minimizing, problems of environmental pollution. Protection could not 'wash off' as a result of rainfall and would not be subject to the additional costs and problems associated with application.
5. Compared with producing a novel chemical insecticide, this approach involves lower development costs.

These advantages are, of course, shared with the introduction of genetic resistance by classical plant breeding. Genetic engineering, however, affords new opportunities to:

1. Transfer a single character in one step into a favourable genetic background without the co-transfer of possibly undesirable linked characteristics.
2. Most importantly, it allows transfer of insect resistance genes across the barriers to conventional plant breeding; allowing transfer between species, genera and even kingdoms.

Selection of genes for transfer

This being so, an issue of prime importance is how to select those genes which are suitable for transfer. The strategy that has been adopted at Durham is grounded in the fact that land plants and insects have co-existed and co-evolved for around 250–400 million years. During this time the plants have evolved many highly effective defence mechanisms against herbivorous insects, while some of the latter have, in turn, evolved mechanisms for overcoming such resistance. In most natural ecosystems these trends appear to be more or less balanced. However, in agricultural systems, particularly those relying on extensive monoculture, crops may be exposed to insects to which they have not evolved an inherent resistance, thereby allowing a catastrophic build up of the insect population. It is in such instances that the insects become 'pests'.

Many of these plant defence mechanisms involve physical barriers to being eaten – spines, hairs, tough surfaces, etc. Another set of defences are based on the enormous 'armoury' of secondary compounds which plants, with their extraordinarily diverse secondary metabolic capabilities, are able to produce (Norris and Kogan, 1980). Such defences are almost certainly the result of the interaction of many genes in complex, well regulated pathways. To transfer such systems by genetic engineering is beyond the capabilities of current technology, though perhaps not of the future (Dawson *et al.*, 1989). However, in some cases resistance is provided by the presence of a specific primary gene product, a protein, in the protected tissue. Examples of such proteins are digestive-enzyme inhibitors and lectins (Gatehouse *et al.*, 1979; 1984; Janzen *et al.*, 1986). Such single-gene products are ideal for genetic engineering. The potential of adopting such an approach is very well illustrated by the cowpea trypsin inhibitor (CpTI) gene story.

Insecticidal properties of CpTI

The cowpea (*Vigna unguiculata* L. (Walp.)) is an important grain legume, providing a major source of dietary protein in West Africa and parts of South America (FAO, 1970), and is widely cultivated and consumed elsewhere. The crop is subject to huge losses in storage due to infestation by the bruchid beetle, *Callosobruchus maculatus* F., which can cause up to 100% damage within 5 months' storage (Singh, 1978). Consequently, a major screening and plant breeding programme was initiated at the International Institute of Tropical Agriculture (IITA) in Nigeria to identify, and introduce into cultivated cowpeas, a source of resistance to this pest.

Out of several thousand cowpea accessions screened at IITA only one, designated TVu2027, showed significant resistance to the larvae of *C. maculatus* (Redden, Dobie and Gatehouse, 1983). This variety provided a potentially useful source of insect resistance genes.

Having established that resistance in TVu2027 did not have a physical basis, seeds of this accession and various susceptible varieties were screened for a range of secondary compounds, including alkaloids, lectins, saponins, non-protein amino acids and protease inhibitors (Gatehouse *et al.*, 1979). Of the toxic and antimetabolic compounds investigated, the only activities detected were inhibitors of the serine proteases, trypsin and chymotrypsin. Bruchid resistance in TVu2027 was associated with elevated levels of trypsin inhibitor (Table 5.2); some two-to-fourfold higher than the susceptible lines. Polyacrylamide gel electrophoresis and isoelectric focusing of the trypsin inhibitor fraction from these varieties suggested that the differences were purely quantitative, rather than qualitative (Gatehouse *et al.*, 1979).

The antimetabolic properties of CpTI were confirmed in bioassays with *C. maculatus* on artificial seeds incorporating various fractions extracted from cowpea

Table 5.2 DEVELOPMENT OF *C. MACULATUS* TO ADULTHOOD IN RELATION TO TRYPSIN INHIBITOR (TI) CONTENT OF SEEDS FROM DIFFERENT COWPEA LINES

Accession	Ti Content (% w/w)	Adult emergence (%)
TVu2027	0.92	0
TVu4557	0.44	95.1 ± 6.2
TVu76	0.34	90.0 ± 2.5
TVu3629	0.30	90.6 ± 9.5
TVu37	0.26	86.6 ± 4.3
TVu57	0.25	91.7 ± 6.7
TVu1109E	0.23	89.0 ± 6.7
TVu1502-1D	0.19	92.0 ± 6.0

meal (Gatehouse *et al.*, 1979). Only those fractions containing active CpTI at levels approaching those which naturally occurred in the seeds of TVu2027 were effectively insecticidal to the bruchid (Table 5.3). Thus, the elevated levels of CpTI play a major role in the resistance of TVu2027 to bruchids. The trypsin inhibitors extracted from cowpeas were more effective antimetabolites than those extracted from a number of other legumes (Gatehouse and Boulter, 1983).

Table 5.3 DEVELOPMENT OF *C. MACULATUS* ON ARTIFICAL SEEDS SUPPLEMENTED WITH VARIOUS COWPEA SEED MEAL FRACTIONS

'Seed' composition	Adult emergence (% control)
Control	100
+10% albumin	42
+10% albumin − CpTI*	96
+10% globulin	97
+0.1% CpTI	102
+0.5% CpTI	26
+0.8% CpTI	0

* CpTI removed from the albumin fraction by affinity chromatography on trypsin conjugated sepharose.

Of prime interest in terms of a wider use of CpTI was the observation that it was antimetabolic to a wide range of insects in feeding trials on artificial diets. The range of susceptible insects covered members of the order Lepidoptera as well as Coleopterans like *Callosobruchus*. Among the insects which are susceptible are some pests of major economic importance, against which there is a high expenditure on insecticides (Table 5.4). These include the Lepidopterans *Heliothis* (tobacco budworms and corn earworms) and *Spodoptera* (armyworms) and the

Table 5.4 INSECT PESTS AGAINST WHICH CpTI IS EFFECTIVE IN ARTIFICIAL DIETS

Insect	Common name	Crops attacked*
Lepidoptera		
Heliothis virescens	Tobacco budworm	Tobacco, cotton
H. zea	Corn earworm	Maize, cotton, beans, tobacco
H. armigera	Bollworm	Cotton, beans, maize, sorghum
Spodoptera littoralis	Armyworm	Maize, rice, cotton, tobacco
Chilo partellus	Stalk-borer	Maize, sorghum, sugarcane, rice
Coleoptera		
Anthonomus grandis	Boll weevil	Cotton
Diabrotica undecimpunctata	Corn rootworm	Maize
Callosobruchus maculatus	Cowpea seed weevil	Cowpea, soyabean
Tribolium confusum	Flour beetle	Most flours
Costelytra zealandica†	Grass grub	Grasses, clover

* *see* Hill (1983).
† Determined by Dr O.R.W. Sutherland, DSIR.

Coleopterans *Diabrotica* (corn rootworms) and *Anthonomus* (cotton boll weevil). This broad range of effectiveness is typical of plant-based protection mechanisms.

It is assumed that the primary mode of action of CpTI is to inhibit essential digestive proteases resulting in abnormal development and death due to a deficiency of essential amino acids. There is reason to believe that it may also affect other insect systems involving proteases. Because its site of action is the actual catalytic site of an enzyme, and because it probably affects more than one enzyme, the ability of the insects to evolve resistance to CpTI by a single, or even a few, mutational events should be minimal. Resistance might be achieved by detoxifying CpTI before it could inhibit its target enzymes, but this would probably require a major alteration to the biochemistry and physiology of the insect gut.

Although CpTI inhibits mammalian trypsin, it is not toxic to mammals. Food is consumed by organisms, not by enzymes, and CpTI is degraded by pepsin in the very acid conditions of the stomach before it encounters the serine proteases of the small intestine in the mammalian gut.

Characterization of CpTI

Biochemical studies showed that the trypsin inhibitors in cowpea seeds belonged to the Bowman–Birk type protease inhibitor group (Gatehouse, Gatehouse and Boulter, 1980). The Bowman–Birk type protease inhibitors are small polypeptides of around 80 amino acids and an exceptionally high degree of disulphide cross-linking (*see* Richardson, 1977; Ryan, 1981). They are typically found in legume seeds, but related proteins have been identified in cereals (Odani, Koide and Ono, 1986). They are double-headed inhibitors – i.e. each inhibitor molecule can bind to, and thereby inhibit, two enzyme molecules, which in some cases may be different types of serine protease at the two inhibitory sites (Figure 5.2).

The cowpea inhibitors comprise a small family of four major isoinhibitors. They are encoded by a larger gene family, although it may be that there are only four

Figure 5.2   Model of the interaction between Bowman–Birk type protease inhibitors and their target enzymes. Trypsin is represented by *Tyrannosaurus*, chymotrypsin by *Stegosaurus*. Above is a trypsin/trypsin inhibitor, below a trypsin/chymotrypsin inhibitor. The inhibitor model is based on X-ray crystallographic data of Suzuki *et al.* (1987). The models of the enzymes are not strictly accurate!

active genes. Three of the isoinhibitors are trypsin/trypsin inhibitors, the fourth a trypsin/chymotrypsin inhibitor. The primary sequence of the latter was established by the manual DABITC method (Chang, Brauer and Whittman-Liebold, 1978), confirming their identity as Bowman–Birk type inhibitors (Hilder *et al.*, 1989a).

Transfer of CpTI genes

The determined protein sequence allowed us to synthesize mixed sequence oligonucleotide probes complementary to all the coding possibilities of short regions of the polypeptide. These provided the basis for selecting CpTI encoding clones from a cowpea cotyledon cDNA library. DNA sequencing has established that these include representatives of trypsin/trypsin and trypsin/chymotrypsin inhibitors (Hilder *et al.*, 1989a).

A full length cDNA clone encoding a trypsin/trypsin inhibitor (pUSSRc3/2) was selected for transfer into tobacco plants. A 550 bp *Sca*1 – *Alu*1 restriction fragment was blunt-end ligated into the *Sma*1 site of the binary vector pRok2, kindly provided by Drs T. Kavanagh and M. Bevan. This placed the inserted DNA under the control of the 'strong', constitutive cauliflower mosaic virus (CaMV) 35S gene promoter (Guilley *et al.*, 1982) and the nopaline synthase gene transcription terminator (Bevan, Barnes and Chilton, 1983). The blunt-end ligation allowed selection of constructs in which the CpTI encoding sequence was in the correct orientation with respect to the promoter and terminator (pRokCpTI+5) or reversed, so that the CpTI message could not be produced (pRokCpTI-2). The constructs contained the NOS–NEO gene to provide a selectable marker (kanamycin resistance) for transformed plants (Hilder *et al.*, 1987).

These constructs were mobilized into *Agrobacterium tumefaciens* LBA4404 and used to transform tobacco leaf discs. Transgenic plants were regenerated from these by standard procedures (Fraley *et al.*, 1983; Bevan, 1984; Horsch *et al.*, 1985).

Characterization of transgenic plants

Various molecular biological characteristics of the transgenic tobacco plants containing the CpTI gene constructs have been determined (Hilder *et al.*, 1987; 1989b). Southern blots and dot-blots of transgenic tobacco plant DNA reveal that all of the plants investigated contain multiple, unrearranged copies of the construct in tandem, head-to-tail repeats (Figure 5.3). Segregation analysis of kanamycin resistance in the S_1 seed of the original transformants showed that insertion had occurred at a single locus in most plants, at two independent loci in the others.

Expression of the CaMV promoter – CpTI – NOS terminator gene was measured by dot immunobinding assays of rabbit anti-CpTI antiserum binding to protein extracts from young leaves. A wide range, from below the limit of detection to around 0.9% of total soluble protein was detected in different plants transformed with the +5 construct (no binding was observed with the −2 plants). Western blots of protein extracted from the CpTI expressing plants suggested that correct processing occurs; the product in them is identical to that in cowpea cotyledons (Figure 5.4). Expression of CpTI was not correlated with the number of copies of the gene inserted.

Estimates of the relative levels of expression of the NOS–NEO gene were obtained from the levels of kanamycin required to kill seedlings. Expression of this gene appeared to be similarly variable between different individual transformants, although since selection for kanamycin resistance had taken place there were no non-expressers remaining in the population of plants studied. With this gene there was also no correlation between level of expression and copy-number. Neither was there any correlation between the levels of expression of the two co-introduced genes within individual transformed plants.

There has been some controversy in the literature concerning the relative importance of the chromosomal location of insertion ('position effect') and the number of copies inserted on the level of expression of inserted genes (Jones, Dunsmuir and Bedbrook, 1985; Nagy *et al.*, 1985; Eckes *et al.*, 1986; Stockhaus *et al.*, 1987; Baumlein *et al.*, 1988). Our results indicate that with these promoters the chromosomal environment exerts an overriding influence on expression, but that the response of particular promoters to any given chromosomal environment may

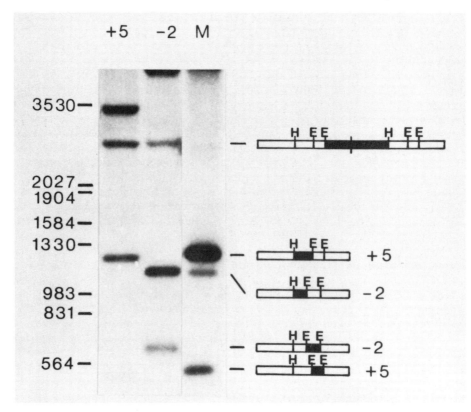

Figure 5.3 Southern blot of HindIII/EcoR1 digested genomic DNA from transgenic tobacco plants transformed with the +5 and −2 constructs, probed with pRokCpTI. The origin of the bands is indicated on the right. M, Hind/Eco digest of a 5:1 mixture of pRokCpTi+5:pRokCpTI-2

be different. This may account for the frequently observed lack of coordinate expression of co-introduced, linked genes (Jones, Dunsmuir and Bedbrook, 1985; An, 1986). It may be that certain promoters, for example the potato ST–LS1 (Stockhaus *et al.*, 1987), are insensitive to position effects, in which case expression levels are related to copy number. The relative importance of chromosomal position and gene copy-number is further clarified in the case of the CaMV–CpTI chimaeric gene by comparison of the levels of CpTI expression in plants which are hemizygous for this insertion and their homozygous S_1 progeny. The latter contains twice the number of copies of the gene as the former, but in an identical chromosomal environment. In this case the homozygotes express approximately twice the quantity of CpTI as the hemizygotes. Thus, when position effects can be discounted, expression appears to be proportional to copy-number.

These position effects must operate over quite considerable distances. From the number of insertion loci and the insert copy number we can calculate that the majority of these constructs are in an identical local sequence environment composed of tandem repeats of the construct itself. Most copies of the genes are at least 10 kb from the nearest host DNA. The long range over which position effects

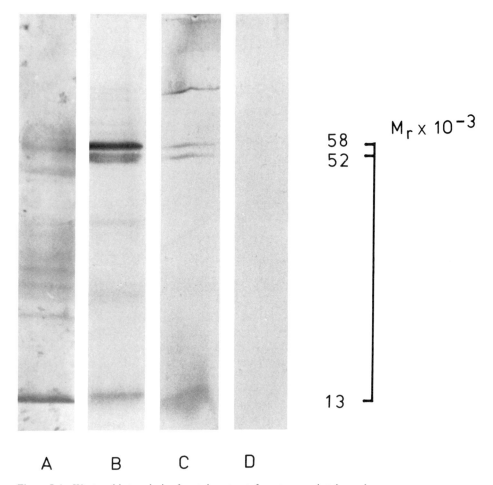

$$M_r \times 10^{-3}$$

58
52

13

A   B   C   D

Figure 5.4   Western blot analysis of protein extracts from transgenic tobacco leaves, probed with polyclonal anti-CpTI antiserum. A, affinity purified CpTI; B, extract of cowpea TVu2027 seed meal; C, +5/5 leaf tissue; D, −2/8 leaf tissue

operate has also been indicated by the failure of long flanking sequences of host DNA to buffer position effects on the expression of a foreign gene (Dean *et al.*, 1988).

Plants which express high levels of CpTI in the leaves also express the protein in roots, stems and floral tissues at levels which should be insecticidally effective.

Insect resistance to transgenic plants

The critical question regarding these plants was whether expression of CpTI in them led to enhanced resistance to insect pests. The population of transformed plants was sealed into individual plantaria along with eight first instar larvae of *Heliothis virescens*, the tobacco budworm (Figure 5.5). *H. virescens* was selected as

Figure 5.5   Insect bioassay on transgenic tobacco plants. Young regenerated plants sealed into plantaria with eight first instar larvae of *H. virescens*

the test organism since it is classified as a serious pest which readily eats tobacco plants.

Insect survival and damage to the plants was assessed after 7 days. Approximately one in five of the +5 transgenic plants appeared to be more resistant to insect attack. Plants which survived the first trials were subjected to further rounds of infestation with *H. virescens*. Plants which still appeared to have enhanced resistance, plus some controls, were clonally replicated as stem cuttings to allow statistically meaningful trials to be performed. It was not until this stage that levels of CpTI expression were actually measured. The plants which we had selected as 'potentially resistant' were those which were expressing CpTI to the highest levels. The somewhat arbitrary cut-off point was about 0.5% of soluble

Figure 5.6   Transgenic plants expressing (+5/5, left) or not expressing (−2/8, right) CpTI 7 days after infestation with *H. virescens* larvae

protein as CpTI. The 'best' plant [+5/5] was the highest CpTI expresser at around 0.9% of total soluble protein (Hilder *et al.*, 1987).

Subsequent insect bioassays were carried out on sets of at least 10 clonal replicates of +5/5 or on seed grown plants from the homozygous line +5/51-A20 which was derived from selfing +5/5. Resistance was measured from the point of view of the insects, as survival and insect biomass; and of the plants, as percentage leaf area destroyed – estimated by image analysis of the leaves. These bioassays have conclusively established that:

1. CpTI expressing plants have significantly enhanced resistance to *Heliothis* (Figure 5.6).
2. Resistance extends to a broad range of pests that are capable of attacking tobacco (Table 5.5).
3. Resistance is stably inherited, so far to the seventh generation.

Those Coleopteran pests which are of major economic interest unfortunately will not feed on tobacco plants. We have, therefore, had to rely on extended studies in artificial diets to determine insecticidal effectiveness against Coleopterans. It has been our experience with Lepidopterans tested on both artificial diets and transgenic plants that the former tend to underestimate the effectiveness of CpTI. This is no doubt because the artificial diets are optimized for insect survival – a condition which very few plants achieve! Nevertheless, the results suggest that the levels of CpTI expression in the appropriate parts of transgenic plants would also be effective against a range of Coleopteran pests.

Table 5.5 SUMMARY OF INSECT BIOASSAY RESULTS ON CpTI-EXPRESSING TRANSGENIC PLANTS. FIGURES ARE PRESENTED AS PERCENTAGE OF CONTROLS, FROM WHICH ALL WERE SIGNIFICANTLY DIFFERENT AT $P<0.05$

Insect	Leaf area eaten	Insect biomass
Heliothis virescens	40.3	27.0
H. zea	32.0	52.8
Spodoptera littoralis	49.3	62.7
Manduca sexta	43.6	17.6
*Autographa gamma**	ND	0

* = *Plusia gamma*; determined by E. Macaulay, Rothamstead Experimental Station.
ND: not determined.

Comparison with B.t.t. expressing transgenic plants

To date, the only other published works on genetically engineering foreign genes into plants to enhance resistance to insects have employed modified endotoxin proteins (B.t.t.) from the parasporal crystals of the bacterium *Bacillus thuringiensis*. Preparations of B.t.t. have been in limited use as insecticidal sprays for many years (Dulmage, 1980). A number of groups have recently reported enhanced insect resistance in transgenic tobacco and tomato plants expressing toxic fragments of B.t.t. (Barton, Whiteley and Yang, 1987; Fischhoff *et al.*, 1987; Vaeck *et al.*, 1987). It is inevitable that comparisons have been made between B.t.t. and CpTI as potential protective agents in transgenic crops.

Tobacco and tomato plants expressing modified B.t.t.s display impressive resistance to B.t.t.-susceptible Lepidopteran pests. However, a feature of B.t.t. is the specificity of activity of any particular strain against a very limited range of insects. Different strains of B.t.t. have been identified which together cover a wide range of Lepidopterans (Wilcox *et al.*, 1986) and a few have been reported with activity against some Coleopterans (Hernstadt *et al.*, 1986). It is possible that there are situations in which specific resistance to a single pest species might be advantageous, but in order to obtain field protection against a broad spectrum of insect pests using B.t.t. alone the introduction of many different bacterial genes would probably be required.

The toxicity of the bacterial protein to susceptible insects is considerably higher than that of the plant protein, CpTI. However, the perceived importance of absolute toxicity with this type of compound is, we believe, erroneously based on a false analogy with synthetic chemical insecticides. The latter tend to be generally toxic and persistent; since they are not naturally present in the environment they are not readily metabolized to harmless products. The types of compound it is proposed to introduce into plants are intrinsically biodegradable, being proteins. The levels at which CpTI would need to be presented match those at which it is already present in the consumed part of a food crop. We are not aware of the wider toxicological significance of expressing B.t.t. at effective levels in crops, but would expect it to have an equally low persistance in the environment. Furthermore, genetically engineered insect resistance involves no application risks or expenditure no matter what 'weight per hectare' is required for effectiveness. Provided that these compounds can be *efficiently* expressed at *effective* levels within the plant, what those actual levels are is largely irrelevant.

That effective levels of expression of B.t.t. and CpTI *can* be achieved in transgenic plants is clearly demonstrated by the respective insect feeding trials. The higher toxicity of B.t.t. means that much lower levels of expression are required for effectiveness than is the case with CpTI. Interestingly, correspondingly lower levels of expression are achievable from the bacterial genes, even using the same 'strong' CaMV promoter. It has been observed in Professor Boulter's laboratory that other genes which are naturally expressed in legume seeds are expressed to similarly high levels as CpTI in transgenic plants. Whether the differences in apparent 'acceptability' to the host plant's synthetic systems of these plant and bacterial genes is a general feature of the origin of the genes, or some special characteristic of B.t.t. or the B.t.t. genes, is yet to be established.

It has been suggested that the expression of these foreign proteins, particularly at the levels required for effectiveness of CpTI, might impose some 'yield penalty' on the host plant. We have attempted to address this question systematically by measuring various phenotypic characteristics and 'yield' parameters on large populations of CpTI-expressing transgenic plants, non-expressing transformed plants and untransformed controls, under a range of controlled growth conditions (Hilder *et al.*, unpublished data). Some small differences were observed between non-transformed and transformed plants, but not between CpTI-expressing and non-expressing transformants. Thus, there may be some minor 'penalty' resulting from the transformation process – which could probably be reduced or eliminated in a subsequent breeding programme – but expression of 1–2% of a foreign plant protein appears to be well within the plants 'spare' synthetic capacity and imposes no additional yield penalty.

Both B.t.t. and CpTI afford promising, novel approaches to the problem of protecting crops from insects. They are not exclusive. It is our belief that the performance of each needs to be assessed where it really matters, in the farmers' fields, before we can draw any conclusions about the relative benefits of the two systems.

Future developments

The next step in the development of CpTI is its transfer to major crop plants. Transformation/regeneration systems either are, or should soon be, available for most of the major crops (Umbeck *et al.,* 1987; Toriyama *et al.,* 1988; Rhodes *et al.,* 1988; Shimamoto *et al.,* 1989) and for many locally important ones. AGC Ltd, who have the proprietary rights over the CpTI gene, have concluded that it is only worthwhile introducing the gene into the very best currently available genetic backgrounds (Barfoot and Connett, 1989). They have, therefore, licensed it to a number of companies who have the technology and premium germplasm for transformation of major crops. More details on the proposed commercialization of CpTI may be found in Barfoot and Connett (1989).

The aim at Durham is to identify and obtain other plant genes involved in field resistance to insects. These should act on quite different targets within the insects' metabolism. We shall then be in a position to 'pyramid' a number of different resistance factors into crops; building up a multi-mechanistic resistance package. The advantages of using such combinations of genes are seen as:

1. Increased insecticidal effect. The insects would be subject to 'attack' on more than one front. The effects of 'pyramiding' different mechanisms might be synergistic rather than simply additive. CpTI might be particularly valuable in such a strategy as, not only is it intrinsically antimetabolic, but it might also protect other introduced proteins from premature proteolysis in the insect gut.
2. Broader range of effect. Insects which tolerate one resistance mechanism might be susceptible to another.
3. Reduction in the likelihood of the insects developing resistance – since this would require concurrrent, multiple favourable mutational events.

In this way we hope to contribute to the development of crop plants with inherent, durable resistance to the range of insect pests which they are likely to meet in an 'integrated pest management' agricultural system. It is important to recognize that future development depends on the goodwill of the primary sources of resistance, frequently research institutes in the Third World; and on fundamental research providing the scientific ground from which these developments spring.

Acknowledgements

This work forms part of the Agricultural Genetics Co. Ltd. insect resistance programme. We are grateful to the following for intellectual, technical or material input: Dr R. A. Barker, Dr M. Bevan, Dr R. A. Connett, Ms G. Davison, Dr J. A. Gatehouse, Mr S. Hughes, Mr H. Minney, Ms S. Sheerman.

References

An, G. (1986) Development of plant promoter expression vectors and their use for analysis of differential activity of nopaline synthase promoter in transformed tobacco cells. *Plant Physiology,* **81**, 86–91

Barfoot, P. D. and Connett, R. A. (1989) AGC's cowpea enzyme inhibitor gene and its potential market opportunity. *Ag Biotech. News and Information,* **1**, 177–182

Barton, K. A., Whiteley, H. R. and Yang, N-S. (1987) *Bacillus thuringiensis* delta endotoxin expressed in *Nicotiana tabacum* provides resistance to lepidopteran insects. *Plant Physiology,* **85**, 1103–1109

Baumlein, H., Muller, A. J., Schiemann, J., Helbing, D., Manteuffel, R. and Wobus, U. (1988) Expression of *Vicia faba* legumin gene in transgenic tobacco plants. Gene dosage dependent protein accumulation. *Biochemie und Physiologie der Pflanzen,* **183**, 205–210

Bevan, M. (1984) Binary *Agrobacterium* vectors for plant transformation. *Nucleic Acids Research,* **12**, 8711–8721

Bevan, M., Barnes, W. M. and Chilton, M. D. (1983) Structure and transcription of the nopaline synthase region of T-DNA. *Nucleic Acids Research,* **11**, 369–385

Chang, J. Y., Brauer, D. and Whittman-Liebold, B. (1978) Microsequence analysis of peptides and proteins using 4NN-dimethylaminoazobenzene 4'-isothiocyanate/phenylisothiocyanate double coupling method. *FEBS Letters,* **93**, 205–214

Dawson, G. W., Hallahan, D. L., Mudd, A., Patel, M. M., Pickett, J. A., Ladhans, L. A. *et al.* (1989) Secondary plant metabolites as targets for genetically modifying crops for pest resistance. *Pesticide Science,* **27**, (in press)

Dean, C., Favreau, M., Tanakis, S., Bond-Nutter, D., Dunsmuir, P. and Bedbrook, J. (1988) Expression of tandem gene fusions in transgenic tobacco plants. *Nucleic Acids Research,* **16**, 7601–7617

Dulmage, H. T. (1981) Insecticidal activity of isolates of *Bacillus thuringiensis* and their potential for pest control. In *Microbial control of pests and plant diseases 1970–1980* (ed. H. D. Burges), Academic Press, New York, pp. 193–222

Eckes, P., Rosahl, S., Schell, J. and Willmitzer, L. (1986) Isolation and characterization of a light-inducible, organ specific gene from potato and analysis of its expression after tagging and transfer into tobacco and potato shoots. *Molecular General Genetics,* **205**, 14–22

FAO (1970) *The State of Food and Agriculture.* FAO, Rome, p. 274

Fischhoff, D. A., Bowdish, K. S., Perlak, F. J., Marone, P. G., McCoormick, S. M., Niedermeyer, J. G. *et al.* (1987) Insect tolerant transgenic tomato plants. *Biotechnology,* **5**, 807–813

Fraley, R. T., Rogers, S. G., Horsch, R. B., Sanders, P., Flick, J., Adams, S. *et al.* (1983) Expression of bacterial genes in plant cells. *Proceedings of the National Academy of Sciences, USA,* **80**, 4803–4807

Gatehouse, A. M. R. and Boulter, D. (1983) Assessment of the antimetabolic effects of trypsin inhibitors from cowpea (*Vigna unguiculata*) and other legumes on development of the bruchid beetle *Callosobruchus maculatus. Journal of the Science of Food and Agriculture,* **34**, 345–350

Gatehouse, A. M. R., Dewey, F. M., Dove, J., Fenton, K. A. and Pusztai, A. (1984) Effect of seed lectin from *Phaseolus vulgaris* on the development of larvae of *Callosobruchus maculatus*: mechanism of toxicity. *Journal of the Science of Food and Agriculture*, **35**, 373–380

Gatehouse, A. M. R., Gatehouse, J. A. and Boulter, D. (1980) Isolation and characterisation of trypsin inhibitors from cowpea (Vigna unguiculata). *Phytochemistry*, **19**, 751–756

Gatehouse, A. M. R., Gatehouse, J. A., Dobie, P., Kilminster, A. M. and Boulter, D. (1979) Biochemical basis of insect resistance in *Vigna unguiculata*. *Journal of the Science of Food and Agriculture*, **30**, 948–958

Gatehouse, A. M. R. and Hilder, V. A. (1988) Introduction of genes conferring insect resistance. In *Brighton Crop Protection Conference – Pests and Diseases, 1988*, vol. 3. The Lavenham Press, Lavenham, pp. 1245–1254

Guilley, H., Dudley, R. K., Jonard, G., Balazs, E. and Richards, K. E. (1982) Transcription of cauliflower mosaic virus DNA: detection of promoter sequences and characterisation of transcripts. *Cell*, **30**, 763–773

Hernstadt, C., Soares, G. G., Wilcox, E. R. and Edwards, D. L. (1986). A new strain of *Bacillus thuringiensis* with activity against coleopteran insects. *Biotechnology*, **4**, 305–308

Hilder, V. A., Barker, R. F., Samour, R. A., Gatehouse, A. M. R., Gatehouse, J. A. and Boulter, D. (1989a) Protein and cDNA sequences of Bowman-Birk protease inhibitors from the cowpea (*Vigna unguiculata*). *Plant Molecular Biology* (in press)

Hilder, V. A., Gatehouse, A. M. R. and Boulter, D. (1989b) Potential for exploiting plant genes to genetically engineer insect resistance exemplified by the cowpea trypsin inhibitor. *Pesticide Science*, (in press)

Hilder, V. A., Gatehouse, A. M. R., Sheerman, S. E., Barker, R. F. and Boulter, D. (1987) A novel mechanism of insect resistance engineered into tobacco. *Nature*, **330**, 160–163

Hill, D. S. (1983) *Agricultural Insect Pests of the Tropics and their Control*, 2nd ed. Cambridge University Press, Cambridge

Horsch, R. B., Fry, J. E., Hoffman, N. L., Eichholtz, D., Rogers, S. G. and Fraley, R. T. (1985) A simple and general method for transforming genes into plants. *Science*, **227**, 1229–1231

Janzen, D. H., Ryan, C. A., Liener, I. E. and Pearce, G. (1986) Potentially defensive proteins in mature seeds of 59 species of tropical leguminosae. *Journal of Chemical Ecology*, **12**, 1469–1480

Jones, J., Dunsmuir, P. and Bedbrook, J. (1985) High level expression of introduced chimeric genes in regenerated transformed plants. *EMBO Journal*, **4**, 2411–2418

Meeusen, R. L. (1986) Reported in *Agrow*, **29**, 9

Nagy, F., Morelli, G., Fraley, R. T., Rogers, S. G. and Chua, N. H. (1985) Photoregulated expression of pea rbcS gene in leaves of transgenic plants. *EMBO Journal*, **4**, 3063–3068

Norris, D. M. and Kogan, M. (1980) Biochemical and morphological bases of resistance. In *Breeding Plants Resistant to Insects*, (eds F. G. Maxwell and P. R. Jennings). John Wiley & Sons, London, pp. 23–60

Odani, S., Koide, T. and Ono, T. (1986) Wheatgerm trypsin inhibitors. Isolation and structural characterisation of single-headed and double-headed inhibitors of the Bowman–Birk type. *Journal of Biochemistry*, **100**, 975–983

Redden, R. J., Dobie, P. and Gatehouse, A. M. R. (1983) The inheritance of seed resistance to *Callosobruchus maculatus* F. in cowpea (*Vigna unguiculata* L.

Walp). I: Analysis of parental, F_1, F_2, F_3 and backcross seed generations. *Australian Journal of Agricultural Research,* **34**, 681–695

Rhodes, C. A., Pierce, D. A., Mettler, I. J., Mascaranhas, D. and Detmer, J. T. (1988) Genetically transformed maize plants from protoplasts. *Science,* **240**, 204–207

Richardson, M. (1977) The proteinase inhibitors of plants and microorganisms. *Phytochemistry,* **16**, 159–169

Ryan, C. A. (1981) Proteinase inhibitors. In *The Biochemistry of Plants,* vol. IV, (ed. A. Marcus). Academic Press, New York, pp. 351–370

Shimamoto, K., Terada, R., Izawa, T. and Fujimoto, H. (1989) Fertile transgenic rice plants regenerated from transformed protoplasts. *Nature,* **338**, 274–276

Singh, S. R. (1978) Resistance to insect pests of cowpea in Nigeria.In *Pests of Grain Legumes and their Control in Nigeria,* (eds S. R. Singh, H. F. Van Emden and T. A. Taylor). Academic Press, New York, pp. 267–297

Stockhaus, J., Eckes, P., Blau, A., Schell, J. and Willmitzer, L. (1987) Organ specific and dosage dependent expression of a leaf/stem specific gene from potato after tagging and transfer into potato and tobacco plants. *Nucleic Acids Research,* **15**, 3479–3491

Suzuki, A., Tsunogae, Y., Tanaka, I., Yamane, T., Ashida, T., Norioka, S. *et al.* (1987) The structure of Bowman–Birk type protease inhibitor A-II from peanut (*Arachis hypogaea*) at 3.3A resolution. *Journal of Biochemistry,* **101**, 267–274

Toriyama, K., Arimoto, Y., Uchimaya, H. and Hinata, K. (1988) Transgenic rice plants after direct gene transfer into protoplasts. *Biotechnology,* **6**, 1072–1075

Umbeck, P., Johnston, G., Barton, K. and Swain, W. (1987) Genetically transformed cotton (*Gossypium hirsutum*) plants. *Biotechnology,* **5**, 263–266

Vaeck, M., Reynaerts, A., Hofte, H., Jansens, S., De Beukleer, M. D., Dean, C. *et al.* (1987) Transgenic plants protected from insect attack. *Nature,* **328**, 33–37

Wilcox, D. R., Shivakuma, A. G., Melin, B. E., Miller, M. F., Benson, T. A., Schopp, C. W. *et al.* (1986) Genetic engineering of bioinsecticides. In *Protein Engineering,* (eds M. Inouye and R. Sarma). Academic Press, New York, pp. 395–413

6

GENETIC ENGINEERING OF PLANTS FOR RESISTANCE TO THE HERBICIDE 2,4-D

DANNY LLEWELLYN, BRUCE R. LYON, YVONNE COUSINS,
JOHN HUPPATZ, ELIZABETH S. DENNIS and W. JAMES PEACOCK
CSIRO Division of Plant Industry, Canberra City, ACT, Australia

Introduction

Cotton is currently the third largest export crop in Australia and its importance is increasing annually. CSIRO supports the cotton industry through an on-going plant breeding and agronomy programme based at its research station at Narrabri in northern New South Wales. Cotton varieties produced by CSIRO (Siokra and Sicala) have been successfully integrated into the Australian Industry and in 1988 constituted over 70% of the seed planted. CSIRO continues to improve its varieties by traditional breeding techniques, but we have also embarked on a number of programmes aimed at improving existing genotypes using genetic engineering techniques. Our first programme was to develop a tissue culture regeneration system for Australian cottons and then to adapt this to gene transfer systems based on *Agrobacterium tumefaciens*. The initial choice of an agronomically useful character to introduce into cotton was partially instigated by the Industry itself, when a number of growers suggested that we look into the annually recurring problem of spray drift damage caused to cotton crops by the herbicide 2,4-D (2,4-dichlorophenoxyacetic acid).

2,4-D is commonly used on cereal crops in the same valleys that cotton is grown or on fallow fields in wheat/fallow/cotton rotations to control broadleaf summer weeds. It cannot be used on cotton itself once the seedlings have emerged because of the extreme sensitivity of this crop to the herbicide. The majority of spraying of 2,4-D is undertaken from the air in light aeroplanes, so spray drift, even on relatively calm days, away from the cereals into the cotton growing areas is almost inevitable. There are also instances where the tanks on the planes have not been washed completely clear of 2,4-D and in subsequent spraying of an unrelated herbicide, miticide or insecticide on to a cotton crop, considerable damage has been caused by the residual 2,4-D. Although these problems may often be covered by insurance it would be preferable to most farmers that they should not occur at all. We therefore began to investigate the feasibility of developing a synthetic resistance gene against 2,4-D that could eventually be engineered into cotton and perhaps other sensitive summer crops to protect them against accidental damage.

Weed control in cotton is generally an expensive business costing over $34 million annually, about two-thirds of it for chemical weedicides alone. Cotton is sensitive to many herbicides and only a few of the more expensive ones can be used post-emergence, and then often only with complicated application procedures. Most weed control is therefore done pre-emergence and any weeds that survive through to the appearance of the cotton crop must be removed mechanically by hand chipping (costing around $10 million annually). If we can increase not only the tolerance of cotton to low spray drift levels of 2,4-D, but also to directly applied 2,4-D at effective field application levels, then it may be possible to reduce the total need for mechanical weed chipping by using this relatively cheap compound as an effective post-emergence herbicide.

Genetic engineering options for synthetic herbicide resistances

Genetically engineered herbicide resistances are no longer new. Effective resistances have been developed to a variety of potent 'knockdown' herbicides such as glyphosate and the sulphonylureas (Comai *et al.*, 1985; Shah *et al.*, 1986; Haughn *et al.*, 1988). In these cases the molecular target of the herbicide has been well characterized and resistances have been developed through the manipulation or replacement of the target gene. Comai *et al.* (1985) selected a strain of *Salmonella typhimurium* that was resistant to glyphosate because of a mutation in the 5-enolpyruvylshikimate 3-phosphate synthase (EPSP synthase) gene. This enzyme is the target gene for glyphosate, so when expressed in plants, the mutant enzyme could substitute for the endogenous plant enzyme that had been inactivated by the herbicide. Shah *et al.* (1986) used a different approach to achieve the same result. They selected a petunia cell line that was resistant to glyphosate through the over-production of the EPSP synthase enzyme. This petunia gene was cloned and then re-introduced into petunia with a new, much stronger promoter. The new chimaeric gene resulted in the over-production of the enzyme and hence resistance at a whole plant level. Haughn *et al.* (1988) have achieved a similar result with an acetolactate synthase gene from *Arabidopsis* which confers sulphonylurea herbicide resistance on transgenic tobacco plants. While all of these resistance genes are certainly effective, they probably have some deleterious effects, especially in the absence of the herbicide.

Similar principles to those described above could probably not be applied to developing a resistance gene against 2,4-D because its primary target is still unknown. 2,4-D is an analogue of the plant hormone group known as auxins and at low levels is often used as a substitute for natural auxins in plant tissue culture systems. Some of the symptoms of 2,4-D damage to cotton, e.g. twisting of shoots are suggestive of a hormone-induced mechanism such as perturbed rates of cell division, but it is likely that there are other more drastic and undetermined phytotoxic effects.

A second approach that has been used to synthesize herbicide resistance genes has been the engineering of detoxification pathways in susceptible plants. A number of herbicides are selective, i.e. are toxic to some plant species but not others. The difference in sensitivity is often not associated with a different in susceptibility of the molecular target, but to an ability of the resistant plant to detoxify or break down the herbicide before it has a chance to inactivate the susceptible target gene product. Two examples now exist where a microbial gene encoding an enzyme that detoxifies a herbicide has been modified for expression in plants and introduced into a model species by transformation (DeBlock *et al.*, 1987; Stalker, McBride and Malyj, 1988). The novel detoxification enzymes were expressed and conferred resistance to relatively high levels of phosphinothricin (Basta) and 3,5-dibromo-4-hydroxybenzonitrile (bromoxynil), respectively. These artificial systems mimic detoxification systems found in plants and are likely to place the least genetic, physiological or yield burden on the plant, both in the presence or absence of the herbicide. Furthermore, the mode of action of the herbicide need not be known provided that the appropriate detoxification genes can be found. This second strategy was considered the most promising for 2,4-D since a variety of microorganisms were known that naturally degraded the herbicide in the environment.

A source of 2,4-D degradation genes

2,4,-D has a relatively short residual life in soils, decreasing by about half every 14 days or so. The breakdown in the soil is performed by a variety of microorganisms including bacteria, yeasts and fungi from several taxonomic groups. The most well characterized organisms are strains of *Alcaligenes eutrophus*, which are relatively simple Gram-negative rod bacteria found in most aerated soils. Several different strains have been isolated from soils in fields commonly sprayed with 2,4-D and were all shown to grow on synthetic media with 2,4-D as the sole source of carbon (Don and Pemberton, 1981). The common denominator between all the strains is a large 75 kb plasmid which has subsequently been shown to encode many of the enzymes necessary for the breakdown of 2,4-D to easily consumed metabolites. The overall pathway of 2,4-D degradation (Don *et al.*, 1985) in *A. eutrophus* is shown in Figure 6.1 along with the spatial organization of the plasmic encoded genes involved in the pathway (Don *et al.*, 1985; Streber, Timmis and Zenk, 1987). The prospect of engineering all six genes to function in plants was somewhat daunting, but we expected that possibly only the first one or two genes in the pathway would be essential, and that the plant would provide some pathway for the degradation or removal of intermediate products of degradation. Seed germination studies carried out in the presence of 2,4-D or the first enzymatic degradation product, 2,4-dichlorophenol (DCP) indicated that, in tobacco at least, DCP was some 50 to 100-fold less toxic than 2,4-D and perhaps only the first enzyme, 2,4-D mono-oxygenase (encoded by *tfdA*) need be expressed in plants to detoxify 2,4-D sufficiently. When we initiated our programme, the *tfdA* gene was the only gene not yet localized on the *tfd* plasmid, and we began by attempting to clone the gene by complementation and growth on 2,4-D analogues such as phenoxyacetic acid. Shortly thereafter, the gene was identified and sequenced by Streber, Timmis and Zenk (1987), and we were able to progress quickly towards the genetic engineering of the *tfdA* sequence for expression in plants.

Genetic engineering of the *tfd*A gene

Using the sequence data of Streber, Timmis and Zenk (1987) we were able to localize the coding region of the *tfdA* gene to a 980 bp *Xba* I-*Nco* I fragment as indicated in Figure 6.2. The *Xba* I site occurs 102 bp upstream of the start of translation and the *Nco* I site 6 bp downstream of the stop codon of the gene. Both sites were filled in with the Klenow fragment of DNA polymerase and synthetic *Bam*H1 linkers ligated to the ends. The coding region fragment was then cloned

OCH₂COOH ... tfd A → ... OH ... tfd B → ... OH

2,4-Dichlorophenoxyacetic acid   2,4-Dichlorophenol   3,5-Dichlorocatechol

tfd C

HOOC ... C=O ... ← tfd F ← COOH C=O ← tfd D ← COOH COOH

cis-2-dichlorodiene-lactone   trans--2-chlorodiene-lactone   2,4-Dichloromuconate

tfd E

COOH
COOH

chloromaleylacetic acid

(a)

2,4-D DEGRADATION GENES OF pJP4

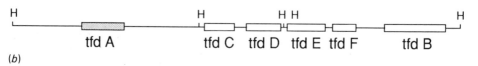

tfd A   tfd C   tfd D   tfd E   tfd F   tfd B

(b)

Figure 6.1   (a) Proposed pathway for the degradation of 2,4-D in *Alcaligenes eutrophus* strain JMP134. Intermediate degradation products are indicated and the plasmid encoded genes responsible for each step are shown adjacent to the arrows. (b) Structural arrangement of the 2,4-D degradation genes on the plasmid pJP4 isolated from JMP134. *Hind*III restriction sites (H) around the gene cluster are indicated. The fragment carrying the *tfdA* (stippled) is *Hind* fragment B (*see* Don *et al.*, 1985). Fragments and genes are only approximately to scale

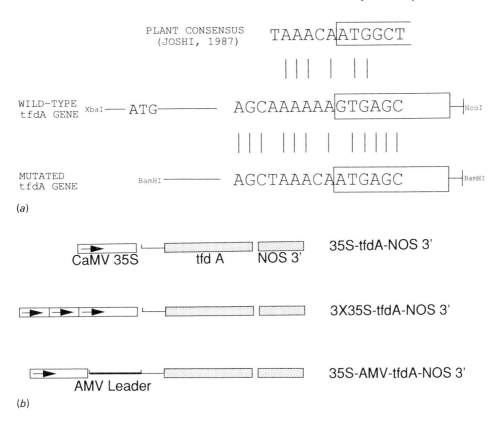

Figure 6.2 Genetic manipulations of the *tfdA* gene for expression in plant cells. (a) Construction of a coding region cassette. Part of the sequence around the translation start site of the *tfdA* gene from *A. eutrophus* is shown in the centre line. The out-of-frame ATG in the leader region is indicated. The upper line shows the consensus sequence around the translation start sites of 79 different plant genes (Joshi, 1987), with similarities to the wild type *tfdA* gene indicated. The lower line shows the completed coding region cassette after *Bal* 31 digestion and oligonucleotide directed mutagenesis. Coding regions are boxed. (b) Schematic diagram of the three gene constructions tested in transgenic tobacco plants. The *tfdA* coding cassette is stippled and the nopaline synthase 3' end hatched. The arrows represent the upstream enhancer region of the cauliflower mosaic virus (CaMV) 35S promoter. The thick line is a 45 bp fragment containing the leader region of the alfalfa mosaic virus (AMV) coat protein gene. Construct designations are shown on the right

into the polylinker of the pUC19 plasmid at the *Bam*HI site. The orientation of the gene was such that the *Eco*RI site of pUC19 was at the 5' end of the coding region fragment. Inspection of the sequence of the *tfdA* gene indicated the presence of an out-of-frame ATG translation start codon in what would be the 5' non-translated leader region of the gene when it was introduced into plant cells. This sequence was removed by *Bal* 31 exonuclease digestion which deleted approximately 20 bp down from the original *Xba* I site. Further examination of the sequence of the *tfdA* gene revealed an unusual start codon. This bacterial coding region begins with a GTG codon rather than the usual ATG (Figure 6.2). We were unsure whether this novel codon would be correctly recognized in plant cells and decided to modify the sequence to an ATG by oligo-mutagenesis. At the same time we also modified two other bases just upstream of the ATG so that the sequence around the start of translation would conform more closely to a consensus sequence compiled from a large number of plant genes (Joshi, 1987). It was reasoned that such a modified sequence (Figure 6.2) would allow the gene to be expressed well in plant cells and hence give higher levels of tolerance to 2,4-D. In support of this assertion, the mutated sequence, when expressed in *Escherichia coli* from the *lacZ* promoter appeared to be more active than the non-mutated GTG-containing sequence.

In order to express the mono-oxygenase gene in plants it was necessary to provide a strong plant promoter and transcription termination and polyadenylation signals. The latter were provided by a fragment of the nopaline synthase gene originally from the T-DNA of *Agrobacterium tumefaciens* (Herrera-Estrella *et al.,* 1983). Promoter sequences were provided by the 35S promoter of the cauliflower mosaic virus (CaMV), a strong and essentially constitutive promoter in many dicot plants (Odell, Nagy and Chua, 1985). Two additional modified 35S promoters were used in an attempt to improve on the expression of the viral promoter alone. In the first, two additional copies of the so-called upstream enhancer region of the 35S promoter (the region between −90 and −340) were ligated on to the 35S promoter at −340. All three enhancer regions were in the same orientation. Kay *et al.* (1987) have suggested that multiple copies of this enhancer increase the level of expression of the promoter more than arithmetically. In the second construction an additional 45 bp was added to the small leader region of the 35S promoter fragment (essentially only two base pairs). The addition consisted of the leader region of another viral gene, the coat protein gene of the alfalfa mosaic virus (AMV). This AMV leader has been suggested to improve the translational competitiveness of viral genes, at least *in vitro*, and perhaps also heterologous genes to which it is attached (Jobling and Gerhke, 1987).

All three chimaeric *tfdA* genes, the 35S-*tfdA*-Nos, 3x-*tfdA*-Nos and 35S-AMV-*tfdA*-Nos (*see* Figure 6.2b), as well as a control construction with the *tfdA* gene in an orientation not compatible with correct expression, were cloned into the binary plant transformation vector, pGA470, developed by An *et al.* (1987). This plasmid carries left and right T-DNA borders that direct integration into the plant genome and a selectable kanamycin resistance gene that allows the recovery of transformed plant cells. The recombinant plasmids were introduced into the disarmed *Agrobacterium* strain LBA4404 (Hoekema *et al.,* 1983) by triparental mating. Correctness of the mobilization was determined by Southern blot analysis.

Generation and analysis of transgenic plants

Although we were ultimately interested in introducing these genes into Australian cotton cultivars, we were uncertain that simply expressing the *tfdA* product would provide any significant level of tolerance to 2,4-D. To test the gene constructions and the basic hypothesis behind them, we produced transgenic tobacco plants containing all four chimaeric genes as a model system before going on to put them into cotton. Transgenic tobacco plants were produced by *Agrobacterium* infection of leaf pieces using relatively standard protocols. Over 20 independent rooted plants were produced for each construction, although, to date, we have only completed the analysis on the plants containing the 35S-AMV-*tfdA*-Nos chimaeric gene. In order to reduce the number of plants that we would eventually have to analyse in the glasshouse we screened the rooted plants using an enzymatic assay for the 2,4-D mono-oxygenase to find individuals that were expressing the highest levels of enzyme. These individuals were likely to be the ones exhibiting the highest levels of tolerance to the herbicide. Since no existing *in vitro* assay existed for the mono-oxygenase we developed one using ring labelled [^{14}C]-2,4-D and oxygen as substrates and NADH as a cofactor. ^{14}C-labelled DCP produced by the reaction was separated from the 2,4-D substrate by thin layer chromatography and visualized by autoradiography after fluorographic enhancement. The assay was optimized using induced cell free extracts of *E. coli* containing the mutated *tfdA* gene under the expression of the *lacZ* promoter. The activity in plant extracts appeared to be considerably lower than in bacteria, but an identifiable DCP product was detected in extracts from some of those plants containing chimaeric genes. Some DCP was produced by control plants, presumably by endogenous mono-oxygenase enzymes with relaxed substrate specificities. These enzymatic activities are, however, insufficient to convert sufficient of the 2,4-D to provide effective protection against the herbicide. Other radiolabelled products were also observed in both control and transgenic plants and these are presumably other chemical modifications of either 2,4-D or DCP caused by endogenous enzymes. A number of the plants showing the highest level of production of DCP were chosen for further study.

Leaf pieces from the primary regenerants and control plants were produced with a sterile cork borer and placed on our standard medium for tobacco shoot regeneration, but with the addition of various concentrations of 2,4-D from 0 up to 12.8 mg/l. Control plants, or plants containing the *tfdA* gene in the wrong orientation for expression, were more sensitive to the added 2,4-D and after 8 weeks there was very little evidence of shoot formation at 0.4 mg/l, although some callus had formed. The leaf pieces were chlorotic and senescing. Some of the transgenic plants, on the other hand, still produced green shoots on up to 12.8 mg/l. The increased resistance in the transgenic plants varied from 10- to 30-fold over the control plants. In most cases the resistance in the leaf disc assay correlated well with the intensity of the DCP band in the enzyme assay and not to any of the other radiolabelled products. Genomic Southern analysis on some of the plants indicated that most contained single copy, or at most two copy, insertions of the 35S-AMV-*tfdA*-Nos chimaeric gene. A selection of the transgenic plants were grown to maturity in a containment glasshouse and allowed to self-pollinate. Progeny seed was collected and assayed further.

F_1 progeny seeds were surface sterilized and plated on a suitable germination medium containing from 0 to 3.2 mg/l 2,4-D. Seeds of control plants failed to germinate or, if cotyledons emerged, failed to elongate on as little as 0.2 mg/l 2,4-D. Transgenic plants expressing 2,4-D mono-oxygenase activity, on the other hand, germinated and elongated on as much as 3.2 mg/l 2,4-D. In some progeny sets over 90% of the seeds germinated and continued to grow, suggesting multiple gene insertions, while in others there seemed to be the expected three to one segregation of a single copy insertion of a dominant character. At higher levels of 2,4-D there were some seeds obviously more resistant than the rest and it seems likely that these were the homozygotes. We are currently propagating these plants through another generation to test this hypothesis.

Randomly selected progeny seedlings from one 2,4-D mono-oxygenase expressing transformant and a control plant were planted in small pots (four seedlings per pot) and grown in a containment glasshouse until they were 6 weeks old. Groups of 16 plants each were sprayed with a commercial preparation of 2,4-D isopropyl ester in acetone at concentrations from 0 to 1000 p.p.m. Control plants sprayed with as little as 30 p.p.m. began to show symptoms after only 3 days when they began to wilt and show abnormal twisting of the stems. Many of the transgenic plants only began to show these same symptoms at 1000 p.p.m. Some of the transgenic plants were more affected than others, presumably because of segregation of the 2,4-D resistance gene among the F_1 progeny. After about 3 weeks control plants sprayed with over 100 p.p.m. were dead or dying while many of the clearly resistant transgenic plants had grown out of the effects of the herbicide, even when sprayed at 1000 p.p.m. (Figure 6.3). We are currently testing seedlings selected for an ability to germinate on high levels of 2,4-D to see whether they are resistant to much higher level of sprayed herbicide as well as screening transgenic plants containing our other two gene constructions. While it is difficult to estimate the equivalent field application rate corresponding to our experimental sprayings we estimate that resistance to 1000 p.p.m. in our system corresponds to at least 1–2 kg/hect which is at least four- to eightfold above normal field application rates for 2,4-D. We hope to be able to field-test these transgenic plants in the coming year (1990) to confirm this under proper field conditions, but it is clear that

Figure 6.3   Phenotypic 2,4-D resistance in the progeny of transgenic tobacco plants. The plants on the left are untransformed tobacco seedlings while those on the right are transgenic plants expressing the 35S-AMV-*tfdA*-Nos 3' chimaeric gene construct. Both groups of seedlings were sprayed, until dripping, with 1000 p.p.m. 2,4-D ester and photographed 3 weeks after spraying

the bacterial 2,4-D mono-oxygenase is effective in protecting transgenic tobacco plants from the effects of 2,4-D at levels above the normal field application rate. Since the enzyme assays and preliminary Northern analyses indicate that considerable improvement in the expression of the *tfdA* gene is possible, we will continue to optimize the gene constructions in tobacco, but we now have enough confidence that the gene will function in cotton to concentrate our efforts on introducing these gene constructs into this commercial species.

Current status of our cotton transformation programme

The regeneration of cotton plants from tissue culture, let alone transformation, has been a difficult task. Regenerability is highly genotype dependent and it was not until recently that reliable regeneration via somatic embryogenesis could be achieved (Gawel, Rao and Robacker, 1986; Trolinder and Goodin, 1987), and then predominantly with 'Coker' varieties of cotton. Transformation has recently been reported for Coker cottons using *Agrobacterium*-mediated gene transfer and regeneration from either seedling cotyledon (Firoozabady *et al.*, 1987) or hypocotyl (Umbeck *et al.*, 1987) explants. Transformed callus tissue was selected on media containing kanamycin at relatively low levels (25–35 µg/ml) and somatic embryos induced from the callus. These embryos were germinated on appropriate media and grown to maturity in soil. The frequency of transformation even with Coker cotton is very low but still manageable. We screened a number of Australian varieties for regenerability essentially using the procedures of Trolinder and Goodin (1988a,b) and were able to find two varieties with acceptable responses. Embryogenic callus has been generated for Siokra 1–3 and Siokra 1–4, and plants regenerated. At least with Siokra 1–3 we can obtain embryogenic suspension cultures, and hence plants, from between 10 and 20% of seedlings put into culture, but it still takes between 12 and 18 months to produce rooted plants ready for transfer to the glasshouse. The long lag time for cotton regeneration has made optimization difficult but we are now using our established regeneration protocols to attempt transformation with *Agrobacterium* vectors.

Australian cultivars appear to be much more resistant to kanamycin than American Coker varieties and we have had to use 100 µg/ml to prevent the growth of non-transformed tissues. We have used a variety of *Agrobacterium* strains, including both co-integrate and binary type vectors carrying selectable kanamycin resistance genes driven by either the nopaline synthase promoter or the 35S promoter. Assayable marker genes such as nopaline synthase itself and the versatile 35S-β glucuronidase (GUS) gene (Jefferson, Kavanagh and Bevan, 1987) have been included so that any kanamycin resistant callus produced could be quickly assessed for transformation. Hypocotyl explants proved to be difficult and many false positive, slow growing calli grew under selection. None of these calli expressed detectable levels of any of the marker genes. Cotyledon explants on the other hand have recently given us a relatively high frequency of transformed callus using a binary vector system (*nos*-kanamycin as selectable gene and 35S-GUS as the assayable marker) with a wild type Ti plasmid as the helper system. This callus grows rapidly on 100 µg/ml kanamycin and assays positively for GUS enzyme activity by both histochemical and fluorometric assays. The callus type has the look of callus with somatic embryogenic potential and we are currently in the process of trying to induce embryogenesis. Concurrently with this programme we have repeated the transformations with identical strains of *Agrobacterium* carrying the three 2,4-D resistance gene constructs and have reasonable expectations that we can recover transformed plants over the next year.

Concluding remarks

The herbicide detoxification system that we have described has proved to be extremely effective in at least one plant species and we expect that it will also be useful in other crops that are sensitive to 2,4-D. The plants containing the engineered resistance gene appear to grow and flower normally, producing just as many viable seeds as non-transformed plants. In other words, there appears to be no yield penalty in the absence of any applied herbicide. When sprayed at relatively high levels of 2,4-D there is a slight retardation of growth but the plants recover quickly. Since most commonly used herbicides are broken down by soil microorganisms it would seem feasible to apply the same principles to those herbicides and quickly to develop a whole range of herbicide resistance genes that could, once transformation procedures become routine, revolutionize farming practices in high input crops such as cotton.

Acknowledgements

This research has been supported in part by Cotton Seed Distributors Pty Ltd (Wee Waa, NSW) and the Australian Cotton Research Council.

References

An, G., Watson, B., Stachel, S., Gordon, M. and Nester, E. (1985) New cloning vehicles for transformation of higher plants. *EMBO Journal*, **4**, 277–284

Comai, L., Facciotti, D., Hiatt, W., Thompson, G., Rose, R. and Stalker, D. (1985) Expression in plants of a mutant *aroA* gene from *Salmonella typhimurium* confers tolerance to glyphosate. *Nature, (London)*, **317**, 741–744

DeBlock, M., Botterman, J., Vandewiele, M., Dockyx, J., Thoen, A., Gossele, V. *et al.* (1987) Engineering herbicide resistance in plants by expression of a detoxifying enzyme. *EMBO Journal*, **6**, 2513–2518

Don, R. and Pemberton, J. (1981) Properties of six pesticide degradation plasmids isolated from *Alcaligenes eutrophus* and *Alcaligenes paradoxus*. *Journal of Bacteriology*, **145**, 681–686

Don, R., Weightman, A., Knackmuss, H. and Timmis, K. (1985) Transposon mutagenesis and cloning analysis of the pathways for degradation of 2,4-dichlorophenoxyacetic acid and 3-chlorobenzoate in *Alcaligenes eutrophus* JMP134(pJP4). *Journal of Bacteriology*, **161**, 85–90

Firoozabady, E., DeBoer, D., Merlo, D., Halk, E., Amerson, L., Rashka, K. *et al.* (1987) Transformation of cotton (*Gossypium hirsutum* L.) by *Agrobacterium tumefaciens* and regeneration of transgenic plants. *Plant Molecular Biology*, **10**, 105–116

Gawel, N., Rao, A. and Robacker, C. (1986) Somatic embryogenesis from leaf and petiole callus cultures of *Gossypium hirsutum*. *Plant Cell Reports*, **5**, 457–459

Haughn, G., Smith, J., Mazur, B. and Sommerville, C. (1988) An *Arabidopsis* acetolactate synthase gene in tobacco confers resistance to sulfonylurea herbicides. *Molecular and General Genetics*, (in press)

Herrera-Estrella, L., DeBlock, M., Messens, E., Hernalsteens, J., Van Montagu, M. and Schell, J. (1983) Chimeric genes as dominant selectable markers in plant cells. *EMBO Journal*, **2**, 987–995

Hoekema, A., Hirsch, P., Hooykaas, P. and Schilperoort, R. (1983) A binary plant vector strategy based on separation of *vir-* and T-region of *Agrobacterium tumefaciens* Ti-plasmid. *Nature (London)*, **303**, 179–180

Jefferson, R., Kavanagh, T. and Bevan, M. (1987) GUS fusions: β-glucuronidase as a sensitive and versatile gene fusion marker in higher plants. *EMBO Journal*, **6**, 3901–3907

Jobling, S. and Gerhke, L. (1987) Enhanced translation of chimeric mRNAs containing plant viral untranslated leader sequences. *Nature (London)*, **325**, 622–625

Joshi, C. (1987) An inspection of the domain between putative TATA box and translation start sites in 79 plant genes. *Nucleic Acids Research*, **15**, 6643–6653

Kay, R., Chan, A., Daly, M. and McPherson, J. (1987) Duplication of CaMV 35S promoter sequences creates a strong enhancer for plant genes. *Science*, **236**, 1299–1302

Odell, J., Nagy, F. and Chua, N. H. (1985) Identification of DNA sequences required for activity of a plant promoter: the CaMV promoter. *Nature (London)*, **313**, 810–812

Shah, D., Horsch, R., Klee, H., Kishmore, G., Winter, J., Tumer, N. *et al.* (1986) Engineering herbicide tolerance in transgenic plants. *Science*, **233**, 478–481

Stalker, D., McBride, K. and Malyj, L. (1988) Expression in plants of a bromoxynil-specific bacterial nitrilase that confers herbicide resistance. In *Genetic Improvements of Agriculturally Important Crops: Progress and Issues*, (eds R. T. Fraley, N. M. Frey and J. Schell). Cold Spring Harbour Laboratory, Cold Spring Harbour, pp. 37–40

Streber, W., Timmis, K. and Zenk, M. (1987) Analysis, cloning and high-level expression of 2,4-dichlorophenoxyacetate monooxygenase gene *tfdA* of *Alcaligenes eutrophus* JMP134. *Journal of Bacteriology*, **169**, 2950–2955

Trolinder, N. and Goodin, J. (1987) Somatic embryogenesis and plant regeneration in cotton (*Gossypium hirsutum* L.). *Plant Cell Reports*, **6**, 231–234

Trolinder, N. and Goodin, J. (1988a) Somatic embryogenesis in cotton (*Gossypium*): I. Effects of source of explant and hormone regime. *Plant Cell, Tissue and Organ Culture*, **12**, 31–42

Trolinder, N. and Goodin, J. (1988b) Somatic embryogenesis in cotton (*Gossypium*): II. Requirements for embryo development and plant regeneration. *Plant Cell, Tissue and Organ Culture*, **12**, 43–53

Umbeck, P., Johnson, G., Barton, K. and Swain, W. (1987) Genetically transformed cotton (*Gossypium hirsutum* L.) plants. *Bio/technology*, **5**, 263–266

7

FUNCTIONAL ANALYSIS OF SEQUENCES REGULATING THE EXPRESSION OF HEAT SHOCK GENES IN TRANSGENIC PLANTS

FRITZ SCHÖFFL, MECHTHILD RIEPING and EBERHARD RASCHKE

Universität Bielefeld, Fakultät für Biologie (Genetik), Bielefeld, West Germany

Introduction

A sudden increase in temperature, or heat shock (hs), induces rapid changes of gene expression in all organisms. During the hs response, the rate of transcription of the hs genes is increased dramatically and these new mRNAs are preferentially translated into the hs proteins (hsps). The translation of the pre-existing mRNAs for most of the non-hs proteins pauses during the hs period. The hsps are thought to protect the cells from detrimental effects of thermal stress, thus causing acquired thermotolerance. The nature of the hs response suggests it is homeostatic. (For review *see* Nover *et al.*, 1984; Lindquist, 1986; Schöffl *et al.*, 1986, 1988; Lindquist and Craig, 1988.)

The activation of hs genes has attracted much interest as a model for studying the coordinate regulation of gene expression. In plants, most investigations have concentrated on the transcriptional regulation of genes for small hsps. The small hsps are a very diverse group of proteins. Different organisms have different numbers of these proteins ranging from one in yeast (26 kD), seven in *Drosophila* (22–27 kD) to more than 30 in plants (17 to approximately 30 kD) (Lindquist and Craig, 1988). In soyabean, the most prevalent small hsps (17–18 kD) belong to two different but related gene families (Schöffl *et al.*, 1986; Raschke, Baumann and Schöffl, 1988). Small hsps from different organisms are also related by certain features of the protein structure. It has been shown in many organisms, but not yet in plants, that small hsps are also induced at specific stages in development at normal temperatures (for review *see* Lindquist and Craig, 1988). Nevertheless, the role of these proteins is still unknown.

Transcriptional regulation has been demonstrated for the expression of small hsp genes in soyabean (Schöffl, Rossol and Angermüller, 1987) and this property is functionally related to the occurrence of multiple copies of hs promoter elements (HSEs). HSEs are the binding sites for the activated hs transcription factor (HSF or HSTF) in yeast, *Drosophila* and mammalian cells. HSF pre-exists in sufficient concentrations within the cells in an inactive form, but is rapidly activated in response to hs by post-translational mechanisms (Sorger, Lewis and Pelham, 1987; Wu *et al.*, 1987). In plants, the functional relevance of HSEs is indicated by the faithful transcriptional regulation of plant hs genes (Schöffl and Baumann, 1985; Gurley *et al.*, 1986; Rochester, Winter and Shah, 1986; Schöffl *et al.*, 1986;

Baumann *et al.*, 1987) and by the regulated expression of *Drosophila* hs promoter/reporter gene fusions (Spena and Schell, 1987) in heterologous transgenic plant cells.

This chapter summarizes the progress in studying the regulation of hs gene expression with emphasis on the identification of structural features of hs genes and the delimitations of *cis*-active regulatory sequences required for transcriptional and translational regulation. These results are important for the construction of hs promoter expression cassettes which can be used for highly regulated and efficient expression of chimaeric genes in transgenic plants. Heat shock-driven gene expression of non-hs genes has a wide range of applications, e.g. selection of regulatory mutants that alter the hs response (Schöffl, 1988).

Structural features of heat shock genes

The coordinate expression of small soyabean hsps (17–18 kD) and their possible functional relationship in the hs response prompted the analysis of DNA sequences of the genes by dot matrix comparisons to search for common structural features in regions determining the amino acid sequences and the regulation of the expression of hsps. The genes sequenced to date belong to two families, class I and class VI (Raschke, Baumann and Schöffl, 1988). Members of each family share approximately 90% similarity in the protein coding region at both the nucleic acid and the amino acid levels. Intergenic comparisons between the class I genes *Gmhsp17.3-B* and *Gmhsp18.5-C* show similarities in the entire protein coding region (*see* Figure 7.1d) but, in comparisons between these genes and *Gmhsp17.9-D* (class VI), similarities are confined to the *C*-terminal halves of the hsps (*see* Figure 7.1e,f).

The structural features of the proteins in this conserved region are a hydrophilic domain with the potential for α-helix formation, followed by a hydrophobic region including the highly conserved amino acid sequence GlyValLeuThr and another hydrophilic domain towards the *C*-terminus (Raschke, Baumann and Schöffl, 1988). These domains may be also conserved in small hsps of other plants since hs-induced mRNAs cross-hybridize with soyabean cDNA probes (Schöffl *et al.*, 1986). Similar domains occur also in related hsps in *Drosophila, Caenorhabditis*, humans and in the α-crystallins of the vertebrate eye lens indicating a conservation of structural features in proteins which have a tendency to aggregate under certain conditions (for a review *see* Schöffl, Baumann and Raschke, 1988). A general stress protein of soyabean (*Gmhsp26-A*) which, in contrast to the small hsps does not seem to aggregate during hs, displays hydropathic characteristics similar to that of typical hsps but it lacks the conserved amino acid motif (described above) in its hydrophobic region (Czarnecka *et al.*, 1988). It is still a matter of speculation whether the conserved hydrophobic amino acids of hsps determine hs-dependent hydrophobic interaction and aggregation.

The similarities between nucleic acid sequences of the soyabean hs genes drop below 50% in the 5' and 3'- non-translated regions of the mRNAs. This lack of conservation in these regions is surprising since non-translated sequences of hsp-mRNAs are implicated for preferential translation and RNA stability during thermal stress. Local similarities in flanking regions containing promoter and 5'-upstream sequences indicate the conservation of *cis*-regulatory promoter elements and AT-rich (>75%) repeats. The AT-rich regions are frequently

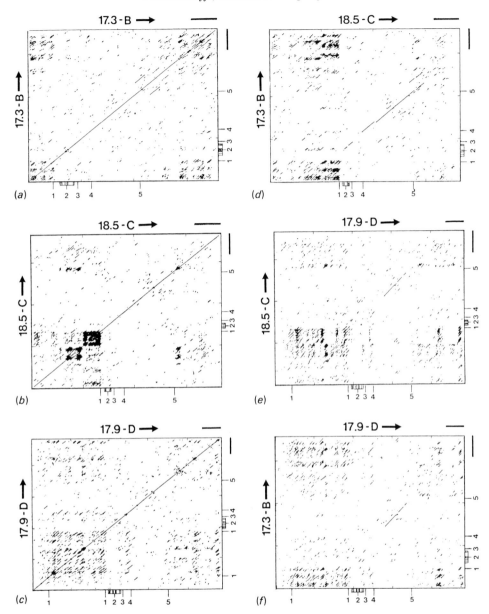

Figure 7.1 DNA similarity search (dot matrix) of soyabean hs genes belonging to two different but related gene families. *Gmhsp17.3-B*, and *−18.5-C* belong to class I and *Gmhsp17.9-D* is a member of class VI (Schöffl *et al.,* 1984; Raschke, Baumann and Schöffl, 1988). Panels (a)–(c): intragenic comparisons revealing repetitive sequences; (d)–(f): intergenic comparisons revealing similarities of sequences in different genes. A match of more than 51% over a span of 41 nucleotides yields an individual dot. Arrows indicate the 5′→3′ orientation of the genes, bars at the upper right hand edges of each panel are length standards for 250 nucleotides. The numbers at the axes of each panel mark important positions in the DNA sequence: (1) runs of simple sequences in the upstream promoter regions, (2) HSE-like promoter sequences, (3) TATA box, (4) translation start and (5) stop codon

preceded by runs of so-called simple sequences (A)$_n$, (T)$_n$ and (AT)$_n$ upstream from the promoter region (Figure 7.2). The AT-rich sequences may represent interspersed intergenic repeats, most of them starting within 300 nucleotides upstream from the transcriptional start site of hs genes. The repeated structure of these sequences is most pronounced in the region upstream from *Gmhsp18.5-C* (*see* Figure 7.1b). The redundancies of two complementary consensus motifs 5'AATTTTT and 5'AAAAATT (Raschke, Baumann and Schöffl, 1988), result respectively in three and two intense signal spots (*see* Figure 7.1b). AT-rich repeats also occur downstream from *Gmhsp17.3-B* (Figure 7.1a). The full length of one intergenic region upstream from *Gmhsp17.9-D* (*see* Figure 7.1c) spans approximately 750 nucleotides. It is still unknown how many of the small soyabean hs-genes are clustered in certain chromosomal regions. Clusters of hs genes are implicated since at least two genes, *Gmhsp17.3-B* and *Gmhsp6834-A*, were localized on one genomic clone (Schöffl and Key, 1983).

The most significant structural similarities between the hs promoter sequences of different soyabean genes are the multiple, frequently interlocked promoter elements termed HSE (Figure 7.2). HSE sequences with similarity (>60%) to the *Drosophila* hs consensus sequence 5' CT-GAA--TTC-AG are also found at other

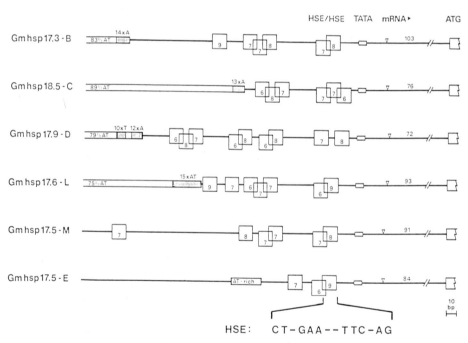

Figure 7.2 Structural features of soyabean hs promoter and upstream sequences. Structural features within approximately 350 nucleotides upstream from the protein coding regions were aligned to the TATA box (after Raschke, Baumann and Schöffl, 1988), all sequence elements and their spacing are drawn to scale: transcript start site (triangles), TATA box (small boxes), HSE-like promoter sequences (open squares), AT-rich intergenic repeats (open bars), A/T/AT runs (hatched bars). Numbers within squares indicate the number of nucleotides identical to the *Drosophila* HSE consensus sequence given below. Numbers above the 5' non-translated mRNA leader sequences indicate their actual length (nucleotides)

locations in hs genes and in many different non-hs genes. Unique to hs promoters is the higher copy number of HSEs within a short region and a special configuration (overlap) of some HSEs. Six to nine copies of HSE-like sequences are spread over a region of approximately 150 nucleotides upstream from the TATA box in five out of six soyabean hs genes. Partially overlapping HSEs occur at least at two locations upstream from every gene except *Gmhsp17.5-E* which contains only three HSE sequences proximal to the TATA box. Five HSEs, located far upstream from this gene (Czarnecka, Key and Gurley, 1989), are not depicted in Figure 7.2. The functional importance of the HSE-containing promoter regions and their 5' flanking sequences for heat-inducible transcription was demonstrated by mutational deletion analysis of *Gmhsp17.3-B* and *Gmhsp17.5-E* and by the construction and usage of synthetic promoter elements in transgenic plants (see below).

Functional analysis of heat shock promoters

The soyabean hs genes *Gmhsp17.3-B* and *Gmhsp17.5-E* are faithfully transcribed in a heat-inducible fashion in heterologous plants (Schöffl and Baumann, 1985; Schöffl *et al.*, 1986; Gurley *et al.*, 1986). A mutational analysis using 5' deletions of promoter and upstream sequences revealed the minimal requirement of *cis*-active sequences for the regulated transcription of *Gmhsp17.3-B* in transgenic tobacco plants (Baumann *et al.*, 1987) and *Gmhsp17.5-E* in sunflower tumours (Czarnecka, Key and Gurley, 1989). It was demonstrated that heat-inducible transcription requires, at least, the two partially overlapping HSE sequences located next to the TATA box. Upstream HSEs imposed a moderate and other upstream sequences a severe modulating effect on the amplitude of transcriptional activation. Full promoter activity was retained with fragments containing 335 bp or 1175 bp upstream from the transcriptional start sites of *Gmhsp17.3-B* and *Gmhsp17.5-E* respectively.

The functional significance of the multiple HSE sequences of *Gmhsp17.3-B* was investigated by 3' promoter deletions linked to a suitable reporter gene for expression studies in transgenic tobacco plants (Schöffl *et al.*, 1989). The deletion of the native TATA box was compensated by a promoter fusion with the otherwise silent ΔCaMV-35S-CATter gene construct (the truncated 35S promoter is devoid of the enhancer sequences required for constitutive transcription). The CAT (chloramphenicol acetyl transferase) activities and the different hs promoter deletions are depicted in Figure 7.3. These results indicate that the first pair of overlapping HSEs (proximal to the TATA box) can be replaced by other native upstream HSEs. A functional redundancy of remote HSEs of the *Drosophila hsp70* gene has been reported by Topol, Ruden and Parker (1985). The different levels of CAT activities (corresponding with equivalent mRNA levels) obtained with the different deletions suggest that the number of HSE sequences and probably their proper spacing from the TATA box is important for efficient gene activation. The manipulated spacing of HSEs in the promoter region of the *hsp70* gene suggested a periodic interaction of HSEs with the *trans*-acting hs binding factor HSF in *Drosophila* (Cohen and Meselson, 1988). The activation of eukaryotic hs genes seems to require multiple interactions of HSE and HSF and possibly also between bound HSF molecules and other TATA box binding proteins and/or RNA polymerase II (Thomas and Elgin, 1988). Other *cis*-active sequences regulating the magnitude of hs gene expression in plants are discussed below.

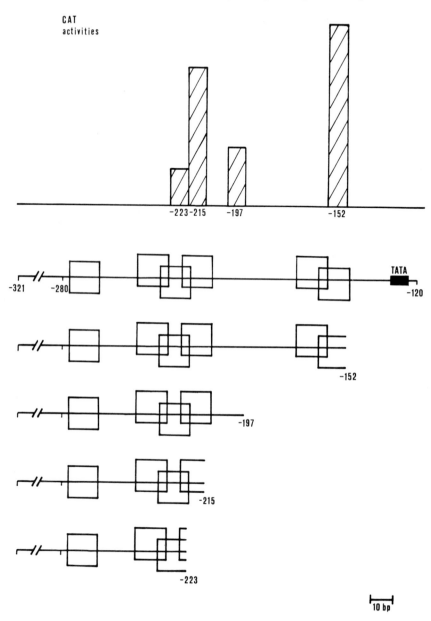

Figure 7.3   Heat shock-induced activities of a reporter gene driven by different 3′-truncated hs promoter fragments. Schematic diagrams of four different promoter fragments of the *Gmhsp17.3-B* hs-gene of soyabean which were linked to the otherwise silent ΔCaMV-35S-CATter gene of a BIN19 derived plant vector (the fusion is not shown). A series of transgenic plants was generated and heat-inducible CAT activities were determined in leaf extracts immediately after heat shock (2 h, 40°C). For details *see* Schöffl *et al.* (1989). The numbers refer to the translational start codon +1. Open squares indicate the position of HSE-like promoter sequences (*see* Figure 7.2), hatched bars represent the relative CAT activities (drawn to scale) induced by the different promoter deletions

Reconstitution of functional hs promoters using synthetic promoter elements

It has been shown that synthetic oligonucleotides with more than 70% homology to the *Drosophila* consensus sequence (HSE) 5' CT-GAA--TTC-AG are able to direct the heat-inducible activity of linked genes in animal cells (Pelham and Bienz, 1982; Voellmy and Rungger, 1982). Later it was found that HSE-like sequences are the binding sites for the *trans*-acting (transcription) factor HSF or HSTF (Parker and Topol, 1984; Wu, 1984). Full promoter activity of the *hsp70* gene in *Drosophila* requires at least two HSE sites (Dudler and Travers, 1984; Topol, Ruden and Parker, 1985), one HSE with a high affinity for HSF and a second site with low affinity for HSF (in the absence of site 1). It was suggested that cooperative binding of HSF at site 2 to an already occupied site 1 is crucial to strong gene activation (Topol, Ruden and Parker, 1985). The overlap of two HSE-like sequences, which seems to be the basic functional unit of a plant hs promoter, would cause steric difficulties to the binding of HSF-like molecules to both sites. It seems more likely that overlapping HSEs represent one single binding site for a HSF-like molecule in plant cells.

Strittmatter and Chua (1987) have shown that a 36mer oligonucleotide containing the first pair of overlapping HSE-sequences and 10 flanking nucleotides of *Gmhsp17.3-B* was sufficient to confer heat inducibility to a light-regulated promoter in transgenic plants. We have synthesized a different sequence consisting of two complementary 32mer oligonucleotides which, after annealing, represented the 26 bp consensus sequence for overlapping HSEs (termed HSE2) flanked by the cohesive ends of a *Kpn*I site (Schöffl *et al.*, 1989). This element differs in seven out of 26 nucleotides in the HSE region from the one used by Strittmatter and Chua (1987). Except for the *Kpn*I site no other flanking nucleotides were added to the HSE2 sequence. One to five copies of these HSE2 sequences were inserted via *Kpn*I into the multiple cloning site of the ΔCaMV-35S-CATter vector (derivative of *BIN19* with a truncated viral promoter as described above).

Transgenic tobacco plants containing these chimaeric genes were examined for heat-inducible CAT activity, CAT mRNA levels, and mRNA levels of tobacco hs genes (Table 7.1). Plants containing constructs with more than one copy of HSE2 showed hs-inducible CAT activity. Individual plants expressed levels as high as those obtained with the most active hs promoter fusion -152 (*see* Figure 7.3 and Schöffl *et al.*, 1989). Neither the orientation (A or B) nor a higher copy number of HSE2 had an elevating effect on CAT activity. Differences in CAT activities correlated with changes in CAT mRNA levels but not with the levels of endogenous tobacco hs mRNAs in these plants (Table 7.1).

It can be concluded that the interlocked HSEs (HSE2) are the essential *cis*-active sequences in native and artificial hs genes. The key features of hs regulatory elements in *Drosophila* are defined as GAA blocks that are arranged in alternating orientations and at two nucleotide intervals (Amin, Anathan and Voellmy, 1988) or as a dimer of a 10 bp sequence nTTCnnGAAn (Xiao and Lis, 1988). Three alternating GAA blocks or one-and-a-half copies of the 10 bp consensus sequence are required for efficient heat-inducible transcription. Nucleotides immediately upstream from the GAA segments and the spacing of HSE copies plays an important role in defining the competence of the regulatory elements. The consensus soyabean HSE2 sequence imperfectly meets the key features of the *Drosophila* hs response element. The 26 nucleotide sequence:

Table 7.1 THE hs RESPONSE IN TRANSGENIC TOBACCO PLANTS CONTAINGIN HSE2-DRIVEN ΔCaMV-35S-CATter CONSTRUCTIONS

Transgenic plants	Orientation of HSE2-sequences	CAT gene copy number	CAT-activities (≈ mRNA)	HS-mRNA levels[3]
1/2	BAA[1]	1	+	++
1/3	ABBBA	2	+	++
1/6	AAA	nd[2]	+	+++
1/7b	BB	nd	+	+
1/7c	BB	nd	+	++++
4/4b	B	nd	(+/−)	++++
2/8	AAA	1	+	+++
1/7a	BB	2	++	++++
1/7d	BB	1	++	++
4/5	AAAAB	2	++	++++
4/6	AAA	nd	++	++
1/5a	AAAAB	2	+++	++
1/8	AAA	2–4	+++	+++
1/5	AAAAB	2	−	++++
Vector controls[4]				
2/9a	−	1	−	+++
2/9b	−	1–2	−	+++

−, +, ++, +++, ++++: semiquantitative differences of hs induced CAT-activities and mRNA concentrations (*see* Schöffl *et al.*, 1989)
[1] Number and orientations (A, B) of synthetic HSE2 sequences present in the ΔCaMV-35-CATter constructs (for details *see* Schöffl *et al.*, 1989).
[2] Not determined.
[3] Levels of hs specific tobacco mRNAs determined by Northern hybridization using a soyabean hs gene as a probe (according to Schöffl, Rieping and Baumann, 1987).
[4] Transgenic plants containing ΔCaMV-CATter (without HSE2 sequences).

caag*GAt*tt*TTC*tg*GAA*ca*TaC*aaga contains 10 nucleotides forming a palindrome (italic letters) in the centre and also two perfect and two imperfect GAA blocks with alternating orientations and 2 bp spacing (capital letters).

It seems rational to assume that the plant HSF binds with high affinity to the centre of HSE2 since sequences in this part are more conserved than its peripheral regions (Schöffl *et al.*, 1989). Full promoter activity seems to require binding of the HSF to a second site (HSE2 or HSE) and may involve cooperative binding. This model is in accordance with the minimal sequences that are necessary to drive heat-inducible transcription from hs promoters of the soyabean *Gmhsp17.3-B* and *Gmhsp17.5-E* genes. Both require at least one HSE2-like sequence in the proximity of the TATA box and in both cases an additional upstream HSE or HSE2-like sequence has an enhancing effect on transcription. Low levels of heat-inducible CAT activity in certain plants containing more than one copy of the synthetic HSE2 sequence (*see* Table 7.1) are probably a consequence of negative position effects.

Too few transgenic plants have been investigated thus far to draw any conclusion about the possible interference of excess HSE or HSE2 sequences with the induction of the hs response. If the level of *trans*-acting HSF is limited in the cell it might be titrated by high affinity binding sites (HSE2?) in competition with the binding sites of the native tobacco hs genes. A several thousandfold excess of the *hsp70* gene 5′ control region was necessary to reduce the activation of other hs genes in Chinese hamster ovary cells (Johnston and Kucey, 1988). Hence, more than a thousand HSF molecules may be present in animal and perhaps also in plant cells.

DNA-protein interaction in the promoter upstream enhancer region

In addition to the proximal promoter sequences, other *cis*-active upstream sequences modulate the amplitude of transcription of hs genes. The functional analysis of 5′ deletions of *Gmhsp17.3-B* in tobacco (Baumann *et al.*, 1987) and *Gmhsp17.5-E* in sunflower (Czarnecka, Key and Gurley, 1989) has revealed an enhancer-like activity of sequences located within 150 bp segments upstream from the HSE containing promoter regions. These sequences stimulate the transcription of the respective hs genes by a factor of 5–10. One common structural feature in these regions is the occurrence of AT-rich sequences (*see* Figure 7.2), a run of $14 \times A$ (*Gmhsp17.3-B*) and an approximately 30 bp segment with more than 80% AT (*Gmhsp17.5-E*). These AT-rich sequences seem to interact with DNA binding factors isolated from soyabean nuclei (Severin, Kliem and Schöffl, 1989; Czarnecka, Key and Gurley, 1989). This interaction is exemplified by a gel retardation assay using a 243 bp DNA fragment of the *Gmhsp17.3-B* gene upstream region and a protein extract containing a binding factor (factor A) (Figure 7.4). This factor is extracted with 250 mM NaCl from nuclei, is sensitive to proteinase treatment and can be partially purified by chromatography (Severin, Kliem and Schöffl, 1989). The binding of the A-factor caused a slight shift of the electrophoretic mobility of the fragment. This shift was successfully competed by excess of identical, non-labelled DNA fragments (not shown), by synthetic poly(dA)-poly(dT) or poly(dAdT) oligonucleotides (Figure 7.4) and by an A-factor-binding upstream fragment of *Gmhsp17.6-L*. Specific binding was not competed by excess DNA from *Escherichia coli* or synthetic poly(dIdC) (Figure 7.4). Similar mobility shifts were obtained with a fragment shortened to 60 bp but still containing the run of $14 \times A$ and with a 106 bp fragment spanning the run of $15 \times AT$ of *Gmhsp17.6-L* (not shown).

The competition experiments indicate that the binding of factor A occurs to AT-rich sites. It cannot be ruled out that this interaction of AT-rich sequences and nuclear proteins is unrelated to the enhancer function, however there is a striking coincidence with enhancer-like properties and protein binding to the AT-rich upstream sequence of *Gmhsp17.5-E*, mentioned by Czarnecka, Key and Gurley (1989). The enhancer-like function seems to apply only to native hs genes; attempts to enhance gene activity of chimaeric genes controlled by the truncated ΔCaMV-35S promoter were unsuccessful (Schöffl and Rieping, unpublished data).

The function of downstream promoter and mRNA leader sequences

The mRNAs of hs genes are preferentially translated in many organisms (for review *see* Nover *et al.*, 1984; Lindquist, 1986). This propensity is attributed to alterations of the translational machinery and to as yet unknown peculiarities of the hs mRNA leader sequences. The inhibition of translation of the bulk of non-degraded normal mRNAs during thermal stress limits the application of hs promoter sequences for the expression of chimaeric genes. Expression cassettes for heat-inducible gene expression should warrant both high inducible transcript levels and efficient translatability of the generated mRNAs. The DNA sequences downstream from the transcriptional start site of *Gmhsp17.3-B* were the prime targets for investigating its relevance for translation.

A

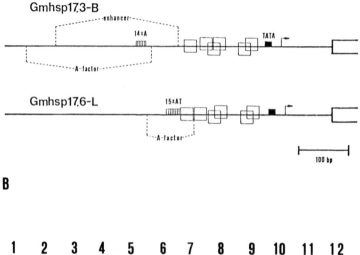

B

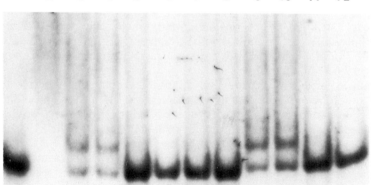

Figure 7.4 Competition of synthetic deoxypolynucleotides with the binding of the nuclear factor A to the promoter upstream fragment of *Gmhsp17.3-B*. (A) Schematic diagrams of the promoter and upstream regions of two soyabean hs genes. Upstream regions delimited for enhancer function and binding of the nuclear factor are indicated. (B) Retardation of a [^{32}P]-labelled DNA fragment (243 bp) from the upstream region of *Gmhsp17.3-B* by factor A. Electrophoresis was carried out on a 4% acrylamide gel, bands were visualized by autoradiography. One ng of the radioactively labelled DNA fragment was preincubated for 30 minutes with 1 μg *E. coli* DNA (1), 2.5 μg protein extract containing factor A (2–12), 1 μg *E. coli* DNA (3), 2 μg *E. coli* DNA (4–8), 0.4 μg [poly(dA)-poly(dT)] (5 and 11), 0.2 μg [poly(dA)-poly(dT)] (6), 0.4 μg [poly(dAdT)] (7 and 12), 0.2 μg [poly(dAdT)] (8), 1 μg [poly(dIdC)] (9), 2 μg [poly(dIdC)] (10–12)

Several *Bal*31-generated 3′ deletions were linked to the CATter reporter gene, transformed into tobacco and examined for heat-inducible CAT activities (Schöffl *et al.*, 1989). All constructs contained the sequences of the functional hs promoter (*see* Figure 7.3), including the native TATA box and the transcriptional start site, but different portions of the 5′ non-translated sequence of the mRNA leader (−12, −86) or the entire leader plus 28 terminal bp from the protein coding region (+28) (Figure 7.5). Heat-inducible CAT activity was found in transgenic tobacco plants containing these constructs, with the exception of the −86 construct (Figure 7.5). The CAT activities correlated with the levels of CAT specific mRNAs and reached approximately equal levels in +28 and −12 plants (Figure 7.5). This demonstrates

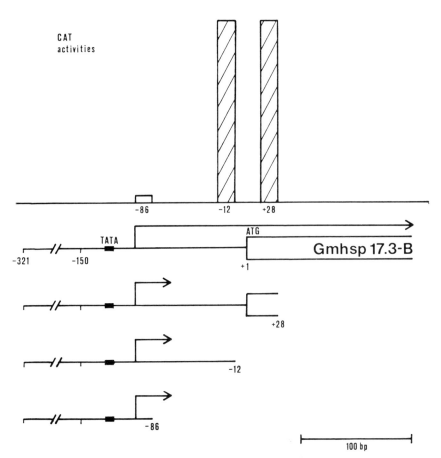

Figure 7.5   Heat shock-induced activities of a reporter gene driven by different transcriptional and translational fusions of a soyabean hs promoter with 3′ deletions. Schematic diagrams of three different deletions of *Gmhsp17.3-B* which were linked to the CATter reporter gene of a *BIN19* derived plant vector. The numbers of the deletions refer to the translational start site +1. Rectangular arrows mark the start site of transcription. A series of transgenic tobacco plants was generated for each construct and heat-inducible CAT activities (hatched bars) were determined in leaf extracts immediately after hs (2 h, 40°C) (for details *see* Schoffl *et al.*, 1989, for symbols *see* Figure 7.2). The CAT activity for the −86 construct represents the low levels obtained only during recovery at 25°C following heat shock

that transcriptional and translational fusions can be used for the regulated expression of foreign genes in plants. The transcriptional fusion construct −86, which is devoid of all but 18 bp of the 5' non-translated RNA region of *Gmhsp17.3-B* produced low levels of heat-induced mRNA but no inducible CAT enzyme activity. The CAT activity recovered in these plants during 1–2 h incubation at 25°C following hs (2 h 40°C). This recovery, typically observed for the bulk of normal mRNAs, is consistent with a translational control of CAT synthesis since the level of its mRNA was not significantly changed during recovery (Schöffl *et al.*, 1989). A recovery of reporter gene activity after hs was also found for neomycin phosphotransferase II expression driven by the *Drosophila hsp70* promoter in tobacco (Spena and Schell, 1987).

The sequences in the 5' non-translated mRNA leader of a hs gene seem to have dual effects, on both transcription and translation. Large deletions of the DNA sequences in the 5' RNA leader region may interfere with the stability of the association between RNA polymerase II and the DNA, thus leading to reduced transcription. It was shown that at normal temperature RNA polymerase II initiates transcription of the *hsp70* gene in *Drosophila* but stalls after approximately 25 nucleotides before the binding of the activated HSF to the promoter region causes, in some way, elongation of transcription (Rougvie and Lis, 1988). It is not known whether structural features at the 5' end of the hs-mRNA or its corresponding DNA coding sequence are required for efficient transcription in plants. The shorter total length of the mRNA leader sequence of the −86 construct (74 bp compared with 103 in the native gene) cannot be the reason for reduced transcription of the chimaeric gene since a 42 bp leader of the ΔCaMV-35S-CATter gene led to much higher hs-induced transcript levels when driven by a plant hs promoter (Schöffl *et al.*, 1989). It cannot be excluded that the manipulations of genes in the 5' RNA leader sequences may have different effects on RNA stability.

The preferential translation of hs-mRNAs could not be attributed to conserved motifs in the leader sequences in *Drosophila* (McGarry and Lindquist, 1985) and sequences in these regions show only a little similarity to the soyabean hs genes (Nagao *et al.*, 1985; Schöffl, Baumann and Raschke, 1988). The low potential for secondary structure formation intrinsic to the hs-mRNA leader sequences may cause unwinding of these RNAs and allow association with components of the translation machinery (Panniers *et al.*, 1986). It can be speculated that certain transcript fusions in the leader region generate secondary structure elements which cannot be unwound for the initiation of translation if the interaction with the Cap binding protein complex is limited. It is known that certain viral RNAs are translated with high efficiency in the absence of active Cap-binding factors in animal cells (Pelletier *et al.*, 1988) or during hs in plants (Dawson and Boyd, 1987). Thus it is not surprising that the ΔCaMV-35S initiated CAT mRNAs escape translational discrimination during hs (Schöffl *et al.*, 1989).

General conclusions

The functional analysis of hs promoters in transgenic plants revealed different *cis*-active sequences which are important for heat-inducibility and enhancement of transcription. The basic units for thermal activation of the transcription are multiple HSE sequences located in the TATA box-proximal upstream region. Most important to the process of activation is the presence of overlapping HSE-like sequences. Artificial hs promoters can be constructed by inserting synthetic HSE2 sequences (consensus sequences for the overlapping plant HSEs) upstream from the TATA box of a truncated viral promoter. Hence, HSE2 is probably the binding site of an as yet, in plants, unidentified hs transcription factor (HSF). The structural feature of the proposed HSF binding site in HSE2 is a central palindrome TTC--GAA flanked by less perfectly conserved blocks of alternating TTC and GAA motifs with 2 bp spacing. In transgenic plants, full promoter activity seems to require more than one HSE2 sequence and/or other upstream elements for enhancement of transcription. The eventual isolation of HSF genes from plants should help to clarify the extent of conservation and possible differences in the mechanism between animal and plant hs-induced transcription. HSF may be also involved in other vital cellular functions since HSF genes are essential in yeast (Sorger and Pelham, 1988; Wiederrecht, Seto and Parker, 1988).

The advantage of using hs promoter elements for heat-inducible expression of linked genes is that the hs reponse is conserved between species and kingdoms. Only a little is known about the mechanism of enhancer-like functions which seem to stimulate transcription activity of soyabean hs genes in tobacco and sunflower. The sequences have not been shown to be involved in thermal induction and it is unknown whether the enhancer-like properties are active in combinations with other promoter/reporter genes. The presence of AT-rich sequences within the putative enhancer regions and the interactions of these sequences with nuclear proteins may represent a key feature of the enhancer function. Several other highly regulated soyabean genes contain protein binding AT-rich upstream sequences (Czarnecka, Key and Gurley, 1989). It is possible to speculate that AT-rich regions may have a role in supercoil dependent expression of hs genes. The importance of torsional stress for hs gene expression is in accordance with the significantly higher activity of the *Gmhsp17.5-E* promoter when assayed in the supercoiled form in a transient expression system (M. Ainley, personal communication). Torsional stress may be either introduced by topoisomerase II or by limited melting of AT-rich regions.

The 5′ non-translated leader sequences of hs genes are important for the translation of mRNAs during hs. A low potential for secondary structure formation intrinsic to hs mRNA leader sequences may be crucial for the association with components of the translation machinery for initiation. Non-perfect transcript fusions between 5′ non-translated hs and reporter gene mRNAs are still recognized but truncation of large portions of the hs-mRNA leader sequence are not tolerated by the translation system during hs. Translation of severely truncated mRNAs resumes during recovery from hs at normal temperatures. The DNA sequences at the 5′ end of the hs-mRNAs seem also to be important for efficient transcription or for the stability of the mRNA. It was shown that the viral sequences, present in the ΔCaMV-35S-CATter constructs, have the capacity to direct high transcript levels and efficient translation of the chimaeric mRNAs in transgenic plants during hs.

Acknowledgements

We thank Heike Behrens and Rolf Hülsewig for technical assistance and we are grateful to Drs Ron Nagao and Mike Ainley (University of Georgia), Eva Czarnecka and Bill Gurley (University of Florida) for sharing unpublished results with us. The work in our laboratory was supported by grants from the Deutsche Forschungsgemeinschaft.

References

Amin, J., Anathan, J. and Voellmy, R. (1988) Key features of heat shock regulatory elements. *Molecular and Cellular Biology,* **8**, 3761–3769

Baumann, G., Raschke, E., Bevan, M. and Schöffl, F. (1987) Functional analysis of sequences required for transcriptional activation of a soybean heat shock gene in transgenic tobacco plants. *EMBO Journal,* **6**, 1161–1166

Cohen, R. S. and Meselson, M. (1988) Periodic interactions of heat shock transcriptional elements. *Nature,* **332**, 856–858

Czarnecka, E., Key, J. L. and Gurley, W. B. (1989) Regulatory domains of the *Gmhsp17.5-E* heat shock promoter of soybean: a mutational analysis. *Molecular and Cellular Biology,* **9**, 3457–3463

Czarnecka, E., Nagao, R. T., Key, J. L. and Gurley, W. B. (1988) Characterization of *Gmhsp26-A*, a stress gene encoding a divergent heat shock protein of soybean: heavy-metal-induced inhibition of intron processing. *Molecular and Cellular Biology,* **8**, 1113–1122

Dawson, W. O. and Boyd, C. (1987) TMV protein synthesis is not translationally regulated by heat shock. *Plant Molecular Biology,* **8**, 145–149

Dudler, R. and Travers, A. (1984) Upstream elements necessary for optimal function of the *hsp70* promoter in transformed flies. *Cell,* **38**, 391–298

Gurley, W. B., Czarnecka, E., Nagao, R. T. and Key, J. L. (1986) Upstream sequences required for efficient expression of a soybean heat shock gene. *Molecular and Cellular Biology,* **6**, 559–565

Johnston, R. N. and Kucey, B. L. (1988) Competitive inhibition of *hsp70* gene expression causes thermosensitivity. *Science,* **242**, 1551–1554

Lindquist, S. (1986) The heat shock response. *Annual Review of Biochemistry,* **55**, 1151–1191

Lindquist, S. and Craig, E. A. (1988) The heat shock proteins. *Annual Review of Genetics,* **22**, 631–677

McGarry, T. J. and Lindquist, S. (1985) The preferential translation of *Drosophila hsp70* mRNA requires sequences in the untranslated leader. *Cell,* **42**, 903–911

Nagao, R. T., Czarnecka, E., Gurley, W. B., Schöffl, F. and Key, J. L. (1985). Genes for low-molecular-weight heat shock proteins of soybeans: sequence analysis of a multigene family. *Molecular and Cellular Biology,* **5**, 3417–3428

Nover, L., Hellmund, D., Neumann, D., Scharf, K.-D. and Serfling, E. (1984) The heat shock response of eukaryotic cells. *Biologisches Zentralblatt,* **103**, 357–435

Panniers, R., Scorsone, K. A., Wolfman, A., Hucul, J. A. and Henshaw, E. C. (1986) Current communications in molecular biology. In *Translational Control,* (ed. M. B. Mathews), Cold Spring Harbor Laboratory Press, New York, pp. 52–57

Parker, C. S. and Topol, J. (1984) A *Drosophila* RNA polymerase II transcription factor binds to the regulatory site of an *hsp70* gene. *Cell,* **37**, 273–283

Pelham, H. R. B. and Bienz, M. (1982). A synthetic heat shock promoter element confers heat-inducibility on the herpes simplex virus thymidine kinase gene. *EMBO Journal,* **1**, 1473–1477

Pelletier, J., Kaplan, G., Racaniello, V. R. and Sonenberg, N. (1988) Cap-independent translation of poliovirus mRNA is conferred by sequence elements within 5′ noncoding region. *Molecular and Cellular Biology,* **8**, 1103–1112

Raschke, E., Baumann, G. and Schöffl, F. (1988). Nucleotide sequence analysis of soybean small heat shock protein genes belonging to two different multigene families. *Journal of Molecular Biology*, **199**, 549–557

Rochester, D. E., Winter, J. A. and Shah, D. M. (1986) The structure and expression of maize genes encoding the major heat shock protein, *hsp70*. *EMBO Journal*, **5**, 451–458

Rougvie, A. E. and Lis, J. T. (1988) The RNA polymerase II molecule at the 5′ end of the uninduced *hsp70* gene of *D. melanogaster* is transcriptionally engaged. *Cell*, **54**, 795–804

Schöffl, F. (1988) Genetic engineering strategies for manipulation of the heat shock response. *Plant Cell and Environment*, **11**, 339–343

Schöffl, F. and Baumann, G. (1985) Thermoinduced transcripts of a soybean heat shock gene after transfer into sunflower using a Ti plasmid vector. *EMBO Journal*, **4**, 1119–1124

Schöffl, F., Baumann, G. and Raschke, E. (1988) In *Plant gene research – temporal and spatial regulation of plant genes* (eds D. P. S. Verma and R. B. Goldberg), Springer Verlag, Vienna, New York, pp. 253–273

Schöffl, F., Baumann, G., Raschke, E. and Bevan, M. (1986) The expression of heat shock genes in higher plants. *Philosophical Transactions of the Royal Society of London, B*, **314**, 453–468

Schöffl, F. and Key, J. L. (1983) Identification of a multigene family for small heat shock proteins in soybean and physical characterization of one individual gene coding region. *Plant Molecular Biology*, **2**, 269–278

Schöffl, F., Rieping, M. and Baumann, G. (1987) Constitutive transcription of a soybean heat shock gene by a cauliflower mosaic virus promoter in transgenic tobacco plants. *Developmental Genetics*, **8**, 365–374

Schöffl, F., Raschke, E. and Nagao, R. T. (1984) The DNA sequence analysis of soybean heat-shock genes and identification of possible regulatory promoter elements. *EMBO Journal*, **3**, 2491–2497

Schöffl, F., Rieping, M., Baumann, G., Bevan, M. and Angermüller, S. (1989) The function of plant heat shock promoter elements in the regulated expression of chimaeric genes in transgenic tobacco. *Molecular and General Genetics*, **217**, 246–253

Schöffl, F., Rossol, I. and Angermüller, S. (1987) Regulation of the transcription of heat shock genes in nuclei from soybean (*Glycine max*) seedlings. *Plant Cell and Environment*, **10**, 113–119

Severin, K., Kliem, M. and Schöffl, F. (1989) In *Proceedings of the Braunschweig Symposium on Applied Plant Molecular Biology* (ed. G. Galling), Braunschweig Technical University, Braunschweig, pp. 179–183

Sorger, P. K., Lewis, M. J. and Pelham, H. R. B. (1987) Heat shock factor is differently regulated in yeast and HeLa cells. *Nature*, **329**, 81–84

Sorger, P. K. and Pelham, H. R. B. (1988) Yeast heat shock factor is an essential DNA binding protein that exhibits temperature-dependent phosphorylation. *Cell*, **54**, 855–864

Spena, A. and Schell, J. (1987) The expression of a heat-inducible chimeric gene in transgenic tobacco plants. *Molecular and General Genetics*, **206**, 436–440

Strittmatter, G. and Chua, N. H. (1987) Artificial combination of two *cis*-regulatory elements generates a unique pattern of expression in transgenic plants. *Proceedings of the National Academy of Sciences, USA*, **84**, 8986–8990

Thomas, G. H. and Elgin, S. C. R. (1988) Protein/DNA architecture of the DNase I hypersensitive region of the *Drosophila hsp26* promoter. *EMBO Journal*, **7**, 2191–2201

Topol, J., Ruden, D. M. and Parker, C. S. (1985) Sequences required for *in vitro* transcriptional activation of a *Drosophila hsp70* hs gene. *Cell,* **42**, 527–537

Voellmy, R. and Rungger, D. (1982) Transcription of a *Drosophila* heat shock gene is heat-induced in *Xenopus* oocytes. *Proceedings of the National Academy of Sciences, USA,* **79**, 1776–1780

Wiederrecht, G., Seto, D. and Parker, C. S. (1988) Isolation of the gene encoding the *S. cerevisiae* heat shock transcription factor. *Cell,* **54**, 841–853

Wu, C. (1984) Activation protein factor binds *in vitro* to upstream control sequences in heat shock gene chromatin. *Nature,* **311**, 81–84

Wu, C., Wilson, S., Walker, B., Dawid, J., Paisley, T., Zimarino, V. *et al.* (1987) Purification and properties of *Drosophila* heat shock activator protein. *Science,* **238**, 1247–1253

Xiao, H. and Lis, J. T. (1988) Germline transformation used to define key features of heat-shock response elements. *Science,* **239**, 1139–1142

8

SPATIAL AND TEMPORAL PATTERNS OF EXPRESSION OF A NOVEL WOUND-INDUCED GENE IN POTATO

ANNE STANFORD, MICHAEL BEVAN
Molecular Genetics Department, Institute of Plant Science Research Cambridge Laboratory, Trumpington, Cambridge, UK

and DON NORTHCOTE
Biochemistry Department, Cambridge University, Cambridge, UK

Introduction

Plants may become wounded in the field either through mechanical stress imposed, e.g. during extreme weather conditions or harvesting, or as a result of injury incurred during insect attack or pathogen invasion. In the laboratory, and in the context of the experiments described below, wounding has been simulated by the excision of tissue from intact and otherwise healthy plants, followed by the incubation of this tissue in a moist and aerobic environment, a process referred to here as ageing. As with other forms of physical stress, wounding brings about extensive changes in the pattern of protein synthesis within the plant. This includes a rapid induction of a number of proteins which are thought to be involved in specific defence mechanisms which can lead to a localized resistance at the lesion and often to an additional systemic protection of tissue distant from the wound. For example, wounding brings about an increased synthesis of enzymes of phenylpropanoid metabolism (Lawton *et al.*, 1983; Cramer *et al.*, 1985a) resulting in the enhanced lignification and suberization of plant tissues (Friend, 1985) and, in some cases, to an accumulation of antimicrobial compounds such as phytoalexins and other phenolics. There is also an accumulation of hydroxyproline-rich glycoproteins within the cell-walls of wounded tissues (Cassab and Varner, 1988). These are postulated to be involved in mechanical strengthening but rather may, as a result of their agglutinating activity, serve to immobilize bacteria upon entry into the cell (Leach, Cantrell and Sequeira, 1982). Wounding also leads to an increase in the concentration of lytic enzymes such as chitinases and glucanases which are able to degrade fungal and bacterial cell walls (Boller, 1985), as well as to a systemic induction of certain proteinase inhibitors which are thought to interfere with the digestive processes of attacking insects (Ryan, 1978).

The elicitation of these defence responses is primarily the result of increased gene transcription. For example, the accumulation of messenger RNAs encoding several enzymes of phenylpropanoid biosynthesis (Chappell and Hahlbrock, 1984; Cramer *et al.*, 1985b) and also those of hydroxyproline-rich glycoproteins (HGRPs) (Lawton and Lamb, 1987) and chitinases (Hedrick *et al.*, 1988) have been shown to reflect a rapid incrase in the transcriptional activity of the corresponding genes.

To investigate the mechanisms by which the transcription of defence-related genes is activated in response to mechanical or biological stress we have characterized two members of a small family of wound-induced genes (*win*1 and *win*2) from potato and used reporter gene fusion to examine the spatial and temporal patterns of wound-induced transcription in transgenic potato plants.

Characterization of wound-induced genes, *win*1 and *win*2

A cDNA clone (clone 5) complementary to a mRNA species which accumulates in potato tubers (King Edward variety) upon wounding (Shirras and Northcote, 1984) was used to isolate a homologous genomic clone from a lambda library of Maris Piper DNA. This was found to contain two related genes which we called wound-induced genes *win*1 and *win*2, arranged in close tandem array (*see* upper part of Figure 8.2) (Stanford, Bevan and Northcote, 1989). These two genes encode almost identical proteins of 200 and 211 amino acids respectively with *win*2 containing a 10 amino-acid extension at the carboxy-terminus. The proteins are rich in glycine, serine and cysteine. A comparison of the predicted amino-acid sequences of *win*1 and *win*2 against all other protein sequences compiled in databases revealed some striking similarities to several plant proteins which have been implicated in plant defence.

First, both genes showed a close homology to hevein, a protein of unknown function from the latex of rubber-trees (Walujono *et al.*, 1975). The same degree of homology was shown for a chitinase from bean (Lucas *et al.*, 1985), and slightly less for a number of plant lectins; wheat-germ agglutinin (Wright, Gavilanes and Peterson, 1984), rice and nettle lectin (Chapot, Peumans and Strosberg, 1986) (Figure 8.1). In all cases, this homology extended over a 43-amino-acid domain which was located at the amino-terminal end of the mature proteins. Likewise, for *win*1 and *win*2, this region was immediately preceded by a 25 amino-acid leader sequence with all the structural characteristics of a signal peptide. This conserved domain is likely to be of some functional or structural significance since mature hevein consists of these 43 amino acids alone and wheat-germ agglutinin consists of four such domains arranged in tandem, of which the third repeat is shown in Figure 8.1. It is notable that all three lectins show a binding specificity for the sugar N-acetyl-glucosamine and its polymers. This includes the polymer chitin which forms a major component of fungal cell walls. Chitinase is also specific to chitin and it is therefore tempting to speculate that this highly-conserved domain is involved in

	residue		residue	homology to win1	win2
win1	26	QQCGRQKGGALQSGNLCCSQFGWCGSTPEFQSPSQGCQSRCTG	68		88%
win2	26	QQCGRQRGGALCGNNLCCSQFGWCGSTPEYQSPSQGCQSQCTG	68	88%	
Hevein	1	EQCGRQAGGKLCPNNLCCSQWGWCGSTDEYCSPDHNCQSNCKD	43	67%	72%
Chitinase	1	EQCGRQAGGALCPGGNCCSQFGWCGSTTDYCGP--GCQSQCGG	41	70%	72%
WGA	88	KCGSQSGGKLCPNNLCCSQWGSCGLGSEFC--GGGCQSGACS	127	55%	55%
Rice Lectin	1	GFLCPNNMCCSQWGYCGLGSEFC--GNTCQTTACS	33	49%	49%
Nettle Lectin	1	QRCGSQGGGTCPALRCCSIWGWCGASSPYC	31	48%	52%

Figure 8.1 Homologies between the deduced amino-acid sequences of *win*1 and *win*2 and the protein sequences of hevein, chitinase, wheat-germ agglutinin and rice and nettle lectins. The positions of the amino-acid residues are numbered in relation to the initiation codon of *win*1 and *win*2, and in relation to the amino-terminus of the mature proteins. Conserved residues are boxed

the recognition or binding of chitin or its subunits. Since chitin is not usually found in plant tissues, it is considered likely that the *win* gene family is involved in the defence of the plant against fungal pathogens.

While the coding sequences of *win*1 and *win*2 are highly conserved, with an overall homology of 81%, the sequences of the 5′ and 3′ flanking-regions and that of the single intron within each gene, were found to be highly divergent with the exception of conserved sequences surrounding putative transcriptional regulatory sequences such as the TATA box and CAAT box. In addition, a conserved 22 bp sequence was localized in the 3′ untranslated region of both genes (Stanford, Bevan and Northcote, 1989). This divergence of the non-coding sequences could indicate that *win*1 and *win*2 are differentially-expressed in response to wounding. To investigate the individual expression of *win*1 and *win*2, gene-specific probes were prepared from the divergent 5′ untranslated regions of each of the two genes and used in both Northern analyses and S1 nuclease protection experiments. The data showed that transcripts complementary to *win*2 accumulate in the leaves, stems, tubers and, to a lesser extent, in the roots of potato in response to wounding, whereas those complementary to *win*1 only accumulate in wounded leaves and stems (Table 8.1). Thus, these two members of the *win* gene family exhibit organ-specific expression in response to wounding.

Table 8.1 THE DIFFERENTIAL ORGAN-SPECIFIC EXPRESSION OF TANDEM GENES *WIN*1 AND *WIN*2 IN RESPONSE TO WOUNDING

	RNA accumulation in wounded tissue	
Organ	win*1*	win*2*
Leaf	+	+
Stem	+	+
Root	−	+
Tuber	−	+

Transcriptional activation of *win*-GUS gene fusions in transgenic potato plants

To examine the timing of induction and cell specificity of *win*2 gene transcription in response to wounding, we constructed transcriptional fusions between the putative regulatory sequences of *win*2 and the coding sequence of the β-glucuronidase reporter gene (Jefferson, Kavanagh and Bevan, 1987). One construct prepared in the Bin19-derived vector pBI101.2 contained the entire 5′ flanking sequences of *win*2, from the carboxy-terminus of *win*1 right to the initiation codon of *win*2 fused to the *GUS* gene, with the polyadenylation signal sequence of the nopaline synthase gene (Bevan, Barnes and Chilton, 1983) (Figure 8.2, Δ *win*2-GUS). In a second construct the *win*2 promoter was truncated up to position-570 from the transcriptional start site (Figure 8.2, Δ*win*2-GUS). These gene fusions were then transferred into the potato variety Desirée via *Agrobacterium tumefaciens* and

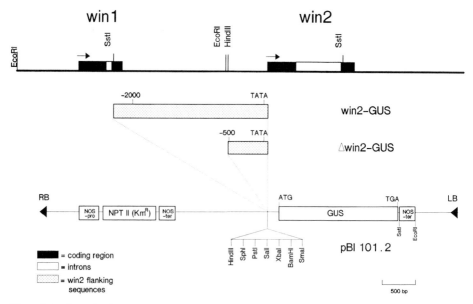

Figure 8.2   Organization of *win1* and *win2* within the genomic clone lambda-St5111, with the structure of the transcriptional fusions *win2*-GUS and *win2*-GUS shown beneath. The 5′ flanking sequences of *win2* were inserted into the polylinker of poly-KozakGUS-NOSter contained within the left (LB) and right (RB) borders of the T-DNA of Bin19. This vector (pBI 101.2) also contained the NPTII gene under the control of the nopaline synthase promoter (NOS-po) and terminator (NOS-ter) sequences

transformants selected for growth on kanamycin. In all cases, plants were regenerated in tissue culture only. This minimized the risk of environmental stresses such as fungal infection, insect attack or mechanical damage which could potentially trigger a wound response prior to excision experiments. For the same reason, mini-tubers were induced *in vitro* on stem-internode cuttings of transformed plants by growth on 7% sucrose.

Chimaeric *win*-GUS gene activity was then investigated in response to wounding in the leaves and stems of potential transformants. Extracts of tissue taken at various times after wounding were assayed fluorometrically for GUS activity using the substrate 4-methyl umbelliferyl glucuronide (MUG). The results of the analysis of a representative *win2*-GUS transformant are shown in Figure 8.3a in which the mean specific GUS activity of three replicate samples taken from the leaves and stems of a single plant has been plotted against time after wounding. Figure 8.3b shows the same samples taken from a potato transformant containing a fusion between the GUS reporter gene and the cauliflower mosaic virus (CaMV) 35S promoter (Odell, Nagy and Chua, 1985). A dramatic induction of GUS activity from barely detectable levels in intact tissue to maximal levels between 48 and 64 h after wounding, with a characteristic lag phase of 16 h, was observed in the leaves and stems of all *win2*-GUS transformants. Averaged over 15 plants derived from five such transformants, GUS activity in leaves and stems increased 430-fold and 450-fold respectively, over the first 48 h of ageing. In contrast, GUS expression under the control of the constitutive CaMV 35S promoter remained at a relatively constant level throughout the ageing time-course. Surprisingly, no β-glucuronidase

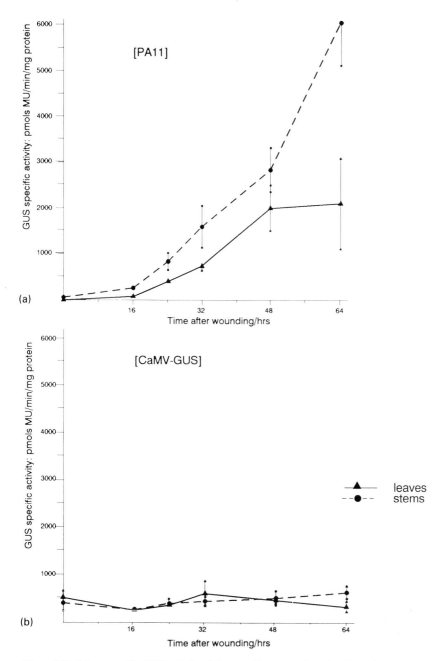

Figure 8.3 Mean specific GUS activity of three replicate samples taken from the wounded leaves (--▲--) and wounded stems (--●--) of (a) a representative potato transformant containing the *win2*-GUS gene fusion (PA11) and (b) a CaMV-GUS transformant (pBI121.1) over a time-course of ageing. The standard error of the means is shown. The GUS activity of tissue homogenates was measured fluorometrically using the substrate methyl umbelliferyl glucuronide and standardized to the protein content of the plant extract

activity was detectable in the leaves of potato plants transformed with *win*2-GUS containing the truncated *win*2 promoter, either before or after wounding (data not shown).

From these data, it could be concluded that sequences present within 2 kb of the 5' flanking region of *win*2 are sufficient and necessary for wound-induced gene expression. Furthermore, it appeared that essential regulatory sequences are located more than 570 bp upstream of the transcriptional start site.

The massive induction of GUS activity in *win*2-GUS potato transformants after wounding made this an ideal system within which to study the spatial pattern of wound-induced transcription using histochemical analysis.

In an initial experiment, a leaf from each of two replicate plants derived from a single independent transformant was wounded aseptically using the pointed ends of a pair of tweezers. One of these leaves was immediately excised and incubated with a histochemical substrate for GUS, 5-bromo-4-chloro-3-indolyl β-D-glucuronide (X-gluc). The second wounded leaf was left to age on the plant for approximately 48 h and was then excised and stained in the same way. Figure 8.4a shows there to be no staining in the unaged leaf, even after a 16-h incubation with the substrate. In contrast, an intense indigo staining indicative of GUS activity was visible around many of the wound sites of the second leaf which had been aged for 48 h (Figure 8.4b). When this same leaf was viewed under different lighting conditions, the blue staining could be seen to extend beyond the wound site, predominantly within the veins of the leaf (Figure 8.4c). To verify that the observed pattern of staining was not merely an artefact due to differential penetration of the substrate into the wound, two leaves taken from replicate plants of a CaMV-GUS transformant were treated in the same way. An even staining was observed over the entire surface of leaves which had been aged for either 0 or 48 h prior to staining, consistent with a constitutive expression of the CaMV-GUS gene. However, less GUS activity was detected around the wound sites where presumably cells had already started to die (Figure 8.4d).

In further experiments to examine the spatial pattern of chimaeric *win*2-GUS gene expression over a time-course of ageing, a single leaf was wounded as before and then divided into quadrants. These leaf segments were then aged for increasing periods of time. Once again, there was no activity in leaf tissue incubated immediately after wounding, but after 48 h of ageing GUS activity was localized in cells immediately adjacent to the cut edges. After 96 h GUS activity was still predominantly localized in tissue bordering the wound, but X-gluc staining could be seen to extend into the veins of the leaf (Figure 8.4c). By 150 h, GUS activity was detectable throughout the entire network of vascular tissue of the leaf, while cells immediately bordering the wound ceased to express GUS. However, the transcriptional activity of the *win*2-GUS gene was not limited to the vascular tissue since at later time-points blue staining had spread to all parts of the leaf.

It is possible that the spread of X-gluc staining away from the original wound site is due to the movement of the GUS protein away from the site of synthesis in the vicinity of the wound. However, no gradient of staining was observed away from the wound site as would have been expected if this were the sole source of the gene product, but rather an induction to similar levels at more distal sites. Furthermore, GUS activity continued to spread throughout the leaf tissue even after the cells immediately bordering the wound had died and ceased to express GUS. It has therefore been concluded that the observed pattern of staining is a direct indication of the spatial pattern of expression of the win2-GUS gene during the wound

Figure 8.4 Histochemical analysis of GUS activity in the leaves of transformed potato plants after wounding. Leaves were vacuum-infiltrated with X-Gluc and incubated at 37°C for 16 hours essentially as described by Jefferson (1987). (a) Leaf from a *win2*-GUS transformant, incubated immediately after wounding. No GUS activity was detectable. (b) Leaf from a replicate *win2*-GUS transformant, stained 48 h after wounding. A blue precipitate indicative of GUS activity was visible around the wound sites. (c) The same leaf as shown in (b) but viewed under different lighting conditions. This shows GUS activity to be predominantly localized around the wound sites but also within the veins of the leaf. (d) Wounded leaf from a CaMV-GUS transformant which has been stained after 48 h of ageing

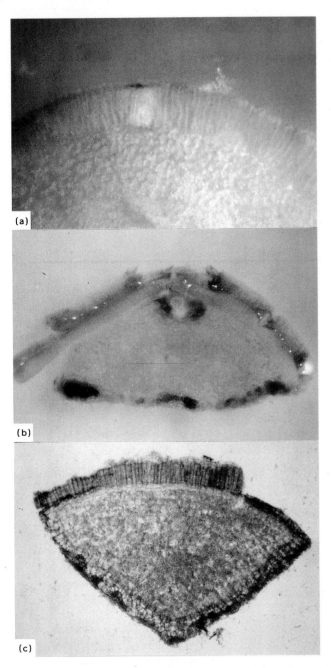

Figure 8.5 Histochemical localization of GUS activity in sections of mini-tubers derived from a *win*2-GUS transformant. Tubers were wounded using the ends of a pair of tweezers, cut in half or quadrants and then either incubated immediately with X-Gluc or aged before staining. (a) Section through a freshly-cut tuber showing localization of blue staining indicative of GUS activity beneath a lenticel within the periderm. (b) Section through half a mini-tuber which had been aged for 48 h prior to staining. GUS activity was localized in cells immediately adjacent to the cut edge. (c) Section through a quartered mini-tuber which had been aged for 150 h after wounding, showing the localization of GUS activity in cells immediately adjacent to the cut edge. The formation of a wound periderm can be observed several cell layers beneath the wound surface

response. Based on this reasoning, the hypothesis has been put forward that chimaeric gene expression under the control of the *win2* promoter is triggered by a putative wound stimulus which is initially generated at the site of the wound, but subsequently spreads to adjacent cells. The observation that gene expression is predominantly localized in the veins of the leaf before being induced systemically strongly suggests that this putative wound stimulus is transported through the vascular system. This would be expected to provide more rapid signalling than via cell-to-cell transfer. Certainly, when we looked at cross-sections of potato stems which had been aged for 0 to 48 h after wounding, we noted a marked induction of GUS activity in both the internal and external phloem and within the xylem (data not shown).

When we looked at cross-sections through mini-tubers, it was surprising to find that there was a background level of GUS activity in two distinct places in the freshly-cut tissue, both beneath the lenticels in the cuticle (Figure 8.5a) and within the developing buds of sprouting tubers. While this localized activity was inconsistent with our earlier RNA studies which had shown there to be no detectable RNA complementary to *win2* in freshly-cut tubers (Stanford, Bevan and Northcote, 1989), it could be easily explained by the fact that RNA was routinely extracted from a central core of tuber tissue following the removal of the periderm and underlying cortical tissue. Figure 8.5b shows a section through a mini-tuber which had been wounded with tweezers, cut in half and aged under sterile conditions for 48 h before sectioning and staining. GUS activity was detected in cells immediately adjacent to the exposed cut edge and surrounding the incision. However, unlike the pattern of expression observed previously in leaves, chimaeric gene activity remained highly localized even after prolonged ageing, by which time the wound periderm could be seen to be well-developed (Figure 8.5c). The pattern of expression was strikingly different in tubers which had started to sprout: a high level of GUS activity was rapidly induced throughout the cortex and vascular system after wounding.

These observations remained consistent with our hypothesis of a wound stimulus which is generated at the wound site and then transported via the vascular system to more distant tissues. In the dormant tuber, little or no flux would be expected through the vascular system of the tuber to other parts of the plant and consequently the wound stimulus would not be readily transmitted away from the cells at the wound surface. Upon sprouting, there is an increasing flux through the vascular system of the tuber to supply the demands of the rapidly-growing bud and the wound stimulus could therefore be relayed throughout the disperse vascular tissue, activating transcription of the *win2*-GUS gene in all parts of the cortex.

The identity of this putative wound stimulus remains unknown. It may possibly be an oligosaccharide, similar to the proteinase inhibitor inducing factor (PIIF) (Bishop *et al.*, 1981), which could be generated from the cell walls of the potato plant as a result of mechanical damage. It has been reported that the acropetal transport of PIIF from wounded leaves is blocked by steaming the petiole, a treatment thought to destroy the phloem while leaving the xylem intact. It has therefore been suggested that transport of PIIF takes place in the phloem (Makus, Zuroske and Ryan, 1981). However, there is as yet no direct evidence for the transport of oligosaccharide elicitors through plants after wounding (Baydoun and Fry, 1985). The spatial pattern of chimaeric gene expression described here certainly suggests that the vascular system provides a major route for signalling the transcriptional activation of defence-related genes.

Future experiments aim to identify this putative wound stimulus and its mode of transport, and also to determine the function of the *win2* gene product in the defence response of the potato.

Acknowledgements

This work was supported by the Gatsby Trust. A.C.S. acknowledges support from a SERC-CASE studentship and the Cambridge Philosophical Society.

References

Baydoun, E. A.-H. and Fry, S. C. (1985) The immobility of pectic substances in injured tomato leaves and its bearing on the identity of the wound hormone. *Planta,* **165**, 269–276

Bevan, M. (1984) Binary *Agrobacterium* vectors for plant transformation. *Nucleic Acids Research,* **12**, 8711–8721

Bevan, M. W., Barnes, W. M. and Chilton, M.-D. (1983) Structure and transcription of the nopaline synthase gene region of T-DNA. *Nucleic Acids Research,* **11**, 369–385

Bishop, P. D., Makus, D. J., Pearce, G. and Ryan, C. A. (1981) Proteinase inhibitor-inducing factor activity in tomato leaves resides in oligosaccharides enzymically released from cell-walls. *Proceedings of the National Academy of Sciences, USA,* **78**, 3536–3540

Boller, T. (1985) Induction of hydrolases as a defence reaction against pathogens. In *Cellular and Molecular Biology of Plant Stress*, UCLA Symposia on Molecular and Cellular Biology (eds J. L. Key and T. Kosuge), New Series. Liss, New York, vol. 22, pp. 247–262

Cassab, G. I. and Varner, J. E. (1988) Cell wall proteins. *Annual Review of Plant Physiology and Plant Molecular Biology,* **39**, 321–353

Chapot, M.-P., Peumans, W. J. and Strosberg, A. D. (1986) Extensive homologies between lectins from non-leguminous plants. *FEBS Letters,* **195**, 231–234

Chappell, J. and Hahlbrock, K. (1984) Transcription of plant defence genes in response to UV light or fungal elicitor. *Nature,* **311**, 76–78

Cramer, C. L., Bell, J. N., Ryder, T. B., Bailey, J. A., Schuch, W., Bolwell, G. P. *et al.* (1985a) Co-ordinated synthesis of phytoalexin biosynthetic enzymes in biologically-stressed cells of bean (*Phaseolus vulgaris* L.). *EMBO Journal,* **4**, 285–289

Cramer, C. L., Ryder, T. B., Bell, J. N. and Lamb, C. J. (1985b) Rapid switching of plant gene expression induced by fungal elicitor. *Science,* **227**, 1240–1243

Friend, J. (1985) Phenolic substances and plant disease. *Annual Proceedings of the Phytochemical Society,* **25**, 367–392

Hedrick, S. A., Bell, J. N., Boller, T. and Lamb, C. J. (1988) Chitinase cDNA cloning and mRNA induction by fungal elicitor, wounding and infection. *Plant Physiology,* **86**, 182–186

Jefferson, R. A. (1987) Assaying chimeric genes in plants: the GUS gene fusion system. *Plant Molecular Biology Reporter,* **5**, 387–405

Jefferson, R. A., Kavanagh, T. A. and Beven, M. W. (1987) GUS fusions: β-glucuronidase as a sensitive and versatile gene fusion marker in higher plants. *EMBO Journal,* **6**, 3901–3907

Lawton, M. A., Dixon, R. A., Hahlbrock, K. and Lamb, C. (1983) Rapid induction of the synthesis of phenylalanine ammonia-lyase and of chalcone synthase in elicitor-treated plant cells. *European Journal of Biochemistry*, **129**, 593–601

Lawton, M. A. and Lamb, C. J. (1987) Transcriptional activation of plant defense genes by fungal elicitor, wounding and infection. *Molecular and Cellular Biology*, **7**, 335–341

Leach, J. E., Cantrell, M. A. and Sequiera, L. (1982) Hydroxyproline-rich bacterial agglutinin from potato. *Plant Physiology*, **70**, 1353–1358

Lucas, J., Henschen, A., Lottspeich, F., Voegeli, U. and Boller, T. (1985) Aminoterminal sequence of ethylene-induced bean leaf chitinase reveals similarities to sugar-binding domains of wheat germ agglutinin. *FEBS Letters*, **193**, 208–210

Makus, D. J., Zuroske, G. and Ryan, C. A. (1981) Evidence that the transport of the proteinase inhibitor inducing factor out of wounded tomato leaves is phloem-mediated. *Horticultural Science*, **16**, 90–91

Odell, J. T., Nagy, F. and Chua, N.-H. (1985) Identification of DNA sequences required for activity of the cauliflower mosaic virus 35S promoter. *Nature*, **313**, 810–812

Ryan, C. A. (1978) Proteinase inhibitors in plant leaves: a biochemical model for pest-induced natural plant protection. *Trends in Biochemical Science*, **3**, 148–150

Shirras, A. D. and Northcote, D. H. (1984) Molecular cloning and characterisation of cDNAs complementary to mRNAs from wounded potato (*Solanum tuberosum*) tuber tissue. *Planta*, **162**, 353–360

Stanford, A. C., Bevan, M. W. and Northcote, D. H. (1989) Differential expression within a family of novel wound-induced genes in potato. *Molecular General Genetics*, **215**, 200–208

Walujono, K., Mariono, A., Hahn, A. M., Scholma, R. A. and Beintema, J. J. (1975) Amino-acid sequence of hevein. In *Proceedings of the International Rubber Conference*, Kuala Lumpur, Malaysia, vol. 2, pp. 518–531

Wright, C. S., Gavilanes, F. and Peterson, D. L. (1984) Primary structure of wheat-germ agglutinin isolectin 2. Peptide order deduced from X-ray structure. *Biochemistry*, **23**, 280–287

9

TUBER-SPECIFIC GENE EXPRESSION IN TRANSGENIC POTATO PLANTS

L. WILLMITZER, A. BASNER, W. FROMMER, R. HÖFGEN, X.-J. LIU,
M. KÖSTER, S. PRAT, M. ROCHA-SOSA, U. SONNEWALD and
G. VANCANNEYT
Institut für Genbiologische Forschung Berlin GmbH, Berlin, W. Germany

Introduction

Potato is one of the important crop plants for which both tissue culture techniques and gene transfer methods have been developed. Cloned genes, the expression of which closely correlates with the development of carbohydrate and protein storage organs, could be used as tools to study the mechanisms that regulate the accumulation of storage components and their possible role in the differentiation process. We have chosen patatin, a protein which is normally restricted to the potato tuber, in order to study its function and possible role in the development of a storage organ.

Tubers represent specific organs of a potato plant since they develop from a stem growing underground by drastic radial expansion to become specialized storage organs (Artschwager, 1924). Under field conditions the induction and growth of tubers is influenced by a variety of environmental factors, which have been extensively analysed in the past. Factors like short day length, low temperature or low nitrogen supply favour tuber formation (for summary *see* Ewing, 1985). Furthermore, certain plant hormones, mainly cytokinins, enhance tuber formation, whereas gibberellic acid inhibits the differentiation into tubers (*see* Melis and van Staden, 1984; Hannapel, Creighton-Miller and Park, 1985). Thus the potato tuber provides an interesting model system for developmental biology.

The patatin protein

The morphological process of tuberization is accompanied by a variety of biochemical changes, the most dramatic ones being the accumulation of starch (which is typical for the storage function of the potato tuber) as well as the appearance of new proteins. Up to 40% of the total soluble protein of potato tubers is represented by a family of immunologically identical glycoproteins with a molecular weight of about 40 kDa which have been given the trivial name 'patatin' (Racusen and Foote, 1980; Park, 1983). Storage proteins accumulate to high levels in seeds or tuber tissues. It is widely accepted that their most important function is to serve as a means of storing nitrogen for germination and early seedling growth. However, in addition to serving as a nitrogen deposit, a number of storage proteins also fulfil other roles which are often connected to protection against pathogens, as in the case of proteinase inhibitors (Richardson, 1976) and certain seed proteins displaying antifungal or antibacterial activity (Ponz *et al.*, 1983; Croy and Gatehouse, 1985).

In the case of patatin it has been speculated that, in addition to representing the main storage protein of potato tubers, it might have a lipid acyl hydrolase (LAH) activity. This proposal was based on the fact that patatin co-purified with the LAH activity in a number of biochemical assays (Racusen, 1984). As this biochemical evidence cannot rule out the possibility that the LAH activity of patatin is due to contamination by a different protein co-purifying with patatin, we decided to set up a genetic test to see whether patatin actually has an LAH activity. As the inactivation of the patatin genes in potato, and thus the construction of a patatin null mutant, is not possible, we decided to transfer the gene encoding patatin to a foreign host devoid of the patatin gene and to express it in a foreign organ.

We therefore expressed the patatin gene in tobacco leaves. For this purpose the protein-coding part as well as the 3'-downstream region of a patatin encoding gene was fused to the 5'-upstream region of ST-LS1, a gene from potato specifically expressed in a leaves and stems. A new esterase activity was found in leaves of transgenic tobacco plants, using polar lipids as substrates (Rosahl, Schell and Willmitzer, 1987). Andrews *et al.* (1988) were also able to demonstrate the appearance of a new LAH activity in a transgenic insect culture. This of course raises the question about the possible function of such massive amounts of LAH activity in potato tubers. One plausible assumption is that the LAH activity of patatin is important for the transition of the tuber from dormancy to vegetative growth, where a high LAH activity could be important for the rapid degradation of cell membranes and thus rapid liberation of certain metabolites. In addition, patatin might also be involved in protection against microbial invasion as the fatty acids liberated by the LAH activity might be used for the formation of wax or certain cytotoxic acid derivatives (Racusen, 1984). By transferring patatin genes into tobacco and obtaining stable expression and accumulation of the patatin protein in tobacco leaves new approaches have become possible which should be helpful in elucidating the possible biological role of this plant storage protein.

Although the cellular function of patatin is unclear, its substrates are part of intracellular membranes; thus it is obvious that the membranes of intact cells must be protected from this enzymatic activity. This could also explain the observed instability and rapid degradation of protoplasts isolated from tubers (Racusen, 1986). Export out of the cell, processing or compartmentation of the protein within vacuoles, where the enzyme should be inactive due to the low pH (Nishimura, 1982), could solve the problem of why intact cells are undamaged. Patatin is synthesized with a signal peptide (Kirschner and Hahn, 1986), which allows the polypeptide to enter the lumen of the endoplasmic reticulum (ER).

During the transport to its final destination the signal sequence is cleaved (Park *et al.*, 1985), the protein becomes *N*-glycosylated and the glycans are further modified to complex glycans (Sonnewald *et al.*, 1989b). Antibodies raised against the glycosylated protein show cross-reactivity with other glycoproteins (phyto-haemagglutinin isolated from Jack-bean; invertase from carrot cell walls), most likely mediated by antibodies reacting with a common glycan epitope present on many plant glycoproteins.

Since we are interested in the glycosylation, stability and activity of the protein, we analysed the immunocytochemical localization of patatin in cells from potato tubers using highly purified antibodies specific for the protein portion of patatin. Two different fixation methods, namely conventional chemical fixation and high pressure freezing followed by freeze-substitution (Müller and Moor, 1983) were used. In both cases the protein was found to be localized in vacuoles (Sonnewald *et al.*, 1989a). Storage of patatin in vacuoles may lead to the inactivation of the esterase activity, since the pH of vacuoles is known to be acidic. The intravacuolar pH of isolated vacuoles from castor bean endosperm was determined to be 5.7–5.9. The isoelectric points for three known patatin proteins are 5.3, 4.8 and 5.2, respectively. Solubility of proteins near to their isoelectric point is minimized and the observation that patatin seems to cluster in vacuoles could indicate that it is aggregated and if so, presumably inactive. Upon wounding, patatin would be released, could become activated and be involved in the rapid wound response, since it is present in such high amounts.

Another speculative possibility is that patatin may be involved in the response to pathogens, via the release of polyunsaturated fatty acids, i.e. arachidonic acid and eicosapentaenoic acid. Esters of arachidonic acid and eicosapentaenoic acid are present in membranes of *Phytophthora infestans* but not in potatoes. These compounds serve as elicitors and are responsible for the hypersensitive response in potatoes (Bostock, Kuc and Laine, 1981).

In order to test whether the compartmentation of patatin is necessary for cells to survive we are constructing chimaeric genes which should allow expression of the protein in the cytosol. In addition, patatin-coding regions of different members of the patatin gene family will be expressed in transgenic plants (i.e. tobacco) using heterologous promoters. This should allow us to study the enzymatic function of single patatin genes which might give us a clue about the biological function that patatin fulfils in higher plant cells.

With regard to the tissue specificity of the patatin expression, the protein was found mainly in parenchyma cells containing high amounts of starch and was not detectable in periderm cells as demonstrated by silver enhanced immunogold localization. The distribution of esterase activity closely followed this pattern. To show this, we performed esterase staining *in situ* with hand sections of tubers using α-naphthyl-acetate, which is known to be a substrate for patatin (Sonnewald *et al.*, 1989a).

The patatin gene family

The study of the patatin gene regulation has been complicated by the fact that the tetraploid genome of potato carries about 40–60 patatin genes, patatin thus being encoded by a gene family of more than 10 members per haploid genome. Several cDNA and genomic clones have been isolated and the complete nucleotide sequence of both the promoter and the coding region has been determined for some of them (Mignery *et al.*, 1984; Bevan *et al.*, 1986; Pikaard *et al.*, 1986; Rosahl *et al.*, 1986b; Twell and Ooms, 1987; Mignery, Pikaard and Park, 1988). The protein coding regions and the promoter region up to position −87 of all genes analysed so far are homologous, whereas upstream from this point the promoters diverge, allowing the genes to be divided into at least two classes. Apart from homologies reaching far into the 5′ sequence, class II genes are also characterized by the presence of a 22 bp sequence in the untranslated leader which is absent in class I genes. Both classes contain a conserved motif which is repeated several times in all promoters analysed so far and which binds a factor isolated from potato nuclei (Prat, personal communication).

One possible explanation for the presence of several different patatin genes in the potato genome is that different members of this multigene family carry different regulatory sequences and therefore react to different environmental, metabolic and developmental signals. Knowing that patatin is under strong transcriptional control, as shown by run off transcription assays (Rosahl *et al.*, 1986a), in principle two approaches can be followed to elucidate the regulation of single members of the family. Gene-specific probes might be the best way to analyse the regulation, but this approach has the disadvantage that prior to the study all genes must have been isolated and sequenced and appropriate regions of diversity must exist.

Gene fusion experiments, using reporter genes such as β-glucuronidase (GUS), have the advantage that single genes can be analysed in a homologous background (assuming appropriate DNA transfer methods for the analysed plant species are available). The GUS system (Jefferson, 1987; Jefferson, Kavanagh and Bevan, 1987) was chosen because of its high sensitivity and the possibility to identify histochemically the tissue- or even cell-specific expression of single genes. This also allows one to differentiate between generally weak promoters and promoters being strongly expressed in only very few cells.

Expression of class I and class II patatin genes in transgenic plants

In potato plants grown under either greenhouse or field conditions expression of patatin is, as a rule, restricted to tubers and to stolons associated with growing tubers (Paiva, Lister and Park, 1982; Rosahl *et al.*, 1986a). In addition, patatin is also expressed in roots, albeit at a 100-fold lower level (Pikaard *et al.*, 1987). Furthermore, expression of patatin is found in the anthers and petals of potato flowers (Vancanneyt *et al.*, 1989). Although tuberization is always accompanied by patatin expression, there are several instances where patatin expression has been observed in non-tuberizing tissues. Thus, patatin is expressed in petioles and stems of potato plants induced for tuberization upon removal of tubers and stolons (Paiva, Lister and Park, 1983). In addition, patatin accumulates to considerable levels in leaves of potato plantlets growing under axenic conditions on media supplied with high levels of sucrose (Rocha-Sosa *et al.*, 1989). In these cases, therefore, induction of patatin expression is independent from the morphological differentiation process of tuberization.

In order to understand the molecular mechanism underlying the complex control of patatin expression, as well as the differential expression of the two classes of patatin genes, we have isolated several members of the patatin gene family representing class I and class II genes. Our particular interest was to determine whether specific *cis*-acting upstream regulatory elements are responsible for the differential expression of these genes. For this purpose chimaeric genes composed of the upstream sequence of class I and class II patatin genes respectively and the β-glucuronidase gene of *Escherichia coli*, were constructed (Figure 9.1) and transferred into potato plants using recombinant *Agrobacterium* strains (Rocha-Sosa *et al.*, 1989; Köster-Töpfer *et al.*, 1989).

By this approach we could show that 1.5 kb of the 5'-upstream promoter region of a class I patatin gene from *Solanum tuberosum* cv. Berolina (B33) is essential and sufficient to direct a high level of tuber-specific gene activity which was on

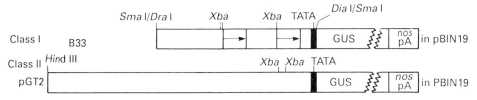

Figure 9.1 The structure of chimaeric patatin/βglucuronidase genes. ■ Homology between class I and class II up to position −87. GUS, βglucuronidase gene; *nos* pA, nopaline synthase 3′-end; TATA, TATA box. (For details *see* Rocha-Sosa *et al., 1989*; Köster-Töpfer *et al., 1989*)

average 100- to 1000-fold higher in tubers compared with leaves, stems and roots of transgenic potato plants. Results of one such experiment are summarized in Table 9.1. Histochemical analysis revealed this activity to be present in parenchymatous tissue but not in the peripheral phellem cells of transgenic tubers.

Table 9.1 GUS-ACTIVITY AVERAGE VALUES FOR POTATO PLANTS TRANSFORMED WITH A CLASS I (B33) OR CLASS II (pGT2) GENE IN pMol 4-MU/mg protein/unit

Gene	Leaf	Stem	Root	Tuber
B33	7	135	76	11 248
pGT2	34	63	1062	337
Untransformed	6	30	37	15

The level of activity observed in tubers with the B33 patatin promoter (2000–15 000 U) (1 U = 1 pmol 4-methylumbelliferol/mg protein/min) is within the same range as that observed with other strong promoters such as the cauliflower mosaic virus (CaMV) 35S promoter (average value in leaves between 2000 and 15 000 U), the photosynthetic ST-LS1 gene (J. Stockhaus, personal communication), or the potato proteinase inhibitor II in potato tubers (Keil, Sanchez-Serrano and Willmitzer, 1989). The B33 gene thus most likely represents an active tuber-specific gene since it carries a strong tuber-specific promoter region and its coding sequence, upon fusion to the CaMV 35S promoter and transfer into tobacco plants, yields a protein which shows immunological cross-reactivity with patatin antibodies and gives rise to a new esterase activity (R. Höfgen, personal communication).

A comparative analysis of the 5′-upstream sequences of the B33 patatin gene reveals a high degree of homology to both class II and class I genes up to position −87, whereas upstream from this up to at least −1.7 kb, sequences are homologous only to class I genes (e.g. Bevan *et al., 1986*; Mignery, Pikaard and Park, 1988). The major difference is the presence of the 208 bp direct repeat in B33. A possible role for this long repeat remains to be shown by means of deletions in expression studies.

Fusion of the 5′-upstream sequence of the class II gene pgT2 from the monohaploid line AM 80/5793 to the glucuronidase reporter and subsequent analysis of transgenic potato plants shows expression in defined cell layers in the tuber and in the root tip. In transgenic tobacco plants the same root specific expression is found (Köster-Töpfer *et al.*, 1989). These data therefore confirm and extend previous observations made for the expression of the endogenous class II patatin genes (Pikaard *et al.*, 1987). It is interesting to note that in this fusion the leader containing the 22 bp insertion was not included, arguing against an important role of this insertion, despite its conservation in class II promoters.

This histochemistry allowed the pgT2 promoter to be classified as being a strong promoter, its activity however being restricted to a minority of cells within the root or tuber (Köster-Töpfer *et al.*, 1989).

Metabolic signals regulating patatin genes

The process of tuberization in potato has been the subject of intensive studies at the physiological and morphological level. Potato tubers originate from stolons which represent lateral shoots normally borne at the basal nodes of the plant. They thus represent a modified underground stem with a radially expanded stem axis (Artschwager, 1924).

Tuber formation results in a drastically altered dry matter distribution within the plant, the tubers representing the sink which attracts most of the photoassimilates formed in the leaves.

Grafting experiments have shown that a tuberization stimulus must be formed in leaves of potato plants kept under conditions favouring tuber formation which can be transmitted from induced to non-induced plants leading to tuber formation in the non-induced ones. The nature of this/these transmissible factor(s) is still unknown (Melis and van Staden, 1984). In earlier studies, carbohydrate levels have been suggested to control tuberization (Wellensiek, 1929). This idea, however, has been rejected and changes in hormone levels are considered as important controlling events (Ewing, 1985). Recent work by several authors (Peterson and Barker, 1979; Forsline and Langille, 1976) has, however, demonstrated that tubers can be induced on potato stem nodes grown *in vitro* on media containing high levels of sucrose alone or in combination with high levels of cytokinin.

In order to try to dissect the morphological differentiation of stem tissue into tubers from the biochemical changes associated with it (i.e. starch formation and expression of new proteins) we tried to find conditions where the expression of patatin can be induced without the morphological changes occurring during tuberization. We therefore decided to test whether or not culturing intact potato plantlets on basic MS media (Murashige and Skoog, 1962) containing different levels of sucrose would result in the activation of patatin genes. We avoided the use of phytohormones (e.g. cytokinin) as previous studies in our laboratory had shown that this might result in the unspecific activation of gene expression (data not shown). Three sets of potato plants kept on MS-medium supplemented with either 2%, 7% or 10% sucrose, were prepared. No gross differences in the morphology of the plants could be detected, although plants kept at high sucrose concentrations grew more slowly than those kept on 2% sucrose. After 3 weeks, leaf, stem and root tissue were tested for the activity of patatin genes at the RNA level.

Patatin RNA accumulates to much higher levels in plants kept on media containing either 7% or 10% sucrose when compared with those kept on media with only 2% sucrose. The expression is highest in leaves, lower in stem and not detectable in roots of potato plantlets. It should be mentioned that this clear induction was observed in about 60–70% of the cases analysed. In the majority of the remaining cases we did not observe any induction, whereas in few cases the level of expression of patatin RNA in leaves of potato plants cultured on 2% sucrose only was already very high and did not show any increase when the plants were put on higher levels of sucrose. Though we do not yet understand why we sometimes failed to induce the expression of these genes, we would like to mention that in all cases where leaves contained high levels of patatin we also observed high levels of starch.

These data prove that the morphological differentiation occuring during tuber formation can be separated from the accompanying biochemical differentiation, at least with regard to patatin accumulation. It should be mentioned that we observed a close correlation between the expression of patatin and a second biochemical activity typical for somatic storage tissue, i.e. starch formation. If this linkage of starch accumulation and patatin expression is not just a coincidence, there could be two explanations for this:

1. The expression of patatin, and perhaps other tuber-specific genes, is controlled by one or several factors present during starch synthesis. If this assumption is correct one has to assume that these factors need to be present at concentrations in excess of a threshold level as low amounts of starch normally present in leaf cells of greenhouse grown plants do not lead to a detectable induction of expression of patatin (Rosahl *et al.*, 1986a).
2. The expression of patatin and the accumulation of large amounts of starch is indicative of the differentiation of the cells into a somatic storage tissue (*see also* Park *et al.*, 1985). The coincidence of both processes could then be explained by assuming that there is a common control.

If the latter assumption turns out to be correct, the system described for the reproducible induction of patatin in leaves of axenically cultured potato plants might represent a favourable system to study the formation of somatic storage tissue. Having regulatory regions of different patatin genes at hand, we therefore tested whether or not the corresponding chimaeric genes containing either a class I or a class II promoter would also react to elevated levels of sucrose in the medium.

Analysis of transgenic potato plants carrying the glucuronidase gene under the promoter of the B33 gene gave the following result. Whereas activity of the B33 gene promoter was very low in leaves from plantlets supplied with 2% sucrose, up to a 100-fold increase was found in leaves from plants supplied with 7% sucrose. This activation is cell specific since it was observed in leaves of mesophyll and epidermal cells but not in vascular tissue and not in guard cells (Rocha-Sosa *et al.*, 1989).

When similar experiments were performed using transgenic potato plants which contained the GUS gene under the control of the class II patatin promoter pgT2, no induction by sucrose could be observed. The 1.5 kb B33 gene promoter, in contrast to the class II promoter, must therefore also carry (a) *cis*-acting element(s) that react to metabolic signals. One and the same member of the gene family can therefore mediate patatin expression in tubers, but also in other organs under defined metabolic conditions. It is conceivable that the availability of starch or one of its precursors is a signal not only to initiate the morphological changes typical for tuberization but also for the activation of 'tuber-specific' genes. If this hypothesis is correct, the B33 glucuronidase gene would provide a convenient marker to study the factors involved in the switching of a somatic tissue into a storage tissue.

Acknowledgements

We thank Carola Recknagel and Carola Mielchen for performing most of the transformation experiments, Beate Küsgen and Regine Breitfeld for the glasshouse work and Anita Bauer for editing the manuscript. This work was supported in part by a grant from the Bundesministerium für Forschung und Technologie (BCT 0389 Molekular und Zellbiologische Untersuchungen an höheren Pflanzen und Pilzen) as well as a grant given by the Alexander-von-Humboldt-Stiftung to M. Rocha-Sosa.

References

Andrews, B., Beames, B., Summers, M. D. and Park, W. D. (1988) Characterization of the lipid acyl hydrolase activity of the major potato (*Solanum tuberosum*) tuber protein, patatin, by cloning and abundant expression in a baculvirus vector. *Biochemistry Journal,* **252**, 199–206

Artschwager, E. F. (1924) Studies on the potato tuber. *Journal of Agricultural Research,* **27**, 809–835

Bevan, M., Barker, R., Goldsbrough, A., Jarvis, M., Kavanagh, T. and Iturriaga, G. (1986) The structure and transcription start site of a major potato tuber protein gene. *Nucleic Acids Research,* **14**, 4625–4638

Bostock, R. M., Kuc, J. and Laine, R. A. (1981) Eicosapentaenoic and arachidonic acids from *Phytophthora infestans* elicit fungitoxic sesquiterpenes in the potato. *Science,* **212**, 67–69

Croy, R. R. D. and Gatehouse, J. A. (1985) Storage proteins. In *Plant Genetic Engineering,* (ed. J. H. Dodds), Cambridge University Press, Cambridge, pp. 143–268

Ewing, E. E. (1985) Cuttings as simplified models of the potato plant. In *Potato Physiology,* (ed. P. H. Li), Academic Press, Orlando, pp. 153–207

Forsline, P. L. and Langille, A. R. (1976) An assessment of the modifying effect of kinetin on *in vitro* tuberization of induced and non-induced tissues of *Solanum tuberosum. Canadian Journal of Botany,* **54**, 2513–2516

Hannapel, J. D., Creighton-Miller, J. C. and Park, W. D. (1985) Regulation of potato tuber protein accumulation by gibberellic acid. *Plant Physiology,* **78**, 700–703

Jefferson, R. A. (1987) Assaying chimaeric genes in plants: the GUS-gene fusion system. *Plant Molecular Biology Reporter,* **5**, 387–405

Jefferson, R. A., Kavanagh, R. A. and Bevan, M. W. (1987). GUS fusions: β-glucuronidase as a sensitive and versatile gene fusion marker in higher plants. *EMBO Journal,* **6**, 3901–3907

Keil, M., Sanchez-Serrano, J. and Willmitzer, L. (1989) Both wound-inducible and tuber-specific expression are mediated by the promoter of a single member of the proteinase inhibitor II gene family. *EMBO Journal,* (in press)

Kirschner, B. and Hahn, H. (1986) Patatin, a major soluble protein of the potato (*Solanum tuberosum* L.) tuber is synthesized as a larger precursor. *Planta,* **168**, 386–389

Köster-Töpfer, M., Frommer, W. B., Rocha-Sosa, M., Rosahl, S., Schell, J. and Willmitzer, L. (1989) A class II patatin promoter is under developmental control in both transgenic potato and tobacco plants. *Molecular and General Genetics,* (in press)

Melis, R. J. M. and Staden, van J. (1984) Tuberization and hormones. *Zeitschrift für Pflanzenphysiologie,* **113,** 271–283

Mignery, G., Pikaard, C., Hannapel, D. and Park, W. (1984) Isolation and sequence analysis of cDNA clones for the major potato tuber protein, patatin. *Nucleic Acids Research,* **12,** 7989–8000

Mignery, G. A., Pikaard, C. S. and Park, W. D. (1988) Molecular characterization of patatin multigene family of potato. *Gene,* **62,** 27–44

Müller, M. and Moor, H. (1983) Cryofixation of thick specimens by high pressure freezing. In *The Science of Biological Specimen Preparation for Microscopy and Microanalysis.* Proceedings of the 2nd Pfefferkorn Conference, (eds J.-P. R. Evel, G. H. Haggis, T. Bernard). SEM Inc., AMF O'Hare, Illinois, pp. 131–138

Murashige, T. and Skoog, F. (1962) A revised medium for rapid growth and bioassays with tobacco tissue cultures. *Physiologia Plantarum,* **15,** 473–497

Nishimura, M. (1982) pH in vacuoles isolated from castor bean endosperm. *Plant Physiology,* **70,** 742–744

Paiva, E., Lister, R. M. and Park, W. D. (1982) Comparison of the protein in axillary bud tubers and underground stolon tubers in potato. *American Potato Journal,* **59,** 425–433

Paiva, E., Lister, R. M. and Park, W. D. (1983) Induction and accumulation of major tuber proteins of potato in stems and petioles. *Plant Physiology,* **71,** 616–618

Park, W. (1983) Tuber proteins of potato – a new and surprising molecular system. *Plant Molecular Biology Reporter,* **1,** 61–66

Park, W. D., Hannapel, D. J., Mignery, G. A. and Pikaard, C. S. (1985) Molecular approaches to the study of the major tuber proteins. In *Potato Physiology,* (ed. P. H. Li), Academic Press Inc., Orlando, pp. 261–278

Peterson, R. L. and Barker, W. G. (1979) Early tuber development from explanted stolon nodes of *Solanum tuberosum* var. Kennebec. *Botanical Gazette,* **140,** 398–406

Pikaard, C. S., Mignery, G. A., Ma, D. P., Stark, V. J. and Park, W. D. (1986) Sequence of two apparent pseudogenes of the major potato tuber protein patatin. *Nucleic Acids Research,* **14,** 5564–5566

Pikaard, C. S., Brusca, J. S., Hannapel, D. J. and Park, W. D. (1987) The two classes of genes for the major potato tuber protein, patatin, are differentially expressed in tubers and roots. *Nucleic Acids Research,* **15,** 1979–1994

Ponz, A., Paz-Ares, J., Hernandez-Lucas, C., Carbonero, P. and Garcia-Olmedo, P. (1983) *EMBO Journal,* **2,** 1035–1040

Racusen, D. (1984) Lipid acyl hydrolase of patatin. *Canadian Journal of Botany,* **62,** 1640–1644

Racusen, D. (1986) Esterase specificity of patatin from two potato cultivars. *Canadian Journal of Botany,* **64,** 2104–2106

Racusen, D. and Foote, M. (1980) A major soluble glycoprotein of potato tubers. *Journal of Food Biochemistry,* **4,** 43–52

Richardson, M. (1976) The proteinase inhibitors of plants and microorganisms. *Phytochemistry,* **16,** 159–169

Rocha-Sosa, M., Sonnewald, U., Frommer, W., Stratmann, M., Schell, J. and Willmitzer, L. (1989) Both developmental and metabolic signals activate the promoter of a class I patatin gene. *EMBO Journal,* **8,** 15–23

Rosahl, S., Eckes, P., Schell, J. and Willmitzer, L. (1986a) Organ-specific gene expression in potato: isolation and characterization of tuber-specific cDNA sequences. *Molecular and General Genetics*, **202**, 368–373

Rosahl, S., Schmidt, R., Schell, J. and Willmitzer, L. (1986b) Isolation and characterization of a gene from *Solanum tuberosum* encoding patatin, the major storage protein of potato tubers. *Molecular and General Genetics*, **203**, 214–220

Rosahl, S., Schell, J. and Willmitzer, L. (1987) Expression of a tuber specific storage protein in transgenic tobacco plants: demonstration of an esterase activity. *EMBO Journal*, **6**, 1155–1159

Sonnewald, U., Studer, D., Rocha-Sosa, M. and Willmitzer, L. (1989a) Immunocytochemical localization of patatin, the major glycoprotein in potato (*Solanum tuberosum*). *Planta*, (in press)

Sonnewald, U., Sturm, V., Chrispeels, M. and Willmitzer, L. (1989b) Targetting and glycosylation of patatin, the major potato tuber protein in leaves of transgenic tobacco plants. *Planta*, (in press)

Twell, D. and Ooms, G. (1987) The 5′ flanking DNA of a patatin gene directs tuber specific expression of a chimaeric gene in potato. *Plant Molecular Biology*, **9**, 365–375

Vancanneyt, G., Sonnewald, U., Höfgen, R. and Willmitzer, L. (1989) Anther and petal specific expression of a patatin like protein in flowers of potato (*Solanum tuberosum*) and pepper (*Capsicum annum*). *The Plant Cell*, **1**, 533–540

Wellensiek, S. (1929) The physiology of tuber formation in *Solanum tuberosum*. *Mededelingen, Landbouwhogeschool Wageningen*, **33**, 6–42

10

REGULATION OF GENE EXPRESSION IN TRANSGENIC TOMATO PLANTS BY ANTISENSE RNA AND RIPENING-SPECIFIC PROMOTERS

D. GRIERSON, C. J. S. SMITH, C. F. WATSON, P. C. MORRIS, J. E. GRAY, K. DAVIES, S. J. PICTON
Department of Physiology and Environmental Science, University of Nottingham School of Agriculture, Sutton Bonington, Loughborough, UK

G. A. TUCKER, G. SEYMOUR
Department of Applied Biochemistry and Food Science, University of Nottingham School of Agriculture, Sutton Bonington, Loughborough UK

W. SCHUCH, C. R. BIRD and J. RAY
ICI Seeds, Plant Biotechnology Section, Jealott's Hill Research Station, Bracknell, Berkshire, UK

Introduction

The ripening of tomato and other fruits involves the regulated expression of specific genes which are believed to encode enzymes that catalyse various aspects of the process. At least 19 different mRNA sequences that accumulate in ripening tomatoes have been cloned (Slater *et al.*, 1985) and several have been sequenced (Mansson, Hsu and Stalker, 1985; Grierson *et al.*, 1986a; Holdsworth *et al.*, 1987; Ray *et al.*, 1987; Sheehy *et al.*, 1987). These mRNAs are presumed to play a role in processes such as the increase in ethylene synthesis, and changes in colour, flavour, and texture that occur during the ripening of tomatoes. However, only one of these mRNAs, that encoding the pectin-hydrolysing enzyme polygalacturonase (PG), has been identified (DellaPenna, Alexander and Bennett, 1986; Grierson *et al.*, 1986a; Sheehy *et al.*, 1987). PG is synthesized *de novo* during ripening (Tucker and Grierson, 1982) and is believed to play a role in the solubilization and depolymerization of the pectin fraction of tomato fruit pericarp cell walls (Themmen, Tucker and Grierson, 1982; Crookes and Grierson, 1983). PG cDNAs have been used to investigate the regulation of gene expression during ripening of normal and mutant tomatoes (DellaPenna, Kates and Bennett, 1987; Maunders *et al.*, 1987; Knapp *et al.*, 1989). Accumulation of the mRNA at the onset of ripening is regulated at the transcriptional level (Sheehy, Kramer and Hiatt, 1988). Initiation and continued production of PG mRNA requires ethylene, which is synthesized by ripening fruit (Davies, Hobson and Grierson, 1988). Expression of the PG gene has also been shown to be switched off at 35°C (Picton and Grierson, 1988). Several groups have identified PG cDNAs (DellaPenna, Alexander and Bennett, 1986; Grierson *et al.*, 1986b; Sheehy *et al.*, 1987). Our clone (pTOM6), which extends 50 bp 5' to the translation start site, was identified by sequence comparison with that obtained from the *N*-terminus of the purified enzyme. PG is synthesized as a pre-protein from which 71 amino acids are cleaved from the

N-terminus to obtain the mature protein. A genomic library was constructed and, using the PG cDNA as a probe, five overlapping clones were identified, from which the sequence of a complete PG gene was obtained (Bird *et al.*, 1988). Southern analysis suggests that there is only one PG gene. It covers approximately 7 kb and contains eight introns ranging in size from 99 to 953 bp.

We have used nucleic acid sequences derived from the PG gene and cDNA to construct chimaeric genes, and expressed them in transgenic plants. These experiments have enabled us to show that ripening-specific gene expression is determined by DNA sequences just upstream from the PG gene and also to develop a new method for interfering with expression of endogenous genes by antisense RNA

Factors affecting expression of the PG gene

Detached mature green tomato fruit may be induced to ripen by incubation in air plus 10 µl/1 ethylene. This induces the accumulation of a number of mRNA sequences, including the PG mRNA (Maunders *et al.*, 1987), which begin to appear before any visible sign of ripening. This is consistent with the suggestion that at least some of these mRNAs are actually involved in causing ripening changes.

Ethylene-controlled processes can be delayed or prevented by supplying silver ions. For tomato, this can be achieved by infiltrating solutions of silver thiosulphate into the vascular tissue of fruit while attached to the plant prior to ripening (Davies, Hobson and Grierson, 1988). It is relatively simple to produce fruit in which parts of the pericarp receive silver, and where ripening is inhibited, whereas other areas remain silver-free and therefore ripen normally. Investigation of the effect of silver on the expression of the PG gene has shown that silver greatly reduces the accumulation of PG mRNA if given prior to ripening, and causes a decline in PG mRNA if given after ripening has begun (Davies, Hobson and Grierson, 1988). The inhibitory effect of silver is specific for PG and a number of other ripening-related mRNAs and many others not involved in ripening remain unaffected. This indicates that ethylene is required both for the initiation of high levels of PG gene expression and also for continued high expression once the PG gene has been switched on. The mechanism of action of ethylene at the molecular level is unknown.

The *rin* mutation, which maps at one end of chromosome 5, causes pleiotropic effects in ripening fruit. Apart from having large sepals, due to the close linkage of the *macrocalyx* gene, *rin* plants are otherwise normal. In ripening *rin* fruit the normal increase in ethylene production associated with ripening does not occur (Herner and Sink, 1973). Lycopene production and PG enzyme activity are also greatly reduced in *rin* fruit, which turn a lemon yellow colour but remain sound and basically unripe for 1 year or more (Herner and Sink, 1973; Grierson *et al.*, 1987). As suggested previously (Grierson *et al.*, 1986a), *rin* plants contain the PG gene but its expression is drastically reduced (DellaPenna, Kates and Bennett, 1987; Grierson *et al.*, 1987; Knapp *at al.*, 1989). Although supplying ethylene to *rin* fruit does not restore normal ripening, the expression of some mRNAs is stimulated (Knapp *et al.*, 1989). This indicates first, that the *rin* phenotype is not due solely to a lack of ethylene production and, second, that *rin* fruit can perceive ethylene but do not respond to it completely. The expression of the PG gene, which maps at a single locus on chromosome 10 (Mutschler *et al.*, 1988) is not enhanced by treating *rin* fruit with ethylene. This indicates that a *trans*-acting factor necessary for the expression of the PG gene is absent or does not function normally in the *rin* mutant (Knapp *et al.*, 1989).

Expression of the PG gene and a number of other unidentified 'ripening genes' is inhibited at 35°C (Picton and Grierson, 1988). A number of ripening processes are affected, not just colour production, and our results indicate that the failure of fruits to ripen normally is due to a reduction in the amounts of a number of mRNAs at elevated temperatures.

Gene constructs used to study the regulation of the PG gene in transgenic plants

The sequences used in constructing chimaeric genes for plant transformation were:

1. the CaMV 35S transcription promoter from cauliflower mosaic virus, a DNA virus, which gives strong expression in many different types of plant cell.
2. the PG promoter, a 1450 bp fragment from the 5′ side of the PG gene.
3. the bacterial chloramphenicol acetyl transferase (CAT) gene
4. a 730 bp fragment from the 5′ half of the PG cDNA, used in the reverse orientation, which includes the translation start site and 50 bp of untranslated sequence.

Each of the gene constructs was terminated with a 3′ sequence derived from the nopaline synthase (*nos*) gene (Figure 10.1).

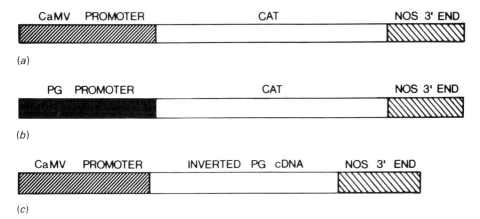

Figure 10.1 Gene constructs incorporated into the tomato genome via Ti plasmids

Chimaeric genes were transferred to *Agrobacterium tumefaciens* Ti plasmids (we used the BIN 19 binary vector system of Bevan, 1984) and thence to plant segments in tissue culture, utilizing the natural gene transfer process that operates in this bacterium. Kanamycin-resistant transgenic shoots were selected, tomato plants were regenerated and gene expression was studied in ripening fruit and other parts of the plant.

Regulation of ripening-specific gene expression

DNA sequences governing the expression of several plant genes have been shown to be located either to the 5' side of the gene (Nagy *et al.*, 1986; Baumann *et al.*, 1987; Ellis *et al.*, 1987), within an intron (Callis, Fromm and Walbot, 1987), or to the 3' side of a gene. We tested the hypothesis that a 1450 bp fragment from the 5' side of the PG gene contains sequences that determine ripening-specific gene expression. This sequence was fused to a CAT reporter gene (*see* Figure 10.1) and transferred to tomato plants (Bird *et al.*, 1988).

Analysis of the DNA of transgenic plants confirmed that chimaeric gene sequences were present. The PG promoter caused the CAT gene to be expressed in ripening fruit. Although the gene was also present in roots, stems, leaves, and green fruit, it was not expressed in these organs (Figure 10.2). This shows that 1450 bp of 5' flanking DNA from the PG gene cause the bacterial CAT gene to behave like a fruit ripening gene. In contrast, when the PG promoter was replaced with the CaMV 35S promoter, expression of CAT activity was observed in all parts of the plant investigated (Figure 10.2). Promoter-deletion experiments are being carried out to define more precisely the location of the ripening-specific control regions in the PG promoter. Quantitative measurements of the amount of PG and CAT mRNA (under control of the PG promoter) that accumulate in transgenic fruit indicate that the CAT transcript is present at a much lower level than anticipated. It is not yet clear whether this is due to instability of the CAT transcript or, alternatively, whether other sequences elsewhere in the PG gene are required

Ripening specific expression of chloramphenicol acetyl transferase in tomatoes

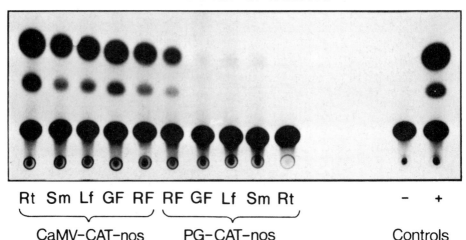

Figure 10.2 Expression of chloramphenicol acetyltransferase (CAT) genes under the control of the CaMV 35S or PG promoter in different organs of transformed tomato plants. CAT assays were carried out by incubating protein extracted from plant organs with ^{14}C-chloramphenicol (Bird *et al.*, 1988). Reaction products were separated by thin layer chromatography and detected by autoradiography. The spots are (in ascending order) chloramphenicol, 1-acetyl chloramphenicol, 3-acetyl chloramphenicol. Rt, root; Sm, stem; Lf, leaf; GF, green fruit; RF, ripe fruit

to give high rates of transcription. A comparison of the transcription rates of the endogenous PG gene and the chimaeric PG promoter-CAT gene in isolated nuclei should help to resolve this question.

Inhibition of PG gene expression by antisense constructs

Tomato plants were transformed with an inverted -PG gene construction ((JR16A) designed to express anti-PG RNA constitutively (Smith *et al.,* 1988). A 730 bp *Hinf* I fragment from the 5′ end of the PG cDNA, including 50 bp of untranslated region, was cloned in the reverse orientation between the CaMV 35S promoter (5′), and nopaline synthase (3′) sequences and transferred to plants. Thirteen

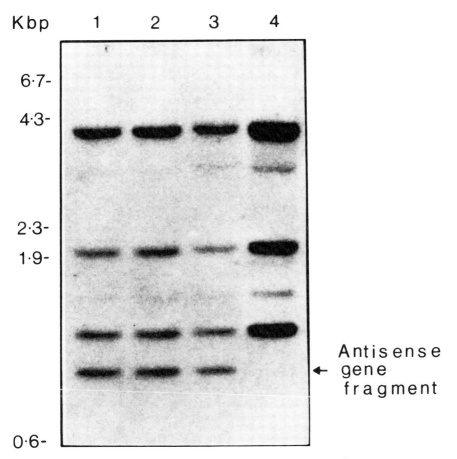

Figure 10.3 Genomic Southern analysis of the endogenous PG gene and added antisense gene in normal and transgenic tomatoes. DNA was extracted from three separate antisense and one untransformed plant, digested with *Eco*RI and *Hind*III, fractionated in a 0.8% agarose gel, blotted on to a nylon membrane, and probed with labelled PG cDNA. Tracks 1–3, transformed plants (containing antisense gene fragment); track 4, untransformed plant (containing only endogenous PG gene fragments)

transformants were identified on the basis of kanamycin resistance and five of these were analysed in detail.

Genomic Southern analysis confirmed the transformed status of these plants and suggested that each contains a single copy of the inserted antisense gene (Figure 10.3). Each line had reduced PG activity during ripening ranging from 5% to 50% of normal. Very similar results have been obtained by Sheehy, Kramer and Hiatt (1988). One of our plants (GR16) has been analysed in detail (Smith *et al.*, 1988). Reduced PG was apparent throughout ripening although other aspects of ripening such as lycopene accumulation and ethylene synthesis were unaffected. Measurement of sense and antisense transcripts by probing Northern blots with strand-specific probes showed that antisense RNA was present only in transformed and not control plants (Figure 10.4). Several different antisense transcripts were detected. These may be due to the fortuitous presence of poly A-addition signals in the antisense RNA sequence (Smith *et al.*, 1988). Antisense RNA under control of the CaMV 35S promoter accumulated in unripe fruit and leaves, as well as ripe fruit. PG mRNA was detected only in ripe fruit (Figure 10.4). In antisense fruit, however, the amount of PG mRNA was substantially reduced. This indicates that antisense RNA does not simply inhibit gene expression by forming duplexes with sense mRNA, thus preventing translation, but actually inhibits the accumulation of sense mRNA.

The inheritance and stability of antisense genes were studied in breeding experiments. GR16 was allowed to self-pollinate and Southern analysis identified

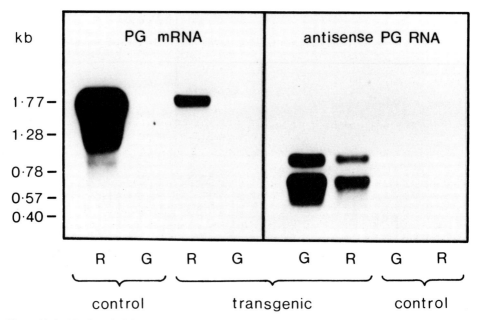

Figure 10.4   Northern hybridization showing expression of the PG sense and antisense transcripts during ripening of normal and transgenic plants. Poly (A)$^+$ RNA was fractionated in a 1.2% agarose gel, blotted on to a nylon membrane, and probed with sense and antisense specific RNA probes for PG. R, RNA extracted from ripe fruit; G, RNA extracted from green fruit

progeny with 0, 1, or 2 copies of the antisense gene. The reduced PG phenotype segregated with the inserted gene. Homozygous antisense F_1 plants had 1% of normal PG and hemizygous plants 20%, while those without the antisense gene had normal PG activity in ripening fruit (Figure 10.5). A representative of each type of progeny was analysed in greater detail and, as with the primary transformants, fruit of antisense plants had reduced PG activity throughout ripening. A similar pattern

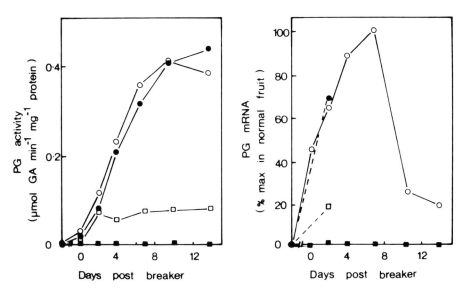

Figure 10.5 PG mRNA and enzyme activity during ripening of normal fruit and F_1 progeny of an antisense plant that have segregated with 0, 1, or 2 copies of the PG antisense gene. Normal, (○); 1 antisense gene, (□); 2 antisense genes, (■); segregated without antisense gene, (●)

was observed when PG mRNA was measured. No deviation from normal ripening was detected in a number of other parameters (ethylene production, lycopene accumulation, invertase activity, pectinesterase activity, and polyurinide solubilization). Although polyuronide solubilization was unaffected, the size of polyuronide fragments extracted from ripe fruit homozygous for the antisense gene was little changed compared with that from normal unripe fruit (Figure 10.6). This suggests that reducing the activity of PG to 1% of normal does not affect the release of soluble polyuronide but its hydrolysis to smaller fragments is largely prevented. The consequences of this in terms of fruit softening and processing quality are still being investigated. Preliminary measurements indicate that in antisense fruit there is no major change in softening, measured by the mechanical force required for fruit deformation (Smith *et al.,* 1988 and unpublished observations).

Discussion

These results indicate that *cis*-acting sequences regulating ripening-specific gene expression lie 5' to the PG gene (*see* Figure 10.2). Further detailed investigation, including promoter-deletion experiments, will be required to identify precisely the appropriate regulatory regions. During this work it will be important to establish whether there are control sequences that govern expression during ripening that are separate from those that give high levels of expression in the presence of ethylene. Furthermore, we need to establish whether other *cis*-acting regions of the chromosome in the vicinity of the PG gene also contribute to the maintenance of high levels of expression.

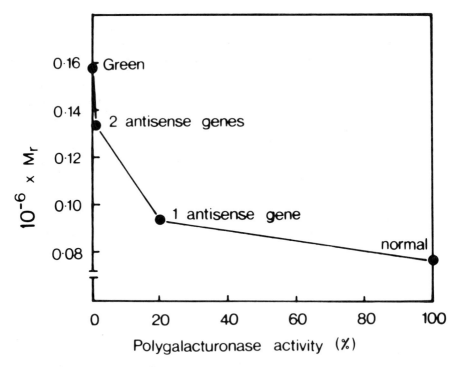

Figure 10.6   Weight-average molecular weight of soluble pectin from normal green and ripe fruit, and ripe fruit with 1 or 2 PG antisense genes. Ripe fruit were collected 14 days after the breaker stage

Studies *in vitro* may help to identify DNA binding proteins with a putative regulatory role in controlling expression of ripening genes. Ultimately, however, the interaction of putative ripening-specific control sequences and *trans*-acting factors needs to be studied *in planta*. Investigations with Ag^+-treated fruit, and fruit held at high temperatures, in which the expression of the PG gene is severely reduced (Grierson *et al.*, 1989) may prove useful in this respect. The *rin* mutation is particularly interesting, and may provide a means of identifying a *trans*-acting factor, since the mutation, on chromosome 5, affects expression of the PG gene on chromosome 10 (Knapp *et al.*, 1989).

PG has been implicated in pectin solubilization and/or softening, and it has also been suggested that it is involved, indirectly, in the control of ethylene production and the initiation of ripening. Using antisense gene constructs we have produced transgenic tomatoes with 1% the normal levels of PG (*see* Figure 10.5). This has enabled us to test the role of PG during tomato ripening. Studies with 1% PG fruit show that fruit develop and ripen at the normal rate. The timing and extent of ethylene and lycopene synthesis are unaltered (data not shown). Furthermore, fruit continue to soften. However, the reduction in pectin molecular weight that occurs during normal ripening is largely prevented (*see* Figure 10.6). Further studies are required to determine whether this has any consequences for fruit storage or processing quality.

The expression of antisense RNA in transgenic plants appears remarkably effective in reducing expression of the endogenous PG gene. Constitutive antisense RNA expression does not appear to have any ill effects on plant growth. The present indications are that antisense RNA works by reducing the amount of sense mRNA. This may well occur in the nucleus, rather than the cytoplasm (Smith *et al.,* 1988) and presumably requires either RNA/RNA or RNA/DNA base pairing. Further work is urgently required to define precisely the mechanism of action. This is important, since the information obtained may aid in the design of even more effective antisense constructs.

Even with the present level of effectiveness of antisense RNA technology, it is clear that it can be used to test the function of identified enzymes, as has been done for PG. It is also clear that the antisense approach will enable the creation of mutants to help assign functions to unidentified genes or cDNAs. In addition, it is now possible to consider altering biochemical pathways, to reduce or eradicate undesirable traits, or to change developmental programmes by the use of antisense RNA. These approaches can be applied not only to fruit ripening, but to many other stages of plant growth and development.

References

Baumann, G., Raschke, E., Bevan, M. and Schöffl, F. (1987) Functional analysis of sequences required for transcriptional activities of a soybean heat shock gene in transgenic tobacco plants. *EMBO Journal,* **6**, 1161–1166

Bevan, M. W. (1984) Binary *Agrobacterium* vectors for plant transformation. *Nucleic Acids Research,* **12**, 8711–8721

Bird, C. R., Smith, C. J. S., Ray, J. A., Moureau, P., Bevan, M. J., Bird, A. S. *et al.* (1988) The tomato polygalacturonase gene and ripening specific expression in transgenic plants. *Plant Molecular Biology,* **11**, 651–662

Callis, J., Fromm, M. and Walbot, V. (1987) Introns increase gene expression in cultured maize cells. *Genes and Development,* **1**, 1183–1200

Crookes, P. R. and Grierson, D. (1983) Ultrastructure of tomato fruit ripening and the role of polygalacturonase isoenzymes in cell wall degradation. *Plant Physiology,* **72**, 1088–1093

Davies, K. M., Hobson, G. E. and Grierson, D. (1988) Silver ions inhibit the ethylene-stimulated production of ripening-related mRNAs in tomato. *Plant, Cell and Environment,* **11**, 729–738

DellaPenna, D., Alexander, D. C. and Bennett, A. B. (1986) Molecular cloning of tomato fruit polygalacturonase: analysis of polygalacturonase mRNA levels during ripening. *Proceedings of the National Academy of Sciences, USA,* **83**, 6420–6424

DellaPenna, D., Kates, D. S. and Bennett, A. B. (1987) Polygalacturonase gene expression in Rutgers, *rin*, nor, and *Nr* tomato fruits. *Plant Physiology,* **85**, 502–507

Ellis, J. G., Llewellyn, D. J., Walker, J. C., Dennis, E. S. and Peacock, W. J. (1987) The *ocs* element: a 16 base pair palindrome essential for activity of the octopene synthase enhancer. *EMBO Journal,* **6**, 3203–3208

Grierson, D., Maunders, M. J., Slater, A., Ray, J., Bird, C. R., Schuch, W. *et al.* (1986a) Gene expression during tomato ripening. *Philosophical Transactions of the Royal Society London*, B, **314**, 399–410

Grierson, D., Purton, M. E., Knapp, J. E. and Bathgate, B. (1987) Tomato ripening mutants. In *Developmental Mutants in Higher Plants*, (eds H. Thomas and D. Grierson), Cambridge University Press, London, pp. 73–94

Grierson, D., Smith, C. J. S., Morris, P. C., Davies, K. M., Picton, S., Knapp, J. E. *et al.* (1989) Manipulating fruit ripening physiology. In *Manipulation of Fruiting*, (ed. C. J. Wright), Butterworths, London, pp. 387–398

Grierson, D., Tucker, G. A., Keen, J., Ray, J., Bird, C. R. and Schuch, W. (1986b) Sequencing and identification of a cDNA clone for tomato polygalacturonase. *Nucleic Acids Research,* **14**, 8595–8603

Herner, R. C. and Sink, K. C. (1973) Ethylene production and respiratory behaviour of the *rin* tomato mutant. *Plant Physiology,* **52**, 38–42

Holdsworth, M. J., Bird, C. R., Ray, J., Schuch, W. and Grierson, D. (1987) Structure and expression of an ethylene related mRNA from tomato. *Nucleic Acids Research,* **15**, 731–739

Knapp, J., Moureau, P., Schuch, W. and Grierson, D. (1989) Organisation and expression of polygalacturonase and other ripening related genes in Ailsa Craig 'Neverripe' and 'Ripening inhibitor' tomato mutants. *Plant Molecular Biology,* **12**, 105–116

Mansson, P., Hsu, D. and Stalker, D. (1985) Characterization of fruit specific cDNAs from tomato. *Molecular and General Genetics,* **200**, 356–361

Maunders, M. J., Holdsworth, M. J., Slater, A., Knapp, J. E., Bird, C. R., Schuch, W. *et al.* (1987) Ethylene stimulates the accumulation of ripening related mRNAs in tomatoes. *Plant, Cell and Environment,* **10**, 177–184

Mutschler, M., Guttieri, M., Kinzer, S., Grierson, D. and Tucker, G. (1988) Changes in ripening-related processes in tomato conditioned by the *alc* mutant. *Theoretical and Applied Genetics,* **76**, 285–292

Nagy, F., Fluhr, R., Kuhlemeier, C., Kay, S., Boutry, M., Green, P. *et al.* (1986) *Cis* acting elements for selective expression of two photosynthetic genes in transgenic plants. *Philosophical Transactions of the Royal Society London*, B, **314**, 493–500

Picton, S. and Grierson, D. (1988) Inhibition of expression of tomato-ripening genes at high temperature. *Plant, Cell and Environment,* **11**, 265–272

Ray, J., Bird, C., Maunders, M., Grierson, D. and Schuch, W. (1987) Sequence of pTOM5, a ripening related cDNA from tomato. *Nucleic Acids Research,* **15**, 10587

Sheehy, R. E., Kramer, M. and Hiatt, W. R. (1988) Reduction of polygalacturonase activity in tomato fruit by antisense RNA. *Proceedings of the National Academy of Sciences, USA,* **85**, 8805–8809

Sheehy, R. E., Pearson, J., Brady, C. J. and Hiatt, W. R. (1987) Molecular characterisation of tomato fruit polygalacturonase. *Molecular and General Genetics,* **208**, 30–36

Slater, A., Maunders, M. J., Edwards, K., Schuch, W. and Grierson, D. (1985) Isolation and characterization of cDNA clones for tomato polygalacturonase and other ripening-related proteins. *Plant Molecular Biology,* **5**, 137–147

Smith, C. J. S., Watson, C. F., Ray, J., Bird, C. R., Morris, P. C., Schuch, W. *et al.* (1988) Antisense RNA inhibition of polygalacturonase gene expression in transgenic tomatoes. *Nature,* **334**, 724–726

Themmen, A. P. N., Tucker, G. A. and Grierson, D. (1982) Degradation of isolated tomato cell walls by purified polygalacturonase *in vitro. Plant Physiology,* **69**, 122–124

Tucker, G. A. and Grierson, D. (1982) Synthesis of polygalacturonase during tomato fruit ripening. *Planta,* **155**, 64–76

11

THE MOLECULAR BIOLOGY OF PEA SEED DEVELOPMENT WITH PARTICULAR REFERENCE TO THE STORAGE PROTEIN GENES

DONALD BOULTER, R. R. D. CROY, I. MARTA EVANS,
J. A. GATEHOUSE, N. HARRIS, A. SHIRSAT and A. THOMPSON
Department of Biological Sciences, University of Durham, Durham, UK

Introduction

Pea has been chosen as an experimental material not only because it is an important legume food source world-wide (Makasheva, 1983), but also because there is a genetic map (Marx, 1985) and both the biochemistry and physiology are amenable to study (Pate, 1985). The developing pea seed in particular, provides an excellent source of experimental material for important problems in cell biology, plant growth regulator physiology, intracellular transport, protein structure–function relationships, molecular evolution, optimization of crop protein quality and yield, as well as in plant developmental biology; peas however, do have the disadvantage for molecular genetics of possessing a large genome ($\sim 5 \times 10^9$ bp).

Developmental biology

The seed is the offspring, normally sexually produced, of the higher land plants, and is the organ of dispersal. It is a more complex propagule than the functionally related dispersal organ, the spore, of the lower plants and considerable development of the fertilized ovule, nourished by the maternal plant takes place before its separation from the latter.

Seed development is normally initiated by fertilization of the ovule which, in the pea, occurs within 24 h of flower opening, and seed development is therefore described in terms of 'days after flowering' (d.a.f) from fertilization to seed maturity. A good description has been established of some of the major biochemical, histological and fine-structural changes which take place during pea seed development and this has led to a model of development consisting of three main developmental sub-programmes which fit into three developmental phases (*see* Boulter, 1981). Thus, the new offspring does not develop continuously, but as a result of internal and external factors, ontogenesis is interrupted by a period of 'developmental arrest' prior to and during seed dispersal.

The first sub-programme is morphogenesis and this phase, which occupies less than one-third of the developmental period, is principally distinguished by cell division activities. These give rise to the embryonic axis at opposite ends of which are the shoot and root meristems, and to the cotyledon cells which will accumulate storage materials. This phase finishes when the number of cells in the various seed compartments, i.e. cotyledons, embryo axis, and shoot has been fixed. During this time the seed remains relatively small and its moisture content remains high at about 85%.

The second sub-programme is to prepare the seed for autonomous growth after seed dispersal but prior to the establishment of the germinating seedling's own metabolic machinery to do so. This phase is one of seed filling and is principally characterized by the enlargement of the cells of the cotyledon and the laying down of reserves. In the pea seed these reserves are principally starch and the storage proteins, vicilin and legumin; by the onset of phase two, cell division has ceased. As well as a metabolism geared to enhance and support starch and storage protein synthesis, which are stored in order to supply nutrients required to establish the seedling on seed germination, other metabolic changes take place which also prepare the seed for subsequent germination and renewed seedling growth. Phase two terminates when the seed reaches physiological maturity and has a moisture content of about 60%; this phase occupies about half the total developmental time period. If immature seeds are taken before this point they may germinate, but do not produce vigorous seedlings unless they are artificially dried over a period of several days to about 60% moisture content (Le Deunff and Rachidan, 1988).

During the first two phases the maternal plant supplies the nutrients for development. Physiological maturity, is followed by a third sub-programme which prepares the seed for its dispersal. During phase three, drying out occurs following the severance of the vascular connections and the moisture content drops to a final level of about 15% (w/w). This is accompanied by a dramatic drop in metabolic activities, after which seed dispersal occurs. 'Stress' proteins are synthesized during this phase also, to protect the metabolic and storage proteins. During seed dispersal and prior to germination, metabolism is quiescent and the water content low. In addition to a relatively impermeable seed coat, some seeds contain inhibitors, which must be leached out before germination is possible, and other factors leading to a dormant state (Bewley and Black, 1982). This is not the case in peas which germinate immediately if conditions are right.

The process of determination, that is the control mechanisms necessary to initiate and stabilize the developmental changes leading to the mature seed, is not understood. The determinants of developmental phase one probably include mitogenic and polarity signals and by analogy with animal systems may involve maternal effect genes whose products activate some zygotic genes. The determinants of phase two may be triggered by the last cell division cycle since this phase starts when cell division has ceased. The drastic drying out in phase three follows severance of water supplies to the seed via the vascular tissues and the increased levels of abscisic acid at this time may be a response to the desiccation. Since many active enzymes have been shown to be potentially active in dry mature seeds (Bewley and Black, 1982), much of the general metabolic machinery of cells (house-keeping enzymes) must survive the dehydration process, possibly mediated by stress proteins.

Changes in protein synthesis

Figure 11.1 shows fresh weight and dry weight values, and Figure 11.2 vicilin and legumin contents measured serologically, at different stages during seed

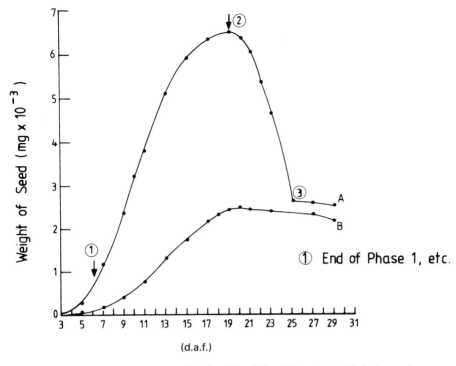

Figure 11.1 Profiles of mean pea seed fresh weights (A) and dry weights (B) during seed development measured in days after flowering (d.a.f.). Numbers 1, 2 and 3 indicate the end of each developmental phase as described in the text

development; the duration of the three developmental phases is also indicated. Figure 11.3 shows the changes taking place in the pattern of accumulated proteins separated on SDS-acrylamide gels, during seed development. These results show the appearance and accumulation of the two major storage protein types, vicilin and legumin, during phase two which correlate with the serological results of Figure 11.2 and the dry weight data of Figure 11.1. Legumin deposition can be seen from 9 d.a.f. The rate is initially quadratic and then from 13–21 d.a.f. approximately, is linear. Vicilin accumulation starts about a day earlier at 7/8 days and is biphasic correlating with vicilin protein deposition predomination up to 14 days, followed by convicilin protein deposition in the latter phase (*see* Figures 11.2 and 11.3).

This description of the changes which take place in protein synthesis during seed development poses a series of questions, the answers to most of which are still largely unknown. For example, why should there be two major classes of storage proteins in peas as well as in many other dicotyledonous seeds. A possible explanation is that, unlike legumin, vicilin contains no sulphur-amino acids. Since legumes often grow in extremely poor soils, many of which, at least before industrialization, were low in sulphur levels, it might be that under such conditions legume seeds could still develop and lay down vicilin storage proteins, since mechanisms exist whereby under sulphur deprivation, legumin synthesis is

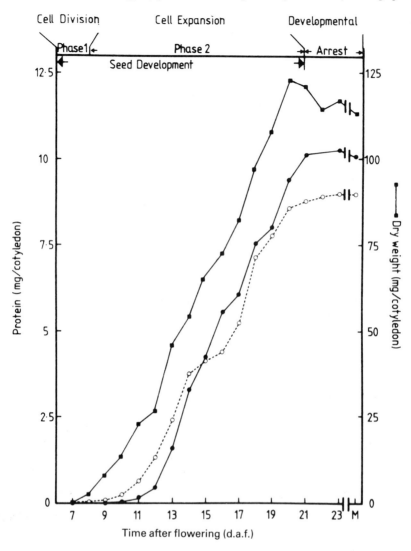

Figure 11.2 Profiles of legumin (●——●) and vicilin (○——○) protein accumulation measured by specific immunoprecipitation assays and total dry weight increase (■——■) during pea seed development expressed as mg/pea cotyledon. Development is measured in d.a.f. and the extent of each developmental phase is indicated. M = mature seed

suppressed whereas vicilin is not (Beach *et al.,* 1985; Evans, Gatehouse and Boulter, 1985). At least in these circumstances the seed would have storage protein reserves which would accommodate most of its amino acid requirements and if then dispersed to a better environment containing some sulphur, could on germination, assimilate the sulphur present in the soil to synthesize the sulphur-amino acids required for seedling growth. This type of biochemical adaptation would reflect the ecological situation of the wild ancestors of present-day peas.

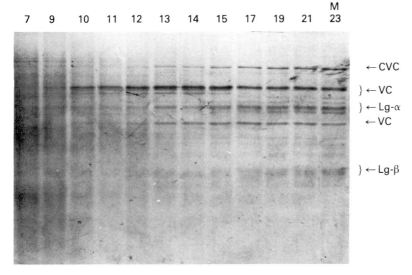

Figure 11.3   Qualitative assessment of the synthesis and accumulation of seed protein polypeptides in pea seeds during development using SDS-PAGE. Developmental stages are indicated in d.a.f. M = mature seed. Polypeptides indicated are, Cvc = convicilin, Vc = vicilin, Lg-α = legumin acidic and Lg-β = legumin basic polypeptides. Reproduced from Gatehouse *et al.* (1982)

STORAGE PROTEIN GENES

The data presented in Figures 11.2 and 11.3 show a greatly increased rate of synthesis (Figure 11.3) and deposition (Figure 11.2) of the storage proteins. Before presenting evidence for the mechanism of this process, it is necessary to describe the genes which encode these storage proteins. Gene copy reconstruction experiments on genomic Southern blots, restriction patterns of isolated genes and gene sequencing (Croy *et al.*, 1982; Casey, Domoney and Ellis, 1986) have all shown that vicilin and legumin are encoded by small multigene families (around 20 and 10 genes respectively). The sequence data shown that vicilin and legumin genes from whatever source, so far as is known, conform to an overall structural pattern (Figure 11.4). The sequences of *legA* and *legJ*, typical of two pea legumin gene subfamilies (*see* Figure 11.6*a*), have been published (Lycett *et al.*, 1984; Gatehouse *et al.*, 1988). Figure 11.5 shows an example of a pea vicilin sequence and the sequence of a pea convicilin gene has been published (Bown, Ellis and Gatehouse, 1988). A considerable amount of protein and nucleic acid sequence data (Gatehouse, Croy and Boulter, 1984; Croy and Gatehouse, 1985; Casey, Domoney and Ellis, 1986) has established the relationship between the various genes encoding the different storage protein polypeptides (Croy *et al.*, 1988) (Figure 11.5 and 11.6*b* for vicilin and Figure 11.6*a* for legumin). About 18 different legumin subunit pairs have been separated on two dimensional (2D) SDS gels ± mercaptoethanol and 2D SDS isofocusing gels (Matta, Gatehouse and Boulter, 1981). Considering the extensive endoproteolytic and co- and post-translational processing which occur with legumin proteins, a total of 10 genes to give rise to

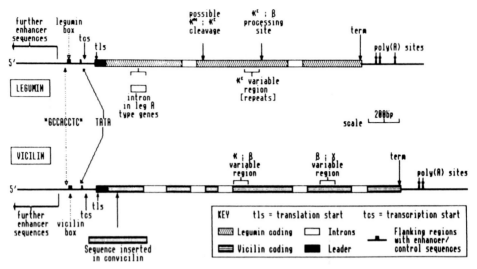

Figure 11.4 Schematic representation of the general features of (*a*) legumin type genes; and (*b*) vicilin type genes. Diagram shows features common and unique to both gene types.

```
vicB ........................................................................TATATATTATATTTTTCTTTTTAATATATAAATAAAGTATAGTATATGTAAAGTAAA -548

vicB CGGATAAATAATAGATAAATAATTAAATGACATATATGTACAATTACATTTTTATATATAAATTGACATATATATGTACAATTACATTTTTATATATTAAGTATAGTATAGTATATATAA -428

vicB AGTAGACGGATAAATAATGATAGATAATTAAATGACGTATATGTACAATTACATTTTTCACATGACAAGTACAAACATATGCACTTCTAAGTGCAAGTTTATGGAGTTATTTGCATGTCT -308

vicB TAGAGCTGGAGCTTGAGTTGTAGGATACAACACTTGTTAAAATTCTCTAGTCAATTCATTAATTCATATACACATGGCCGAAGACAATAATAAAGCATCCTCCTTTTCCATAAGAATGTC -188

vicB CAAAATTCATCAAATTCAAACAAAACTCCACCACCCAAGTAATGTTCTTTTCATTTTGCCACCTCAATTTTGTACATTTTAACACACGTCCATATGCATGGCACAACATGGCCAACTGTTG -68
     ................................................(............VICILIN BOX...............)....................

vicB GTGCATGTTAATTATATAGTTTTATTTTTTATATCTATAAATACACTCATCTCACTGTACTTTATTCATCCAGAGCGACCAAAGTGAGATATTAGTTTCAATCAACAGGCTGCTACTACA 53
     ....................(TATA box)...............(T^S)..................

cDNA ........................................................TTTCTTCTAGGTCTGATCCTCAAAATCCTTTTATCTTCAAGTCTAACAAGTTTCAAACT 59
vicB ATGAAAGCTTCATTTCCACTTTTGATGCTAATGGGAATCTCTTTCCTAGCATCAGTGTGTGTTTCTTCTAGGTCTGATCCTCAAAATCCTTTTATCTTCAAGTCTAACAAGTTTCAAACT 173
     (M  K  A  S  F  P  L  L  M  L  M  G  I  S  F  L  A  S  V  C  V  S  S:R  S  D  P  Q  N  P  F  I  F  K  S  N  K  F  Q  T

cDNA CTTTTTGAGAATGAAAATGGGCACATTCGACTTCTGCAGAAATTTGACCAACGTTCTAAAATTTTCGAGAATCTACAAAACTACCGTCTTTTGGAATATAAGTCCAAACCTCACACAATA 179
vicB CTTTTTGAGAATGAAAATGGGCACATTCGACTTCTGCAGAAATTTGACCAACGTTCTAAAATTTTCGAGAATCTACAAAACTACCGTCTTTTGGAATATAAGTCCAAACCTCACACAATA 293
     L  F  E  N  E  N  G  H  I  R  L  L  Q  K  F  D  Q  R  S  K  I  F  E  N  L  Q  N  Y  R  L  L  E  Y  K  S  K  P  H  T  I

cDNA TTTCTTCCACAGCACACCGATGCCGATTACATCCTTGTTGTACTCAGTG
vicB TTTCTTCCACAGCACACCGATGCCGATTACATCCTTGTTGTACTCAGTGGTAATTTATTATTTATCTAAGTTATTATTTTATTCTACATCTCTCTATGAGCTTCATTCAAATGCGCGGTAT 413
     F  L  P  Q  H  T  D  A  D  Y  I  L  V  V  L  S  (---------------

vicB TTATTATTTGTGAGGAGCTGCGGTAGTAACACTCTCTTTCAACACTCTCAATCACTCACTTTTTATGGTTGAAACATGTGGCCTCGCTTGGAAATAGGCCCACATAAAGTGGTAAGAGCA 533
     --------,---------------------------------------------------------------------------------------------------|IVS-1--------------

vicB CACATATTTCAACAAATGAAAGAGTGTGTGTTTGAGAGAGTGTTGAAAAGAGAGTATTATTAGCATTTCTCTTATTACTTTACATATTTGTTTTGTAAGGATAAATTAAATTCAGTTATA 653
     ------------------------------------------------------------------------------------------------------------

cDNA                       GAAAAGCTATACTCACAGTGTTGAAACCCGATGATAGAAACTCCTTCAACCTTGAGCGCGGAGATACGATAAAACTTCCTGCTGGCACAATTGC 322
vicB AAATCTAATGTCATTGTAATTTCCAGGAAAAGCTATACTCACAGTGTTGAAATCCCGATGATAGAAACTCCTTCAACCTTGAGCGCGGAGATACGATAAAACTTCCTGCTGGCACAATTGC 773
     -----------------------)G  K  A  I  L  T  V  L  */_P  D  D  R  N  S  F  N  L  E  R  G  D  T  I  K  L  P  A  G  T  I  A
```

Figure 11.5 Nucleotide sequences of pea vicilin gene *vicB* (Levasseur, 1988) and the closely homologous cDNA pAD 2.1 (Delauney, 1984). The two sequences combined encode a complete vicilin coding sequence, with 5' flanking sequence and some 3' flanking sequence. IVS = intron; T^S = predicted transcription start; : = *N*-terminus of mature

```
cDNA  TTATTTGGTTAACAGAGATGACAACGAGGAGCTTAGAGTATTAGATCTCGCCATTCCCGTAAATAGACCTGGCCAACTTCAG
vicB  TTATTTGGTTAACAGAGATGACAACGAGGAGCTTAGAGTATTAGATCTCGCCATTGCCGTAAATAGACCTGGCCAACTTCAGGTAATATAACCAATGTTTATCTATTCTCATATCAAATA  893
       Y  L  V  N  R  D  D  N  E  E  L  R  V  L  D  L  A  I  P/A V  N  R  P  G  Q  L  Q <---------------------------
```

```
cDNA  TGCTATGCATTCTAATGTACAAACAAATGTTAGGGGCCTCTACCATAACATCACAACAAAAATTGCGCCTGTACATATTTTTCTGTAATATTTTCCTAATATTTTCTTTATTTTTTTGT  1013
vicB  -------------------------------------------------IVS-2---------------------------------------------------------------------
```

```
cDNA              TCTTTCTTATTGTCTGGAAATCAAAACCAACAAAACTACTTATCTGGGTTCAGTAAGAACATTCTAGAGGCTTCCTTCAAT
vicB  CCTTTTTCAACAGTCTTTCTTATTGTCTGGAAATCAAAACCAACAAAACTACTTATCTGGGTTCAGTAAGAACATTCTAGAGGCTTCCTTCAATGTAAGCATAACACACAATTTTTTTTT  1133
      ------------>  S  F  L  L  S  G  N  Q  N  Q  Q  N  Y  L  S  G  F  S  K  N  I  L  E  A  S  F  N <----------------------
```

```
cDNA                                                                    ACTGATTATGAAGAGATAGAAAAGGTTCTTTTAGAAGAGCATG  538
vicB  CATTTATGTATGTATTAGTTTGGTATTGTATATGTTAATACTCACTTTGTCAATGTATGTATGTGTAAAAAAATATAGACTGATTATGAAGAGATAGAAAAGGTTCTTTTAGAAGAGCATG  1253
      ------------------IVS-3-------------------------------------->  T  D  Y  E  E  I  E  K  V  L  L  E  E  H
```

```
cDNA  AGAAAGAGACACAACACAGAAGAAGCCTTAAGGATAAGAGGGCAGCAAAGTCAAGAAGAGAATGTAATAGTAAAATTATCAAGGGGACAAATTGAGGAATTGAGTAAAAAATGCAAAGTCTA  658
vicB  AGAAAGAGACACAACACAGAAGAAGCCTTAAGGATAAGAGGGCAGCAAAGTCAAGAAGAGAATGTAATAGTAAAATTATCAAGGGGACAAATTGAGGAATTGAGTAAAAAATGCAAAGTCTA  1373
      E  K  E  T  Q  H  R  R  S  L  K  D  K  R  Q  Q  S  Q  E  E  N  V  I  V  K  L  S  R  G  Q  I  E  E  L  S  K  N  A  K  S
```

```
cDNA  CCTCCAAAAAAAGTGTTTCCTCTGAATCTGAACCATTCAACTTGAGAAGTCGCGGTCCTATCTATTCCAACGAGTTTGGAAAATTCTTTGAAATCACCCCAGAGAAAAATCCACAGCTTC  778
vicB  CCTCCAAAAAAGGTGTTTCCTCTGAATCTGAACCATTCAACTTGAGAAGTCGCGGTCCTATCTATTCCAACGAGTTTGGAAAATTCTTTGAAATCACCCCAGAGAAAAATCCACAGCTTC  1493
      T  S  K  K  R/G V  S  S  E  S  E  P  F  N  L  R  S  R  G  P  I  Y  S  N  E  F  G  K  F  F  E  I  T  P  E  K  N  P  Q  L
```

```
cDNA  AAGACTTGGATATATTTGTCAATTCTGTAGAGATTAAGGAG
vicB  AAGACTTGGATATATTTGTCAATTCTGTAGAGATTAAGGAGGTTGATGATAAAATTTATTTTTATAATATAGGAAATTCACCAAATTACACAATGAGATTTCACTTGATCAAATTACAATTGTT  1613
      Q  D  L  D  I  F  V  N  S  V  E  I  K  E <---------------------------------------
```

```
vicB  CTAAATGATTTGATTTTTGTCCTTTGAAGTTATAATGTCAAACTTTTGTTACTAACTTGACATCTCATACACAACAAGTTTTACATACTCAATAACATGTTTTATTTATAGAACATATAT  1733
      -------------------------------------------IVS-4--------------------------------------------------------------------------
```

```
cDNA                                                  GGATCTTTATTGTTGCCACACTACAATTCAAGGGCCATAGTAATAGTAACAGTTAACGAAGGAAAAGGAGATTTTGAAC  898
vicB  CTAATGTATTTATTTAATTATTCTTTCAAATTAAATATTAGGGATCTTTATTGTTGCCACACTACAATTCAAGGGCCATAGTAATAGTAACAGTTAACGAAGGAAAAGGAGATTTTGAAC  1853
      ------------------------------------------->  G  S  L  L  L  P  H  Y  N  S  R  A  I  V  I  V  T  V  N  E  G  K  G  D  F  E
```

```
cDNA  TTGTGGGTCAAAGAAAATGAAAACCAACAAGAGCAGAGAAAAGAAGATGACGAGGAAGAGGAACAAGGAGAAGAGGAGATAAATAAACAAGTGCAAAATTACAAAGCTAAATTATCTTCAG  1018
vicB  TTGTGGGTCAAAGAAAATGAAAACCAACAAGAGCAGAGAAAAGAAGATGACGAGGAAGAGGAACAAGGAGAAGAGGAGATAAATAAACAAGTGCAAAATTACAAAGCTAAATTATCTTCAG  1973
      L  V  G  Q  R  N  E  N  Q  Q  E  Q  R  K  E  D  D  E  E  E  Q  G  E  E  E  I  N  K  Q  V  Q  N  Y  K  A  K  L  S  S
```

```
cDNA  GAGATGTTTTTGTGATTCCAGCAGGCCATCCAGTTGCCCTAAAAGCTTCCTCAAATCTTGATTTGCTTGGGTTTGGTATTAATGCTGAGAACAATCAGAGGAACTTTCTTGCAG  1048
vicB  GAGATGTTTTTGTGATTCCAGCAGGCCATCCAGTTGCCGTAAAAGCAACCTCAAATCTTGATTTGCTTGGGTTTGGTATTAATGCTGAGAACAATCAGAGGAACTTTCTTGCAGGTATAT  2093
      G  D  V  F  V  I  P  A  G  H  P  V  A  V  K  A  S/V S  N  L  D  L  L  G  F  G  I  N  A  E  N  N  Q  R  N  F  L  A <-----
```

```
vicB  TATATTATCACCCAGTCTCTGTCACTATTTATTCATTTTAAGTGTGTATTTTAAAAGTCGACTTCTATTTAAAATCAAGGGGAAAATATTAAGATATGCTTATTATTTTGGTGATTAAAAA  2213
      -----------------------------------------------IVS-5-------------------------------------------------------------------
```

```
                          P  V  K  E  L  A  F  P  G  S  A  Q  E  V  D  R  I  L  E  N  Q  K  Q  S  H  F
cDNA       GCGATGAGGATAATGTGATTAGTCAGATACAGCGACCAGTGAAAGAGCTTGCATTCCCTGGATCAGCTCAAGAGGTTGATAGGATACTAGAGAATCAGAAACAATCCCACTTT  1245
vicB  TTTGAAGGCGATGAGGATAATGTGATTAGTCAGATACACCGA
      ------>  G  D  E  D  N  V  I  S  Q  I P/L R <----
```

```
vicB¹                                            GTATATTCTTGGAGCTGAAACTATCCGTTGCATGTTAGAGCTCTCTGAAACCAAGATTTTTAAGATTCCCTAAGCTAA  2333

       A  D  A  Q  P  Q  Q  R  E  R  G  S  R  E  T  R  D  R  L  S  S  V  I>...............................
cDNA  GCAGATGCTCAACCTCAACAAAGGGAGAGAGGAAGTCGTGAAACAAGAGATCGTCTATCTTCAGTTTGAAGATGTTTCTTAATGAGTGGACAAAATACTATGTATGTATGCTATCAAGAGA  1365
```

```
      ...................<PolyA+>.............................................
cDNA  TATATCTCACGCGGGAGCAATGAATAAAACAATGTTATCTTATAACTATAATTATATATCCACTTTTCTACTATGAATA
```

CONTINUATION OF *vicB¹*

```
vicB¹ CAACAGCCTCCTCTTTATAATCACCAATATACAAGATAACTCCATGCCGGTTCCAATGCAGGTTCCTCTTCTTCAGGTTCATATGCTTCGCAAGCTTACCATCATGATTCGAACACGGT  2453
```

```
vicB¹ AGAGTTCGAAACAATTTTGCCGCTGCCATCTTAGTGATCTCCTGCATGTGCGCTTCACTGAACTTTGAATTCGCTGAAATATAATCAACAGTTCAAGCAAAGTTTGCCTCTCTCCAATG  2573
```

```
vicB¹ ATACGATCCCTGTTCCTTCTACTACCGGCAAGAACTTTTGGTATTTTCCTTTAACTTGTTCTGTTTCTGCAGGTTCGGCGGTTTTATGCATAATGGCTTTTGGTCGTAGTCAAAACCGTG  2693
```

```
vicB¹ CTCTCTCACACCAAGCTTCCTAACAACATCATCAAAACTGTGGCAAC
```

protein. Sequence differences are indicated by bold type; those causing a change in amino acid sequence are indicated as alternative translations. *vicB* is probably inactive, being truncated by a large inserted sequence (*vicB¹*) which is at least 6.5 kb long, and contains repetitive DNA

these subunit pairs appears reasonable (Croy *et al.*, 1988). A comparable degree of heterogeneity has been observed in vicilin polypeptides.

Little is known about the organization within a genetic locus of the storage protein genes for either vicilin or legumin, although the chromosome location of some genes has been established (see Casey, Domoney and Ellis, 1986).

Since equivalent classes of genes have been shown to occur in other legumes, e.g. soyabean (Casey, Domoney and Ellis, 1986), the separation into the two A and J

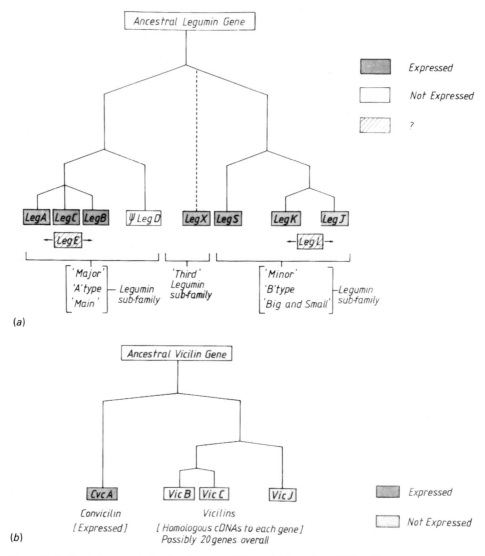

(a)

(b)

Figure 11.6 Evolutionary relationships between different (a) legumin and (b) vicilin gene families based on the available genomic mRNA (cDNA) and protein sequence data. Panels indicate whether genes are expressed (protein or cDNA sequence data), not expressed (pseudogene sequence data), or expression status unknown (gene sequence data)

subclasses probably preceded the evolution of the major legume genera. The similarity between different genes within the same subclass, e.g. A or J, suggests that either the events which were responsible for their duplication were extremely recent, or that genetic convergence has been active. It has sometimes been suggested that the relatively large number of genes for storage proteins is needed to express the large amounts of storage protein (80% of total protein) synthesized at this time (Broekaert and Van Parijis, 1978). However, another explanation has also been suggested.

Many genes, including house-keeping genes, have also been shown to be encoded by small multigene families similar to that of legumin. Furthermore, the different members of such families are very alike. Thus, up to 70% of the genome of many organisms consists of interdispersed repetitive DNA present in gene families from 10 to 100 members (Loomis and Gilpin, 1987). This is in contrast to the middle and highly repeated DNA sequences which are often localized and appear to be generated by unequal crossing-over within localized regions. Loomis and Gilpin (1987) have used computer simulation to show that evolution of most low repetitive sequences, such as the legumin gene family discussed here, could be the unavoidable consequences of duplication/deletion and divergence of inactive portions of genomes generated by relatively recent duplications of single genes and, therefore, highly homogeneous. This repetitive DNA, a by-product of large genomes, has on its own no selective advantage but is a small price to pay to allow rare, selectively advantageous changes in duplicate copies of genes to take place. When divergence occurs at about the same frequency as duplication and deletion, genomes carry repetitive sequences in proportion to their size, thus *Arabidopsis* legumin genes form a smaller family than in pea. This reasoning would not explain however, the presence of different legumin subclasses whose sequences differ considerably, but only those within a subclass. With regard to vicilin, since there is great sequence similarity between the different subclasses these may have themselves arisen in this manner.

As pointed out by Derbyshire, Wright and Boulter (1976) legumin type proteins occur widely in dicotyledonous plants and are also found in monocotyledons. Recently, a legumin type protein has been shown to occur in the seeds of some gymnosperms, e.g. *Ginkgo biloba* (Jensen and Berthold, 1989). The structure of the seed in gymnosperms differs from that of angiosperms in that a large mega-gametophyte persists to nourish the developing embryo, whereas this nutritive function in angiosperms is supplied either by the endosperm or the cotyledons. Thus, the timing of up-regulation of a gene would appear to have changed in the course of evolution from the mega-gametophyte to the embryo, i.e. from before to after fertilization.

Timing, controls and tissue location of up-regulation of storage protein genes

Regulation of the amounts of a gene product present in a cell, i.e. the encoded 'mature' protein (or polypeptide), may occur at the level of:

1. the gene itself (transcription)
2. by processing, transport rate and stability of the primary gene transcript (post-transcription)

3. on the polysomes (translation) and by the rate of co- and post-translational modifications
4. during transport to the site of deposition.

In order to demonstrate that a specific process (e.g. transcription or translation) is solely reponsible for regulation, it is required to demonstrate that the rate of that process is precisely correlated with the rate of deposition of the final gene product, namely the protein. These types of data are usually not available. Some understanding of regulation can be obtained by comparing the rates of transcription, the steady state levels of mRNA and the rates of protein synthesis for a particular gene. However, the results of such experiments are not easy to interpret as certain assumptions have to be made and especially so with the products of a multigene family. Thus, it has not been demonstrated in plants that the 'run off' transcription rates of isolated nuclei are the same as transcription rates

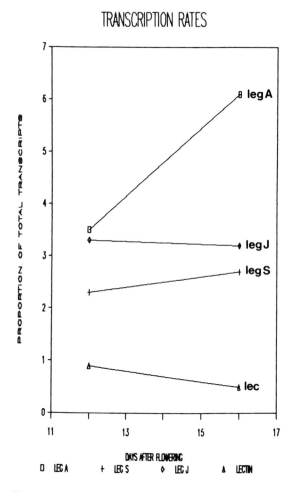

Figure 11.7 Comparison of rates of transcription from different genes at two stages of seed development (12 and 16 d.a.f.), as measured by nuclear transcription run-off assays

in vivo, although this appears to be the case in animals. Also, although there appears to be little turnover of storage proteins, so that rate of deposition can be equated with rate of protein synthesis, some breakdown of storage proteins could occur. Figures 11.2, 11.7 and 11.8 present the results of such assays at the protein, transcription and steady state mRNA levels.

There is very little transcription of the *legA* and *legJ* class genes from nuclei isolated at 8 days (Evans *et al.*, 1984), whereas at 12 days a greatly increased rate of transcription is recorded for both gene classes. This is reflected in the steady state levels of mRNA of the two gene classes and the quadratic rate of increase in legumin protein deposited, establishing that up-regulation of the legumin genes is controlled at the level of gene transcription. Between 12 and 16 days *legA* transcription rates continue to increase, whereas that of *legJ* remains constant and this is reflected in a quadratic rate of increase of *legA* messenger RNA levels and a linear increase in *legJ* mRNA, again suggesting transcriptional regulation of the *legA* genes at the level of mRNA formation. However, this is not matched by legumin protein deposition, which is linear over this period (*see* Figure 11.2), suggesting translational or post-translational level control is now rate limiting. Transcription from the class J genes is constant over the period 12–16 days and this is reflected in a linear increase in mRNA levels. However, if the levels of mRNA and transcription rates for *legA* and *legJ* class genes are compared at 12 and at 16 days, the increased rate of *legA* transcription relative to *legJ* does not fully account for the much greater levels of *legA* message relative to those of *legJ*, strongly

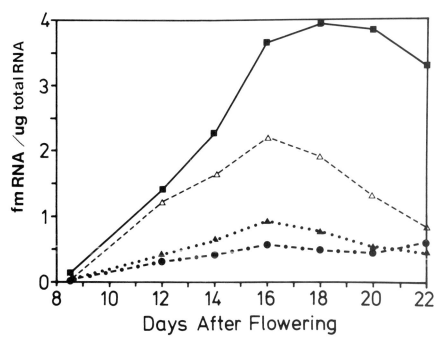

Figure 11.8 Comparison of absolute steady-state levels of specific mRNAs encoding different classes of seed protein polypeptides during pea seed development. RNA levels were assayed by quantitative Northern dot blots using cRNAs for quantitation: *l;egA* (■——058), lectin (△----△), *legS* (▲·····▲) and *legJ* (●—·—·—●)

suggesting that post-transcriptional events modulate the expression of these genes at this stage in development. mRNA levels for *legA* genes are still considerable at 22 d.a.f. (*see* Figure 11.8), while legumin synthesis has by then ceased. Furthermore, polysome density gradient profiles are similar to those at 18 days (data not presented), indicating that translational or post-translational controls are now operating. *legA* and *legJ* class mRNA levels peak at 18 and 16 d.a.f. and decline thereafter, indicating down-regulation of these genes. The data in Table 11.1 show that transcription rates on a per gene basis, are relatively similar for three legumin gene families and the lectin gene. This suggests that the individual

Table 11.1 GENE TRANSCRIPTION

Gene family	Copy no.	Transcription rate per gene family		Transcription rate per gene		Steady state mRNA levels	
		12 d.a.f.	16 d.a.f.	12 d.a.f.	16 d.a.f.	12 d.a.f.	16 d.a.f.
legA	4	1.0	1.0	0.25	0.25	1.0	1.0
legJ	3	0.93	0.51	0.31	0.17	0.20	0.14
legS	2	0.65	0.44	0.33	0.22	0.30	0.25
lec	1	0.27	0.09	0.27	0.09	0.87	0.60

d.a.f., days after flowering.

genes may have promoters of approximately equal activity and the variability in abundance of mRNAs is due to different gene copy numbers in each gene family, together with differential post-transcriptional control. The mRNA levels for vicilin genes peak at 14 days and for convicilin genes at 18 days and subsequently decline, indicating down-regulation in these genes also (Boulter *et al.*, 1987).

It can be concluded that the pattern of gene expression of different storage protein gene families differs and that this is also true for some genes within a gene family. Furthermore, at different stages of seed development, regulatory controls can act principally at different levels.

Run-off transcription assays of nuclei from leaves show that, unlike the control using the light harvesting chlorophyll a/b binding protein genes, the storage protein genes are either not transcribed, e.g. *legA*, or only transcribed at very low levels, e.g. *legJ*, *legS*, and lectin, confirming that transcription of storage protein genes is up- and down-regulated in the course of plant development.

During seed development, prior to its period of up-regulation, legumin mRNA can be demonstrated, albeit in very small amounts, as early as 3 d.a.f. (Harris *et al.*, 1989). This is at a stage when little storage protein can be detected in the seed. Furthermore, this RNA is located in the semi-liquid endosperm of the pea seed, a tissue which is not retained in the mature seed. It is interesting to note that in transgenic tobacco plants transformed with pea legumin genes, pea legumin is deposited in both the endosperm and the embryo of the developing tobacco seed (Croy *et al.*, 1988). In developing pea seeds, low levels of messenger RNA for legumin have been detected in various parts of the developing embryo, but none in the stem apical region. Once up-regulation has been initiated, light and electron microscopic *in situ* messenger RNA analyses for legumin mRNA show that messenger RNA (as well as legumin protein) are mainly found in the parenchymous storage cells of the cotyledons and much later and in lower amounts in the cells of the vascular tissues and the epidermis.

Mechanism of transcriptional control of up-regulation

Transcriptional control of gene expression is common-place in eukaryotes. Most genes have a general promoter consisting of TATA and CAAT sequences 5′ to the transcriptional start site, which work in conjunction with tissue or hormone-specific promoters, 5′ to the gene coding region and are effective in only one of the two possible orientations. In addition, there are enhancer sequences which increase transcription in a position and orientation independent fashion; these may be either tissue-specific or general. Examination of the 5′ regions of the legumin genes show that this same type of regulation using *cis*-nucleotide sequences can be assumed to occur here also. This hypothesis has been tested by transferring the legumin gene to tobacco (Shirsat *et al.*, 1989). Antibodies are available which distinguish the legumin gene products from the endogenous tobacco legumin. These experiments show that a fragment extending 1200 base-pairs 5′ (upstream) of the legumin coding sequence is sufficient to regulate the expression of the pea legumin gene in transgenic tobacco plants, such that it is only expressed in the tobacco seed and is under development control by being up-regulated during phase 2 of seed development. No expression is observed if the legumin box, a putative legumin gene specific regulatory element, is deleted. Sequences between 600 and 1000 base-pairs appear to be needed to 'enhance' the tissue specific expression. Similar results describing the characterization and regulation of the glycinin gene in soyabean, have been reported by Nielsen *et al.* (1989) and for the conglycinin gene family in soyabean by Chen, Pan and Beachy (1988).

It is now essential to generate and identify phenotypes of mutated genes with controlling roles in seed development and to use the techniques of molecular genetics to study their regulation and the function of their products.

References

Beach, L. R., Spencer, D., Randall, P. J. and Higgins, T. J. V. (1985) Transcriptional and post-transcriptional regulation of storage protein gene expression in sulphur deficient pea seeds. *Nucleic Acids Research*, **13**, 999–1013

Bewley, J. D. and Black, H. (1982) *Physiology and Biochemistry of Seeds in Relation to Germination*, vol. 2. Springer Verlag, Berlin

Boulter, D. (1981) Biochemistry of storage protein synthesis and deposition in the developing legume seed. *Advances in Botanical Research 9*, (ed. H. W. Woolhouse) Academic Press, London, pp. 1–31

Boulter, D., Evans, I. M., Ellis, J. R., Shirsat, A., Gatehouse, J. A. and Croy, R. R. D. (1987) Differential gene expression in the development of *Pisum sativum*. *Plant Physiology and Biochemistry*, **25**, 283–289

Bown, D., Ellis, T. N. and Gatehouse, J. A. (1988) The sequence of a gene encoding convicilin from pea (*Pisum sativum* L.) shows that convicilin differs from vicilin by an insertion near the N-terminus. Biochemical Journal, **251**, 717–726

Broekaert, D. and Van Parijis, R. (1978) The relationship between the endomitotic cell cycle and the enhanced capacity for protein synthesis in Leguminosae embryogeny. *Zeitschrift Pflanzenphysiologie*, **86**, 165–175

Casey, R., Domoney, C. and Ellis, N. (1986) Legume storage proteins and their genes. In *Oxford Surveys of Plant Molecular and Cell Biology*, vol. 3, (ed. B. J. Miflin), Oxford University Press, Oxford, pp. 1–95

Chen, Z. L., Pan, N. S. and Beachy, R. N. (1988) A DNA sequence element that confers seed specific enhancement to a constitute promoter. *EMBO Journal*, **7**, 297–302

Croy, R. R. D., Lycett, G. W., Gatehouse, J. A., Yarwood, J. N. and Boulter, D. (1982) Cloning characterisation and sequence analysis of cDNAs encoding *Pisum sativum* L. (pea) storage protein precursors. *Nature*, **295**, 76–79

Croy, R. R. D., Evans, I. M., Yarwood, J. N., Harris, N., Gatehouse, J. A., Shirsat, A. H. *et al.* (1988) Expression of pea legumin sequences in pea, *Nicotiana* and yeast. *Biochemie und Physiologie der Pflanzen*, **183**, 183–179

Croy, R. R. D. and Gatehouse, J. A. (1985) Genetic engineering of seed proteins: current and potential applications. In *Plant Genetic Engineering*, (ed. J. H. Dodds). Cambridge University Press, Cambridge, pp. 143–268

Delauney, A. J. (1984) Cloning and characterisation of cDNAs encoding the major pea storage proteins and expression of vicilin in *E. coli*. *Ph.D. Thesis*, University of Durham

Derbyshire, E., Wright, D. J. and Boulter, D. (1976) Legumin and vicilin, storage proteins of legume seeds. *Phytochemistry*, **15**, 3–24

Evans, I. M., Gatehouse, J. A. and Boulter, D. (1985) Regulation of storage protein synthesis in pea (*Pisum sativum* L.) cotyledons under conditions of sulphur deficiency. *Biochemical Journal*, **232**, 261–265

Evans, I. M., Gatehouse, J. A., Croy, R. R. D. and Boulter, D. (1984) Regulation of the transcription of storage protein mRNA in nuclei isolated from developing pea (*Pisum sativum* L.) cotyledons. *Planta*, **160**, 559–568

Gatehouse, J. A., Bown, D., Gilroy, J. Levasseur, M., Castleton, J. and Ellis, T. H. N. (1988) Two genes encoding 'minor' legumin polypeptides in pea (*Pisum sativum* L.). Characterization and complete sequence of the *legJ* gene. *Biochemical Journal*, **250**, 15–24

Gatehouse, J. A., Croy, R. R. D. and Boulter, D. (1984) The synthesis and structure of pea storage proteins. *CRC Critical Reviews in Plant Sciences*, **1**, 287–314

Gatehouse, J. A., Evans, I. M., Bown, D., Croy, R. R. D. and Boulter, D. (1982) Control of storage protein synthesis during seed development in pea (*Pisum sativum* L.). *Biochemical Journal*, **208**, 119–127

Harris, N., Grindley, H., Mulchrone, J. and Croy, R. R. D. (1989) Correlated *in situ* hybridisation and immunochemical studies of legumin storage protein deposition in pea (*Pisum sativum* L.). *Cell Biology International Reports*, **13**, 23–35

Jensen, U. and Berthold, H. (1989) Legumin-like proteins in gymnosperms. *Phytochemistry*, **28**, 1389–1394

Le Deunff, Y. and Rachidan, Z. (1988) Interruption of water delivery at physiological maturity is essential for seed development, germination and seedling growth in pea (*Pisum sativum* L.). *Journal of Experimental Botany*, **39**, 1221–1230

Levasseur, M. L. (1988) Comparative studies of the nucleotide sequences of pea seed storage protein genes. *Ph.D. Thesis*, University of Durham

Loomis, W. F. and Gilpin, M. E. (1987) Neutral mutations and repetitive DNA. *Bioscience Reports*, **7**, 599–606

Lycett, G. W., Croy, R. R. D., Shirsat, A. H. and Boulter, D. (1984) The complete nucleotide sequence of a legumin gene from pea (*Pisum sativum* L.). *Nucleic Acids Research,* **12**, 4493–4505

Makasheva, R. Kh. (1983) *The Pea.* Oxonian Press Pvt. Ltd., New Delhi, pp. 1–267

Marx, G. A. (1985) The pea genome: a source of immense variation. In *The Pea Crop,* (eds P. D. Hebblethwaite, M. C. Heath and T. C. K. Dawkins. Butterworths, London, pp. 45–54

Matta, N. K., Gatehouse, J. A. and Boulter, D. (1981) Molecular and sub-unit heterogeneity of legumin of *Pisum sativum* (L.) (garden pea) – a multidimensional gel electrophoretic study. *Journal of Experimental Botany,* **32**, 1295–1300

Nielsen, N. C., Dickinson, C. D., Cho, Tai-Ju, Thanh, Vu. H., Scallon, B. J., Fischer, R. L. *et al.* (1989) Characterisation of the glycinin gene family in soybean. *The Plant Cell,* **1**, 313–328

Pate, J. S. (1985) Physiology of pea. In *The Pea Crop*, (eds P. D. Hebblethwaite, M. C. Heath and T. C. K. Dawkins). Butterworths, London, pp. 279–296

Shirsat, A., Wilford, N., Croy, R. R. D. and Boulter, D. (1989). Sequences responsible for the tissue specific promoter activity of a pea legumin gene in tobacco. *Molecular and General Genetics,* **215**, 326–331

12

CONTROLLING ELEMENTS OF THE *PISUM SATIVUM legA* GENE

ANIL T. H. SHIRSAT, NEVILLE WILFORD, RONALD R. D. CROY and
DONALD BOULTER
*Department of Biological Sciences, University of Durham, Science Laboratories,
Durham, UK*

Seed storage proteins are laid down at a specific stage in seed development, to be used as a nitrogen store by the seedling on germination. The genes encoding them are under tissue specific and developmental regulation, and thus provide an excellent system for studying the control of expression of plant genes. The major storage proteins of legume seeds consist of the 11S legumins and the 7S vicilins (Gatehouse, Croy and Boulter, 1985). These proteins are accumulated in the cotyledons of the developing embryo and together constitute between 60 and 80% of the protein of the mature pea seed (Derbyshire, Wright and Boulter, 1976). The transcriptional activities of the legumin and vicilin genes as measured by run off transcription assays in developing pea nuclei, broadly correlate with the rate of legumin and vicilin protein accumulation, showing that these genes are principally controlled at the transcriptional level (Evans *et al.*, 1984; Boulter *et al.*, Chapter 11).

Most of the genes that control the expression of legumin in pea have been isolated, and most work on the functional significance of these genes has concentrated on the major legumin gene, *legA* (Ellis *et al.*, 1988). Although promoter deletion analysis in transgenic tobacco plants of genes coding for the storage proteins of wheat, soyabean and broad bean has been carried out, this has mainly been confined to the 7S vicilin type genes. Seed-specific expression of the legumin type gene from *Vicia faba* (Baumlein *et al.*, 1987) in tobacco has been shown, but a functional deletion analysis of the promoter region was not carried out. We have previously shown (Ellis *et al.*, 1988) that the pea *legA* gene with 1.2 kb of 5′ flanking sequence was able to direct the synthesis of a seed-specific legumin protein when transferred to *Nicotiana plumbaginifolia* via the disarmed binary *Agrobacterium* vector Bin 19 (Bevan, 1984). In order to identify elements within the *legA* promoter which were responsible for tissue-specific promoter activity, a series of promoter deletions was made, and transferred to tobacco via the Bin 19 system (Shirsat *et al.* 1989a).

The four *legA* promoter deletion constructs made are shown in Figure 12.1(*a–c*). After these truncated genes were cloned into *Bin19*, tobacco leaf discs of *N. plumbaginifolia* were infected with agrobacteria containing the various constructs, and shootlets arising from the transformation selected by their resistance to kanamycin. In order to overcome any significant variation in expression which might have arisen due to position effects of the introduced genes in the tobacco genome (*see later*), 15–20 plants were regenerated from each construct, and their

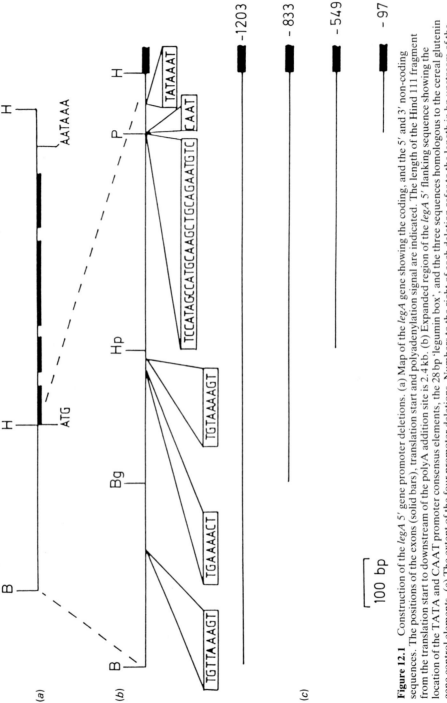

Figure 12.1 Construction of the *legA* 5′ gene promoter deletions. (a) Map of the *legA* gene showing the coding, and the 5′ and 3′ non-coding sequences. The positions of the exons (solid bars), translation start and polyadenylation signal are indicated. The length of the Hind 111 fragment from the translation start to downstream of the polyA addition site is 2.4 kb. (b) Expanded region of the *legA* 5′ flanking sequence showing the location of the TATA and CAAT promoter consensus elements, the 28 bp 'legumin box', and the three sequences homologous to the cereal glutenin gene control elements. (c) The extent of the four promoter deletions. Numbers to the right of each deletion refer to the length in bp upstream of the transcription start site of the *legA* gene. B, *Bam*HI; Bg, *Bgl*II; H, *Hind*III; Hp, *Hpa*II; P, *Pst*I. (Courtesy of Shirsat *et al.*, 1989a)

results individually analysed to lead to a conclusion about levels of expression. Plants were grown to maturity, allowed to flower and self fertilize, and tissue collected for DNA, RNA, and protein analysis. In order to exclude from the experiment transgenic plants which were either transformation 'escapes'. or which had rearranged copies of the introduced genes, all plants regenerated from the four constructs were subjected to Southern blot analysis (Shirsat *et al.*, 1989b). Only those plants having intact unrearranged copies of the introduced genes were selected for further analysis.

The amount of legumin protein synthesized in the leaves and seeds of the plants regeneration from each construct was determined by enzyme-linked immonosorbent assay (ELISA) using a monoclonal anti-legumin antibody (Figure 12.2). Legumin was not detected in the leaves of any of the transgenic plants. Analysis of mature seed samples by ELISA for legumin showed that the plants containing the −97 bp construct did not express the protein. Failure to express was not due to rearrangement or absence of the introduced genes, since Southern analysis showed that the genes were present in an unaltered form. Expression was first seen in a construct containing −549 bp of 5′ *legA* sequence. The presence of increasing lengths of 5′ promoter region from −549 to −1203 bp resulted in an increased in the level of legumin protein synthesized.

It is clear from these results that 97 bp of 5′ flanking sequence, despite the presence of the CAAT and TATA promoter consensus elements was insufficient for expression of the protein. Also, it was shown that the region between −97 and −549 bp has within it a positive control element. This segment of the legumin promoter has been shown to contain a long (28 bp) sequence (5′ TCCATAGC-CATGCAAGCTGCAGAATGTC 3′) which is highly conserved between the legumin genes between different species and has been termed the leg box (Baumlein *et al.*, 1987). The construct with 97 bp of *legA* 5′ sequence only contains the terminal 12 bp from the 3′ end of the leg box sequence, whereas the −549, −833 and −1203 bp constructs contain the whole of the leg box. It is quite likely, therefore, that the leg box sequence is necessary for legumin expression in *Pisum sativum*. Consensus sequences responsible for tissue specificity and promoter activity have also been identified in the lectin, glycinin, and Kunitz trypsin inhibitor genes from soyabean (Goldberg, 1986).

The overall increase in expression seen in transgenic plants containing the −833 and −1203 bp of *legA* 5′ flanking sequence implies the existence of additional positive control elements within this region. Three elements which have been found in this region (*see* Figure 12.1b) are closely related to the consensus sequence TG(T/A/C)AAA(G/A)(G/T) found in the glutenin genes of cereals. This sequence has been implicated in the expression of cereal storage protein genes by promoter deletion analysis (Colot *et al.*, 1987). However, whether this sequence plays any role in pea *legA* gene expression remains to be seen. Although the constructs containing −549 to −1203 bp of 5′ flanking sequence show an increase in expression, it cannot be said that this region contains an enhancer element in the classical sense.

To prove this, the *legA* sequences will have to be shown to confer enhancement of other genes when located in a position and orientation independent manner. It is also equally likely that the far upstream element, if present, is acting in a cooperative fashion with elements in the −549 bp and leg box region to elevate expression, and that the element will not function when removed from its normal gene environment.

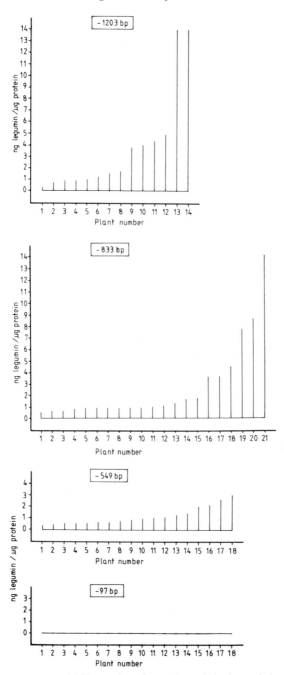

Figure 12.2 ELISA analyses for pea legumin in the seeds from the four sets of transgenic *Nicotiana* plants. The amount of legumin present is expressed in ng/μg of total protein in the seed extracts used. The values for the individual plants containing each construct are arranged in order of increasing legumin expression, with the lowest expressing lines to the left. The boxed numbers over each histogram refer to the *legA* promoter deletions from which the plants were regenerated. (Courtesy of Shirsat *et al.*, 1989a)

In order to exclude the possibility that the constructs were expressing legumin protein in leaf tissue, but that the synthesized legumin was undergoing degradation in a non-seed environment, Northern assays were performed (Figure 12.3). No legumin-specific mRNA was detected in the leaves under conditions where abundant levels were found in seeds, showing that seed specificity was retained by all constructs which expressed legumin. Legumin transcripts were not seen in seeds derived from the -97 bp construct, as expected.

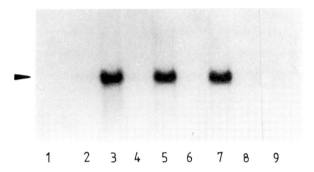

Figure 12.3   Northern blot analysis of leaf and seed total RNA from individual plants containing the four different *legA* constructs. A single-stranded labelled *legA* RNA probe was used in the hybridizations. (1) seed RNA and (2) leaf RNA from plant no. 4 containing the -97 bp construct; (3) seed RNA and (4) leaf RNA from plant no. 17 containing the -549 bp construct; (5) seed RNA and (6) leaf RNA from plant no. 21 containing the -833 bp construct; (7) seed RNA and (8) leaf RNA from plant no. 18 containing the -97 bp construct; (9) control (untransformed) leaf RNA. Arrow shows the location of the 2.1 kb legumin transcript. (Courtesy of Shirsat *et al.*, 1989a)

The possibility that disruption of a hypothetical temporal control element had occurred in one of the constructs was examined. Tobacco seeds from the four sets of transgenic plants were collected at various times after pollination and analysed for the presence of legumin by ELISA. The temporal pattern of legumin expression during seed development remained approximately the same for all the legumin-expressing constructs, and followed the rate of expression of total tobacco protein (Shirsat *et al.*, 1989b). This indicates that the temporal control element, if present, is closely associated with the elements which regulate tissue specificity and promoter activity.

As a large number of independently derived transgenic plants with the *legA* promoter deletions were available, the relationship between copy number and level of expression was examined. Previous reports on the variability of expression have either not analysed the copy number (Chen, Pan and Beachy, 1988), or have dealt with a reporter gene end-product which is susceptible to turnover or breakdown. Two aspects of this work make it more acceptable in reaching a conclusion about the relationship between copy number and levels of expression in transgenic plants: (1) a large sample size (24 plants) was chosen, and many seeds collectively analysed from individual plants to eliminate seed-to-seed variation, and (2) the use of a seed storage protein gene as a reporter and ELISA analysis to detect the final end-product. As there is little or no storage protein turnover in the mature seed, stable accumulation of the legumin protein occurs, enabling a very accurate

Table 12.1 GENE COPY NUMBER AND EXPRESSION LEVELS IN PLANTS
REGENERATED FROM THE TWO CONSTRUCTS. THE GENE COPY
NUMBERS WERE CALCULATED BY SCANNING THE BANDS
CORRESPONDING TO THE INTRODUCED GENES IN A LASER
DENSITOMETER, AND COMPARING THESE VALUES WITH THOSE
OBTAINED FOR THE GENE COPY NUMBER RECONSTRUCTION TRACKS.
TWENTY SEEDS FROM EACH INDIVIDUAL PLANT WERE POOLED AND
ANALYSED FOR LEGUMIN EXPRESSION, AND THE RESULTS
EXPRESSED AS ng LEGUMIN (AS DETERMINED BY ELISA) PER μg OF
TOTAL PROTEIN. THE A SERIES TRANSGENIC PLANTS CONTAIN THE
−1203 bp CONSTRUCT, WHILE THE B SERIES CONTAIN THE −833 bp
CONSTRUCT

Plant no.	Copy no.	Expression ng legumin/μg protein
A12	1	4.9
A9	1	1.7
A5	2	0.9
A7	3	14.0
A3	3	0.05
A11	3	0.0
A1	4	1.0
A4	4	14.4
A6	4	4.0
A2	12	0.68
A10	12	4.3
B3	1	2.3
B6	1	3.5
B7	1	3.5
B8	1	0.5
B9	1	0.9
B5	2	7.6
B1	3	0.0
B4	3	8.5
B10	4	1.6
B11	4	1.5
B2	5	4.3
B12	6	0.7

ELISA: enzyme-linked immunosorbent assay

determination of the level of expression. Southern analysis on plants from the
−1203 bp and −833 bp constructs was performed, and bands corresponding to the
introduced genes scanned with a laser densitometer and compared with gene copy
reconstruction tracks to enable an accurate determination of gene copy number
(Shirsat *et al.*, 1989b). Comparison of the corresponding levels of legumin protein
in pooled seeds from individual plants with the gene copy number (Table 12.1)
showed that there was no relationship between the two. Plant A2 for example from
the −1203 bp construct with 12 gene copies expresses only 0.68 ng/μg of total
protein, whereas plant A7 with only three copies of the introduced gene expresses
14 ng of legumin/μg of total protein – a level 20 times greater than plant A2. Similar
results can be seen for plants regenerated from the −833 bp construct.

The variation of expression is therefore almost certainly due to position effects of
the introduced genes in the tobacco genome.

We have identified sequence elements of the pea *legA* legumin seed storage protein gene which are necessary for its tissue specific expression. It is probable that these elements act to turn on the gene after complexing with specific DNA binding proteins, which are in their turn the product of other control pathways. Much effort is currently being expended to characterize signal transduction pathway genes, and this will ultimately lead to a better overall understanding of the true mechanism of gene regulation.

References

Baumlein, H., Muller, A. J., Schiemann, J., Helberg, D., Manteuffel, R. and Wobus, U. (1987) A legumin B gene of *Vicia faba* is expressed in developing seeds of transgenic tobacco. *Biologisches Zentralblatt,* **106**, 569–575

Bevan, M. (1984) Binary *Agrobacterium* vectors for plant transformation. *Nucleic Acids Research,* **12**, 8711–8721

Chen, Z. L., Pan, N. S. and Beachy, R. N. (1988) A DNA sequence element that confers seed specific enhancement to a constitutive promoter. *EMBO Journal,* **7**, 297–302

Colot, V., Robert, L. S., Kavanagh, T. A., Bevan, M. W. and Thompson, R. D. (1987) Localisation of sequences in wheat endosperm protein genes which confer tissue specific expression in tobacco. *EMBO Journal,* **6**, 3559–3564

Derbyshire, E., Wright, D. J. and Boulter, D. (1976) Legumin and vicilin, storage proteins of legume seeds. *Phytochemistry,* **15**, 3–24

Ellis, J. R., Shirsat, A. H., Hepher, A., Yarwood, J. N., Gatehouse, J. A., Croy, R. R. D. *et al.* (1988) Tissue specific expression of a pea legumin gene in seeds of *Nicotiana plumbaginifolia. Plant Molecular Biology,* **10**, 203–214

Evans, I. M., Gatehouse, J. A., Croy, R. R. D. and Boulter, D. (1984) Regulation of the transcription of storage protein mRNA in nuclei isolated from developing pea (*Pisum sativum*) cotyledons. *Planta,* **160**, 559–568

Gatehouse, J. A., Croy, R. R. D. and Boulter, D. (1985). The synthesis and structure of pea storage proteins. *CRC Critical Reviews in Plant Sciences,* **1**, 287–314

Goldberg, R. B. (1986) Regulation of plant gene expression. *Philosophical Transactions of the Royal Society London, B,* **314**, 343–353

Shirsat, A. H., Wilford, N. W., Croy, R. R. D. and Boulter, D. (1989a) Sequences responsible for the tissue specific promoter activity of a pea legumin gene in tobacco. *Molecular and General Genetics,* **215**, 326–331

Shirsat, A. H., Wilford, N. W., Croy, R. R. D. and Boulter, D. (1989b). Gene copy number and levels of expression in transgenic plants of a seed specific gene. *Plant Science,* **61**, 75–80

13

EXPRESSION OF MODIFIED LEGUME STORAGE PROTEIN GENES IN DIFFERENT SYSTEMS AND STUDIES ON INTRACELLULAR TARGETING OF *VICIA FABA* LEGUMIN IN YEAST

GERHARD SAALBACH, RUDOLF JUNG, GOTTHARD KUNZE, RENATE MANTEUFFEL, ISOLDE SAALBACH and KLAUS MÜNTZ

Zentralinstitut für Genetik und Kulturpflanzenforschung, Akademie der Wissenschaften der DDR, Gatersleben, E. Germany

Introduction

The most abundant storage protein of *Vicia faba* seeds is legumin, an 11S globulin. Two classes of legumin subunits have been found (Bassüner *et al.*, 1983; Horstmann, 1983; Wobus *et al.*, 1986), called A and B. Legumin (Le) A contains up to four methionine (Met) residues per subunit, LeB is completely free of Met. Since both classes occur at about the same level in the seeds of the cultivar Fribo (Müntz, Bassüner and Horstmann, 1984), and the predominating 50 kD subunit of the 7S globulin of *V. faba* is also completely free of Met (Bassüner *et al.*, 1987), the average Met content of *V. faba* storage proteins is extremely low, which considerably impairs the use of faba beans in animal feeds and human nutrition. Recombinant DNA and gene transfer techniques are opening new approaches to the improvement of the nutritional quality of plant seed proteins by direct modification of the respective genes or by transfer of suitable naturally occurring genes. Several attempts at gene modification have been reported (Hoffman, Drong and Donaldson, 1985; Hoffman, Donaldson and Herman, 1988; Radke *et al.*, 1988; Saalbach *et al.*, 1988; Wallace *et al.*, 1988), but in no case so far has the normal functioning in plants of the modified protein been demonstrated. In this chapter we report on the expression of a LeB gene which has been modified to contain four Met codons (Saalbach *et al.*, 1988) *in vitro*, in *Xenopus* oocytes, in yeast, and in tobacco. A second hybrid gene, containing 10 Met codons, constructed from the last exon of the modified LeB gene and a glycinin G2 gene, was also transformed into tobacco. We found that the modification probably interfered with intracellular polypeptide transport, and led to complete failure of accumulation of the corresponding protein in tobacco seeds.

Obviously, the system of biosynthesis, processing, holoprotein formation, intracellular transport, deposition in protein bodies, and degradation during germination of seed storage proteins is highly conserved throughout the plant kingdom, and many aspects of protein structure related to one or the other function might be sensitive to modifications.

Recent progress in yeast and plant systems has revealed that intracellular polypeptide transport into vacuoles and vacuolar protein bodies is not by bulk flow, but requires positive sorting information. In initial attempts, the role of glycans was investigated, since analogy to the mannose-6-phosphate signal of mammalian lysosomal proteins (Sly and Fischer, 1982) was suggested. However, both in yeast (Schwaiger *et al.*, 1982) and in plants (Voelker, Herman and Chrispeels, 1989) it was found that glycans represent no sorting and targeting signal. In yeast, protein sorting information is localized in a short amino acid sequence at the *N*-terminus of a polypeptide (Johnson, Bankaitis and Emr, 1987; Valls *et al.*, 1987; Klionsky, Banta and Emr, 1988). In plants, results on this topic have mainly been reported by Chrispeels and co-workers. They investigated phytohaemagglutinin (PHA) which accumulates in protein storage vacuoles of bean seeds. In yeast, this protein is also efficiently transported to the vacuole (Tague and Chrispeels, 1987). By expressing fusions of different *N*-terminal segments of PHA with the secretory yeast enzyme invertase in yeast, the system which was also used with the above mentioned yeast proteins, it was found that the vacuolar targeting signal also resides in a short polypeptide sequence at the *N*-terminus (Tague and Chrispeels, 1988). In addition, Dorel *et al.* (1989) found that in plants an ER-signal sequence alone is not sufficient to direct a cytosolic protein to the vacuole.

The sequence of legumin, like many other seed storage proteins, does not contain any glycosylation signal, indicating that glycosylation cannot be responsible for sorting. Legumin forms hexameric holoproteins in the protein bodies; trimers have already been found in the endoplasmic reticulum (Chrispeels, Higgins and Spencer, 1982). Although higher structures could form so-called signal patches, it could be suggested that, as with PHA, sorting information is contained in an amino acid sequence stretch at the *N*-terminus. Therefore, we have also used the yeast invertase system for the detection of vacuolar sorting information in the LeB of *V.faba*.

Modification of a legumin gene

A legumin (Le) B gene of *Vicia faba* (Bäumlein *et al.*, 1986) was modified by changing the reading-frame near the 3'-end to introduce Met codons (Saalbach *et al.*, 1988). This was done by removing the four protruding bases of a cleaved unique *Pst*I-restriction site, located 107 bases upstream of the stop codon, with subsequent religation. An early stop codon generated by the frame shift was then removed by *in vitro* mutagenesis. As the final result of the two modification steps, a gene was obtained encoding a modified LeB subunit with a completely different amino acid sequence at the *C*-terminus compared with the original polypeptide: the last 37 amino acid residues of the Met-free LeB subunit are replaced by 52 new amino acid residues including four Met residues (Figure 13.1). A comparison of the hydropathy data according to Hopp and Woods (1981) of the original and modified sequences reveals the latter to be more hydrophobic (Figure 13.2). The last exon of that gene was then used to construct a hybrid gene with a soyabean glycinin G2 gene, where the corresponding last exon was substituted by the modified LeB last exon (Figure 13.3). The resulting hybrid gene contains 10 Met codons.

Expression of the modified genes in different systems

Both genes were at first transformed into tobacco seeds using an *Agrobacterium tumefaciens* Ti-plasmid and the leaf disc transformation procedure. The transformation was proven by Southern blot analysis. In the case of the hybrid

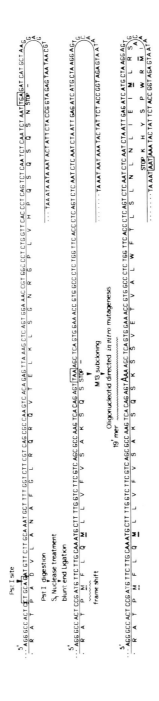

Figure 13.1 Scheme of the two modification steps leading to the modified LeB gene. Comparison of the original and modified gene sequences and the deduced amino acid sequences

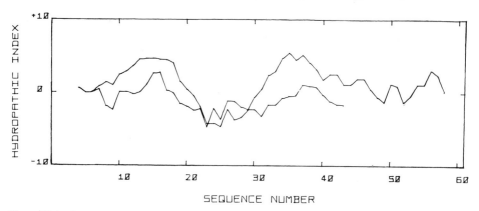

Figure 13.2 Hydropathy plots of the new amino acid sequence (longer line) and the original sequence as depicted in Figure 13.1

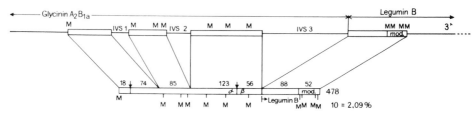

Figure 13.3 Scheme of the hybrid gene constructed from the modified LeB gene and a soyabean glycinin G2 gene

gene, Northern blots revealed the expression of mRNA of the expected size. Immunological techniques (enzyme-linked immunosorbent assay (ELISA) and Western blotting) were used as described (Bäumlein *et al.,* 1987) to detect the modified proteins in the tobacco seeds. After many different experiments we had finally to conclude that the proteins cannot be detected because they do not accumulate in the seeds.

We are now using different expression systems (*in vitro*, in yeast, in *Xenopus* oocytes) to discover the reason for the failure of accumulation of the modified proteins. In yeast, both original and modified Le are accumulated. Original Le is secreted at a level of about 1% of the amount formed but the modified Le is not secreted at all. Preliminary results from experiments with *Xenopus* oocytes indicate that in this system the modified Le is also completely degraded. From the failure of secretion from yeast cells we suggest that the hydrophobic *C*-terminal sequence of the modified Le interferes with intracellular transport.

Studies on intracellular transport signals using legumin-invertase gene fusions

Different *N*-terminal segments of Le were fused to invertase (Figure 13.4) using gene fusion by *in vitro* deletion mutagenesis. The invertase sequence always starts with the first amino acid after the signal sequence cleavage site. The gene fusions

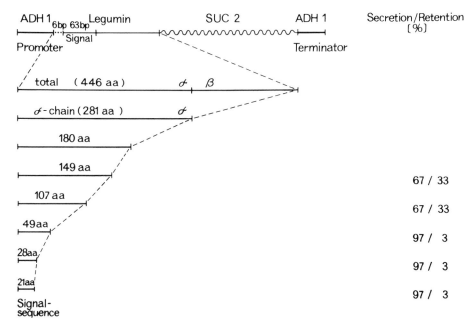

Figure 13.4 Schematic drawings of some LeB-invertase-fusions so far constructed and expressed (if secretion/retention ratios are given) in yeast

were expressed in yeast. The signal sequence of Le efficiently replaced the invertase signal sequence, so that this fusion was efficiently secreted. An additional 28 amino acids of Le did not prevent secretion, but an additional 86 *N*-terminal amino acids of Le caused one-third of the invertase activity to be retained inside the cells. Invertase assays were performed as described (Goldstein and Lampen, 1975). To localize the intracellular activity we isolated vacuoles from the yeast cells according to Stevens, Esmon and Schekman (1982). Indeed, we found that the activity retained inside the cells is completely transported to the vacuole. This result is considered as a first step to the elucidation of the legumin sorting signal. Work is in progress in our laboratory to discover the minimal amino acid sequence necessary for transport to yeast vacuoles and to find out the reason for the fact that only one-third of the activity is retained inside the cells. In addition, we are transforming legumin-invertase and legumin-CAT fusions into tobacco to test different Le segments for potential transport signals in the plant system.

Concluding remarks

Seed storage proteins are an interesting system to study different problems at the gene, the protein, and the cell levels. In addition, these proteins are of great economic importance, since plant seeds are the most important product of agriculture and form the most important source of human nutrition. Therefore, different approaches, including genetic engineering are being used to improve the nutritional quality of these proteins. To date, the testing in plants of two different modified storage proteins has been reported (*see above*). In the first case an oligonucleotide encoding a short polypeptide stretch was introduced into a phaseolin gene, in the latter case a new amino acid sequence was created at the *C*-terminus of LeB. The modified phaseolin was easily degraded in seeds of transgenic plants, the modified LeB could not be detected at all. Although the principle of the modifications is similar, the reasons for the failure of both attempts are obviously different. Bearing in mind the complexity of the storage-protein system, further difficulties have to be expected in modifying such proteins. Different strategies should be applied, including the use of suitable naturally occurring genes, such as the gene encoding the exceptionally sulphur-rich 2S storage protein of the Brazil nut (Altenbach *et al.*, 1987), which has already been shown to be accumulated and correctly processed in transgenic plants (Altenbach *et al.*, 1988).

From the results discussed in this chapter it seems to be clear that signals for intracellular transport are contained in the sequence or structure of seed storage proteins. However, their character is not yet understood. From the few vacuolar sorting sequences so far known a consensus cannot be reached (Klionsky, Banta and Emr, 1988). We compared *N*-terminal regions of many different seed proteins known to be transported to the vacuole. Although in some cases we could observe similarities of such regions of certain protein groups with special properties of the known vacuolar sorting sequences of yeast proteases, e.g. the alternation of negatively charged or hydrophilic amino acids with hydrophobic and positively charged amino acids in the case of the 11S globulins, and the direct sequence similarity in the case of lectins, when compared with the vacuolar sorting sequence of yeast carboxypeptidase Y (Johnson, Bankaitis and Emr, 1987), a striking common feature of the compared sequences could not be found (unpublished data). Divergence of the vacuolar sorting sequences among different proteins could be expected. Structures contributing to such signals should not be modified in attempts to achieve the nutritional improvement of storage proteins.

Acknowledgements

Part of the work was done by G. Saalbach in the laboratory of R. B. Goldberg (Department of Biology, UCLA) on the basis of an exchange programme of the Academy of Sciences (GDR) and the NAS (USA) and on the basis of a salary from the University of California, Los Angeles. The work was supported by the VVB Saat- und Pflanzgut, Quedlinburg (DDR) and the Academy of Sciences of the GDR. We are grateful to Gerhild Jüttner, Ingrid Otto, Brigitte Weiss, and Petra Hoffmeister for technical assistance.

References

Altenbach, S. B., Pearson, K. W., Leung, F. W. and Sun, S. S. M. (1987) Cloning and sequence analysis of a cDNA encoding a Brazil nut protein exceptionally rich in methionine. *Plant Molecular Biology,* **8**, 239–250
Altenbach, S. B., Pearson, K. W., Staraci, L. C., Meeker, G. and Sun, S. S. M. (1988) Processing of Brazil nut sulfur-rich seed protein in transgenic plants. *Journal of Cellular Biochemistry* Suppl. **12C**, L300

Bassüner, R., Manteuffel, R., Müntz, K., Pünchel, M., Schmidt, P. and Weber, E. (1983) Analysis of *in vivo* and *in vitro* globulin formation during cotelydon development of field beans (*Vicia faba* L. var. minor). *Biochemie und Physiologie der Pflanzen*, **178**, 664–684

Bassüner, R., Nong Van, H., Jung, R., Saalbach, G. and Müntz, K. (1987) The primary structure of the predominating vicilin storage protein subunit from field bean seeds (*Vicia faba* L. var. minor cv. Fribo). *Nucleic Acids Research*, **15**, 9609

Bäumlein, H., Müller, A. J., Schiemann, J., Helbing, D., Manteuffel, R. and Wobus, U. (1987) A legumin B gene of *Vicia faba* is expressed in developing seeds of transgenic tobacco, *Biologisches Zentralblatt*, **106**, 569–575

Bäumlein, H., Wobus, U., Pustell, J. and Kafatos, F. C. (1986) The legumin gene family: structure of a B type gene of *Vicia faba* and a possible legumin specific regulatory element. *Nucleic Acids Research*, **14**, 2707–2720

Chrispeels, M. J., Higgins, T. J. V. and Spencer, D. (1982) Assembly of storage protein oligomers in the endoplasmic reticulum and processing of the polypeptides in the protein bodies of developing pea cotelydons. *Journal of Cell Biology*, **93**, 306–313

Dorel, C., Voelker, T., Herman, E. and Chrispeels, M. J. (1989) Transport of proteins to the plant vacuole is not by bulk flow through the secretory system, and requires positive sorting information. *Journal of Cell Biology*, **108**, 327–337

Goldstein, A. and Lampen, J. O. (1975) β-D-Fructofuranoside fructohydrolase from yeast. *Methods in Enzymology*, **42**, 504–511

Hoffman, L. M., Donaldson, D. D. and Herman, E. M. (1988) A modified storage protein is synthesized, processed and degraded in the seeds of transgenic plants. *Plant Molecular Biology*, **11**, 717–730

Hoffman, L. M., Drong, R. and Donaldson, D. (1985) Synthesis and properties of a chimeric high lysine 15 kD zein. *1st International Congress on Plant Molecular Biology*, 56 (abstract)

Hopp, T. P. and Woods, K. R. (1981) Prediction of protein antigenic determinants from amino acid sequences. *Proceedings of the National Academy of Sciences, USA*, **78**, 3824–3828

Horstmann, C. (1983) Specific subunit pairs of legumin from *Vicia faba*. *Phytochemistry*, **22**, 1861–1866

Johnson, L. M., Bankaitis, V. A. and Emr, S. D. (1987) Distinct sequence determinants direct intracellular sorting and modification of a yeast vacuolar protease. *Cell*, **48**, 875–885

Klionsky, D. J., Banta, L. M. and Emr, S. D. (1988) Intracellular sorting and processing of a yeast vacuolar hydrolase: proteinase A propeptide contains vacuolar targeting information. *Molecular and Cellular Biology*, **8**, 2105–2116

Müntz, K., Bassüner, R. and Horstmann, C. (1984) Variability of Met+ and Met− components of the 11S globulins from 3 different grain legumes and within the species of *Vicia faba* L. *Kulturpflanze*, **32**, 5255–5257

Radke, S. E., Andrews, B. M., Moloney, M. M., Crouch, M. L., Kridl, J. C. and Knauf, V. C. (1988) Transformation of *Brassica napus* L. using *Agrobacterium tumefaciens*: developmentally regulated expression of reintroduced napin gene. *Theoretical and Applied Genetics*, **75**, 685–694

Saalbach, G., Jung, R., Saalbach, I. and Müntz, K. (1988) Construction of storage protein genes with increased number of methionine codons and their use in transformation experiments. *Biochemie und Physiologie der Pflanzen*, **183**, 211–218

Schwaiger, H., Hasilik, A., von Figura, K., Wiemken, A. and Tanner, W. (1982) Carbohydrate-free carboxypeptidase Y is transferred into the lysosome-like yeast vacuole. *Biochemical and Biophysical Research Communications*, **104**, 950–956

Sly, W. S. and Fischer, H. D. (1982) The phosphomannosyl recognition system for intracellular and intercellular transport of lysosomal enzymes. *Journal of Cellular Biochemistry*, **18**, 67–85

Stevens, T. H., Esmon, B. and Schekman, R. (1982) Early stages in the yeast secretory pathway are required for transport of carboxypeptidase Y to the vacuole. *Cell*, **30**, 439–448

Tague, B. W. and Chrispeels, M. J. (1987) The plant vacuolar protein, phytohemagglutinin, is transported to the vacuole of transgenic yeast. *Journal of Cell Biology*, **105**, 1971–1979

Tague, B. W. and Chrispeels, M. J. (1988) Identification of a plant vacuolar protein targeting signal in yeast. *Plant Physiology*, **86** (suppl.), 84

Valls, L. A., Hunter, C. P., Rothman, J. H. and Stevens, T. H. (1987) Protein sorting in yeast: the localisation determinant of yeast vacuolar carboxypeptidase Y residues in the propeptide. *Cell*, **48**, 887–897

Voelker, T. A., Herman, E. M. and Chrispeels, M. J. (1989) *In vitro* mutated phytohemagglutinin genes expressed in tobacco seeds: role of glycans in protein targeting and stability. *The Plant Cell*, **1**, 95–104

Wallace, J. C., Galili, G., Kawata, E. E., Cuellar, R. E., Shotwell, M. A. and Larkins, B. A. (1988) Aggregation of lysine-containing zeins into protein bodies in *Xenopus* oocytes. *Science*, **240**, 662–664

Wobus, U., Bäumlein, H., Bassüner, R., Heim, U., Jung, R., Müntz, K. *et al.* (1986) Characteristics of two types of legumin genes in the field bean (*Vicia faba* L. var. minor) genome as revealed by cDNA analysis. *FEBS Letters*, **201**, 74–80

14

REGULATORY ELEMENTS OF MAIZE STORAGE PROTEIN GENES

G. FEIX, M. SCHWALL, M. HAASS, U. MAIER, T. QUAYLE, L. SCHMITZ
and A. TRUMPP
Institute of Biology III, University of Freiburg, Freiburg, W. Germany

Cis-acting regulatory elements have been recognized in many organisms as essential components of regulated genes. These elements occur primarily upstream of the coding region, but can also be part of the 3'-flanking region or of intron sequences. Structures such as the TATA or the CAAT box and their essential roles in the correct initiation of RNA synthesis have been firmly established in many cases. More recently, a growing number of additional 'upstream regulatory elements' have been identified and great efforts are currently being undertaken to uncover the functional significance of such elements and their involvement in regulatory *cis-trans* interactions (Wasylyk, 1988).

The molecular study of regulatory elements has recently also been applied to plant genes, confirming the functional relevance of TATA and CAAT boxes (or the comparable AGGA box) (Heidecker and Messing, 1986). The search for further 'upstream regulatory elements' was then approached, first by looking for sequence homologies to established elements of animal or yeast gene systems, but has now been extended by direct experiments towards an understanding of the functional role of these elements (Willmitzer, 1988). These attempts are, however, severely hampered by the absence of a functional *in vitro* transcription system from plants and, in the case of monocotyledonous plants, by the additional lack of a stable homologous transformation system to be used for the analysis of appropriate mutants. Therefore, work with other methods has become of particular importance for the analysis of plant genes (i.e. the *in vitro* analysis of specific complexes formed between nuclear proteins and regulatory elements and the use of transient transformation assays). Work dealing with these aspects is described here using the zein genes of maize as an example.

Zein genes

Zein proteins constitute the major storage protein of maize and are synthesized exclusively and in great amounts at defined times in the developing endosperm. They are coded by different classes of zein genes (the monogenic 10, 15, and 27 kDa genes and the multigenic 19 and 21 kDa genes) and are processed after their translation, before being packaged into protein bodies. The majority of the zein protein is synthesized from the 19 and 21 kDa genes which together may be present in up to 100 copies per genome. It is this class of genes that has been primarily studied in the past and with which this chapter also deals. The analysis of several representative genomic clones of the 19 and 21 kDa size classes has led to the following picture of these genes (Brown *et al.*, 1988).

The *coding region* is free of introns and contains a blockwise arrangement of amino acid sequences which presumably allows the synthesized proteins to assume secondary and tertiary structures important for the packaging in the protein bodies (Argos *et al.*, 1982).

The *5′ flanking region* contains the two promoters P1 and P2, with P2 lying directly in front of the coding region and P1 lying one kb further upstream (Langridge and Feix, 1983). Both promoters are preceded by respective TATA and CAAT boxes, which are almost the only regions of maintained homology in the 5′-flanking region of different classes of zein genes (19 and 21 kDa classes) (Brown, Wandelt and Feix, 1986). One significant additional region of perfect homology is a 15 bp consensus sequence at the −300 region. Furthermore, the upstream regions display a number of sequence segments with similarity to enhancer-like elements from other organisms. These elements, in addition to the well-defined promoters and the consensus sequence, are of interest for the interaction with regulatory proteins.

The *3′-flanking region* of zein genes contains two polyadenylation sites and is expected to harbour further functional sequence elements which have not yet been analysed.

As zein genes appear to be tandemly arranged, it will be of importance for a correct assignment of regulatory sites to delineate the respective 5′ and 3′ borders of individual gene structures. In the case of a cloned 13 kb DNA fragment (Langridge *et al.*, 1985), only one intact gene could be identified, while in another example two coding regions, 3 kb apart, were found on a single genomic fragment (Spena, Viotti and Pirrotta, 1983). Furthermore, it has been reported that two overlapping cosmid clones covering a total of 100 kb contain 10 separate zein specific hybridizing regions (Messing and Rubenstein, personal communication, 1986), but more information is needed for a complete picture of zein gene arrangements and the length of their flanking sequences.

Despite our structural knowledge of zein genes, attempts to correlate presumed regulatory elements of the 5′ flanking regions with functions are in a very early stage, except for the promoters which have been both structurally and functionally mapped (Brown *et al.*, 1988). Recessive, *trans*-acting regulatory mutants of the opaque type indicate the occurrence of *cis-trans*-interactions (Nelson and Burr, 1973), but corresponding molecular studies have, as yet, not been performed. It was this gap in our knowledge of zein gene expression that initiated the following experiments described here, which deal with the identification and functional assignment of regulatory elements occurring in the 5′-flanking region of zein genes.

Specific interactions of nuclear proteins with the 5′-flanking region of zein genes

The study *in vitro* of specific protein DNA interactions was greatly stimulated by the recent establishment of efficient analytical methods such as nitrocellulose filter binding, electrophoretic band shift, South Western and footprint techniques. In these tests, the gene regions of interest are isolated as small fragments which are then individually reacted with crude or enriched nuclear extracts of particular tissues. The specific protein DNA complexes formed *in vitro* are then analysed with one of the above methods. This type of analysis was performed on the 5'-flanking region of the zein genes pMS1 and pMS2 (19 kDa class) and pML1 (21 kDa class) with the aim of identifying the formation of protein DNA complexes specific for zein promoters, endosperm tissue and developmental stages of this tissue. For a first assessment of the protein binding capacity of the complete 5'-flanking region, nitrocellulose filter binding assays were performed with subfragments in order to detect the formation of specific DNA protein complexes with nuclear proteins from seedlings, endosperm tissue or from a maize cell culture (Maier *et al.*, 1988). This and other *in vitro* techniques will of course detect only relatively stable complexes and will miss those interactions which form transiently *in vivo* and/or are of low stability. The number of specific complexes identified with the nitrocellulose binding assay was, however, surprisingly high and included cases of tissue and developmental stage dependent interactions. The analysis of such interactions is of particular interest as they represent some of the major regulatory features of zein gene activation.

Several of these interactions were selected for further study and the results are summarized in Figure 14.1. This figure shows the binding specificities that were obtained by the interaction of the radioactively labelled fragments A, B, C, D and E with various nuclear protein preparations, which were previously enriched by heparin–agarose chromatography. The indicated results are based on the analysis, of the complexes formed *in vitro* by band shift analyses in polyacrylamide gels. The observed specificities were always confirmed by competition experiments with

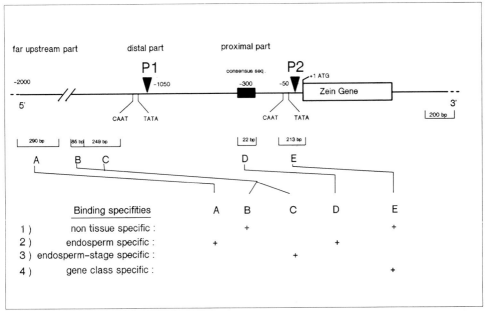

Figure 14.1 Binding of nuclear proteins to upstream regions of zein genes. P1 and P2 indicate the dual promoters of zein genes. A, B, C, D and E are restriction enzyme fragments used for the protein binding studies.

increasing amounts of specific and unspecific DNAs. The reactions often displayed complex band patterns, not just single retarded bands. Such complexities are already known from other gene systems where they resulted from protein DNA interactions of varying stabilities and from additional protein–protein interactions.

In the case of the promoter P2 *fragment E* of pMS1, the observed binding complexity could be further resolved utilizing a protein preparation purified by fragment E coupled affinity column chromatography. This led to the identification of five specific binding proteins from endosperm tissue. Some of the binding proteins display zinc-finger properties as determined by inclusion of the Zn-chelator 1,10 *o*-phenanthroline in the binding reaction (Hooft *et al.*, 1987). Binding of nuclear proteins to fragment E could also be observed with nuclear proteins derived from seedling tissue, indicating a tissue independent protein DNA interaction in the region of the promoter. Slight differences in the binding pattern observed with the P2 fragment E were evident if a comparable P2 promoter fragment of the gene pML1 (21 kDa class) was used in the binding assay. This finding may reflect differences seen in the expression of zein genes of different classes in opaque type mutants (Nelson and Burr, 1973).

The specific interaction of the -300 consensus sequence contained in *fragment D* of pMS1 is endosperm specific (Maier *et al.*, 1987). Since in this case, a DNAse I footprint was obtained with crude nuclear extracts, it is believed that the interacting DNA binding protein occurs in the endosperm tissue in rather high amounts and that a particularly stable complex is formed. This protein DNA interaction at the '-300 box' is of special interest as the observed binding site coincides with very similar sequences occurring at the same position in storage protein genes from other cereals (Forde *et al.*, 1985), indicating a particular importance of this sequence with relevance to the gene and tissue type.

The protein DNA interactions seen with the far upstream *fragment C* of pML1 displayed, besides tissue specific components, an additional specificity for the developmental stage of the endosperm material used. Band shift experiments performed with proteins from endosperm tissue either 7 or 14 days old (representing developmental stages before and after induction of zein gene expression) showed a specific appearance and disappearance of certain bands, while other bands of the complex band shift pattern remained unchanged. Differences in the protein pattern interacting with fragment C could also be demonstrated in South Western experiments. We consider this finding of stage specificity of particular interest which will allow further experiments towards an identification of developmentally relevant regulatory proteins and their respective genes.

Specific binding of *fragment B* of pML1 proved to be more difficult to analyse with respect to nuclear proteins from endosperm, but extracts from seedlings led to a clear retarded band in polyacrylamide gels. This reflects the difficulties often encountered in preparing active nuclear extract. *Fragment B* is of particular interest because it displays a high enhancing activity on the 35S promoter in transient transformation assays (*see* next section).

Fragment A of pMS2 is interesting due to its unique sequence which contains four closely related elements of dyad symmetry which could allow the formation of several stable stem loop structures (Quayle, Brown and Feix, 1989). This fragment exhibits specific binding of nuclear proteins from endosperm tissue, however, the functional relevance of this far upstream region is not known.

In conclusion, evidence is presented for the presence of several different binding specificities over extended regions of the 5'-flanking region of zein genes. It can, however, be expected that more specific interactions take place which escape detection by the methods used. Such further interactions will be more difficult to recognize because of their probable dynamic and flexible nature and their multicomponent composition. A great help towards a better understanding of the regulatory structures will come from transformation experiments with subfragments and mutants from the flanking regions. It can be imagined that the intact environment of the living cell will provide the conditions necessary for the regulated enhancement of gene activity. Some initial experiments towards this goal are described in the following section.

Transient transformation assays in maize cell derived protoplasts

The choice of cell type used in the preparation of protoplasts, as well as the transformation method, are crucial points for successful transient transformation tests. The use of cell cultures as starting material is appealing for its simplicity, but potential genetic changes of cultured cells occurring during the extended periods of maintenance must be considered. The embryonic maize 'Black Mexican Sweet' (BMS) suspension cell culture (Chourey and Zurawski, 1981) used in this study seems, however, to be stabilized as it was successfully used in transient transformation assays of other plant genes. However, the preparation of active protoplasts directly from plant tissue has been difficult, and only recently have we succeeded in obtaining active protoplasts from growing endosperm (Schwall and Feix, 1988), the natural zein gene expressing tissue. This isolation was only possible from material up to the eleventh day post pollination, which marks (under our growing conditions) only the very beginning of zein gene expression. In spite of this disadvantage, such protoplasts presently represent the closest system to the homologous situation *in vivo* of zein gene induction.

Protoplasts of both types – BMS and endosperm cell derived – were used for activity measurements of *zein promoter structures*. For this purpose, promoter structures were cloned into the basic construct pUCAT which contains as an expression box the CAT gene together with the 35S terminator sequence (Figure 14.2). Plasmid pMS1CAT was constructed by inserting the complete 5'-flanking sequence of pMS1 containing the two promoters P1 and P2 into pUCAT. pMS1P2CAT was similarly constructed by inserting a truncated fragment with only the P2 promoter. The additional insertion of a synthetic DNA segment with the sequence of the octopine synthase enhancer (ocs) (Ellis *et al.*, 1987) generated the plasmids pMS1OCS and pMS1P2OCS (Figure 14.2). Insertion of the 35S promoter from the cauliflower mosaic virus (CaMV) resulted in the plasmid pDW2. The various plasmids were electroporated into both types of protoplasts. The newly synthesized CAT enzyme was assayed with [^{14}C]-labelled chloramphenicol as substrate. The results are depicted in Figure 14.3. The complete promoter fragment containing P1 and P2 is inactive in BMS protoplasts, but shows a low but significant activity in the endosperm protoplasts. This result indicates that the activity of the complete zein promoter structure is repressed in BMS cells and gets a slight activation in the 10-day-old endosperm cells which are in the very early stage of endogenous zein gene induction.

Attempts to increase this weak activity with the help of the ocs enhancer sequence (Ellis *et al.*, 1987) were unsuccessful, indicating further requirements for

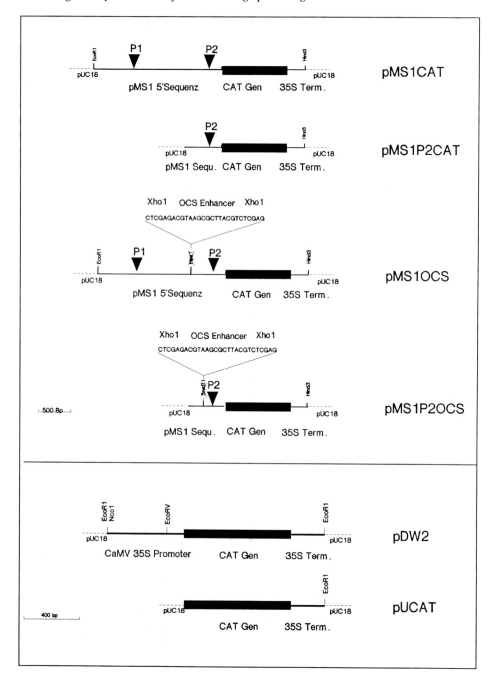

Figure 14.2 Promoter constructs used in transient transformation assays. The zein clone pMS1 is described in Brown, Wandelt and Feix (1986). Plasmid pDW2 was obtained from T. Hohn/Friedrich Miescher Institute, Basel, Switzerland. OCS stands for octopin synthetase (as described by Ellis *et al.*, 1987). Plasmid constructions were performed by established methods

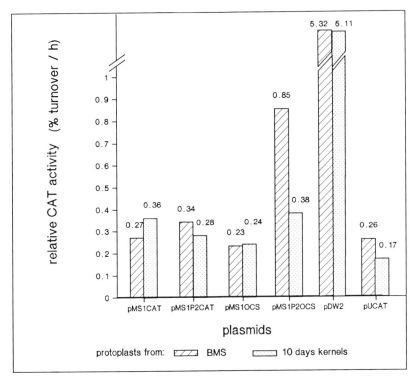

Figure 14.3 Measurement of promoter activity in maize protoplasts. Isolation of protoplast from BMS cells and 10-day-old endosperm tissue as well as transformation assays and measurement of CAT activity were performed as described by Schwall and Feix, 1988

the functioning of this enhancer in these cells. The shortened promoter element (plasmic pMS1P2CAT), however, showed a slight activity in BMS protoplasts, suggesting that sequences of regulatory importance lying further upstream have been deleted in this plasmid. It is interesting to note that the ocs enhancer does function in this case in BMS protoplasts. The endosperm protoplasts, however, seem to be more restrictive in this regard, since they do not respond to the ocs enhancer element or to the truncated promoter fragment. The 35S promoter construct pDW2 used as a positive control was very active in both protoplast types.

In another set of experiments *far upstream sequences* of pML1 were tested in BMS protoplasts for their influence on the activity of the constitutive 35S promoter (Ow, Jacobs and Howell, 1987) after inserting them into an enhancer/silencer trap plasmid. The 35S promoter was shown in pilot experiments to be very active in BMS cells and would therefore easily allow the detection of both enhancing and silencing effects. The 1000 bp region preceding the promoter P1 of pLM1 was chosen for the testing *in vivo* in order to confirm for zein genes the potential importance of far upstream sequences for the regulation of their corresponding genes. An indication of the importance of these sequences had already been obtained by the interaction *in vitro* with nuclear proteins (*see* Figure 14.1). The upstream region was subdivided into the four adjacent fragments A, B, C, and D

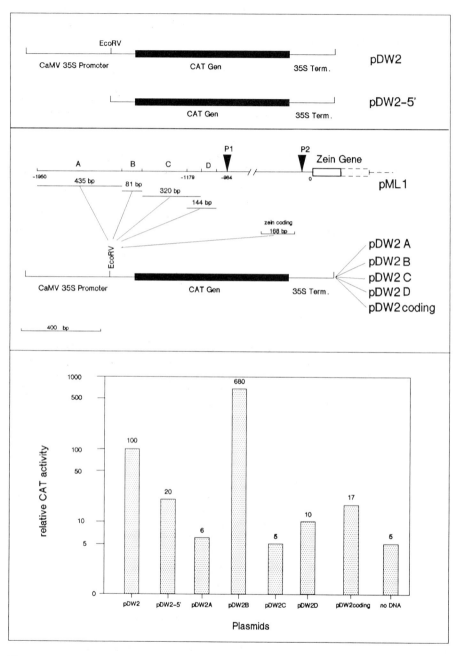

Figure 14.4   Influence of 'upstream' regions of pML1 on the CaMV 35S promoter. pML1 is described in (Pintor-Toro, Langridge and Feix, 1982). Plasmid constructions were by established methods. Isolation of protoplast from BMS cells, transformation assays and measurement of CAT activity as described by Schwall and Feix, 1988. The *y*-axis is logarithmic

which were inserted independently into the Eco RV site of the CaMV 35S promoter leading to the constructs shown in Figure 14.4. In the plasmid pDW2 coding, a zein coding fragment was integrated which had previously been shown not to interact with nuclear proteins and was assumed not to contain any regulatory sequences. The various constructs were electroporated into BMS protoplasts, which in turn were analysed for newly synthesized CAT activity. The result obtained with the plasmid pDW2 containing the 35S promoter together with its own enhancer was arbitrarily set to represent 100 units (Figure 14.4). The plasmid pDW2-5′ in which the majority of the enhancer had been deleted showed a reduced activity of 20 units as also seen in another system (Ow, Jacobs and Howell, 1987). A similar result was obtained with the control plasmid pDW2 coding.

These two plasmids were included in the experiments for an estimate of the probable activity loss of the complete 35S promoter resulting from the insertions at the Eco RV site which separate the enhancer part of the 35S promoter from the promoter core. Most surprising of the various transformation assays was the dramatic enhancing effect on the 35S promoter of fragment B, which exceeded by far the influence of the 35S enhancer itself. Fragments A and C display, however, an opposite effect on the 35S promoter by reducing its activity to background levels. Fragment D acts only marginally as a silencer element under these conditions. Taken together, the results clearly indicate a close package of regulatory elements with different activities in the far upstream region of the zein clone pML1, reminiscent of the situation seen with other regulated genes. Of course, the analysis of the influence of these elements on zein promoters within the context of endosperm cells will be even more interesting and conclusive. For such analyses, appropriate expression systems *in vivo* are greatly needed in maize. Work with a zein expressing cell culture to be used for transient transformation assays as well as attempts to achieve stable transformations are in progress. In particular, the latter approach will eventually allow the testing of regulatory elements under homologous conditions in the growing maize seed itself.

The zein genetic system and applied aspects

The studies described in this chapter as well as other recent experiments from our and other laboratories indicate that for their functioning zein genes need a more versatile set-up than previously anticipated. Some of these concepts are summarized in Table 14.1. The multigenic nature of the zein gene system has led to sets of variable genes with some being very similar to each other and others

Table 14.1 FEATURES OF THE ZEIN MULTIGENE SYSTEM

Line dependent variable number of active and inactive genes
Extended gene specific flanking sequences with complex arrangement of promoters and *cis*-acting regulatory elements
Cell type, stage and gene class dependent interactions with *trans*-acting nuclear proteins
Endosperm specific and efficient gene expression requires extended upstream regions
Flanking sequences with further signals for other cellular functions
Coding regions must retain specific sequences for processing, folding and packaging of the proteins

showing quite substantial differences. The variability is further increased by line dependent differences in the number of active genes as demonstrated by an analysis of zein proteins (Righetti and Gianazza, 1977). More final conclusions, however, must await the further isolation and characterization of additional zein genes, since only a limited number have been analysed to date. A great part of this variability may lie in the potential of individual genes to respond optimally to slightly different cellular conditions occurring in the multicellular endosperm tissue. It is evident that a great number of interactions with other cellular components have to occur at the zein gene structures for an ordered and specific gene expression. Some of the interactions have been identified, but it is expected that interactions of a more complex nature are involved which act in a concerted fashion. This view is in line with previous findings in other gene systems. Furthermore, signal structures for other purposes will reside in the flanking sequences of zein genes and potentially overlap with the regulatory elements discussed so far. One example is the occurrence of splice sites for the processing of the RNA derived from the promoter P1 (Brown and Feix, 1990). Further structures necessary for DNA replication and chromosome structures can be easily envisaged to occur in the flanking regions spanning several kb.

The eludication of regulatory aspects will also be of relevance for the envisaged use of the zein genetic system in genetic engineering. Presently the focus is primarily on three application topics. Thus, the 5'-flanking region is considered to be very useful for the construction of seed specific expression modules. Furthermore, the coding region is considered to be susceptible to an improvement in its content of essential amino acids and also for the integration of alien coding sequences of interest with the aim of their synthesis and storage in maize endosperm. Finally, the use of regulatory loci involved in the highly specific synthesis of zein proteins may be of potential interest for an application with other gene systems. However, before achieving substantial success in any of these or other applications there is clearly a need for an even better understanding of the functioning of zein genes.

Acknowledgements

The experiments described in this chapter were supported by funds from the Deutsche Forschungsgemeinschaft and the Fonds der Chemischen Industrie.

References

Argos, P., Pedersen, K., Marks, M. D. and Larkins, B. A. (1982) A structural model for maize zein proteins. *Journal of Biological Chemistry,* **257**, 9984–9990
Brown, J. W. S. and Feix, G. (1990) A functional splice site in the 5' untranslated region of a zein gene. *Nucleic Acids Research*, (in press)
Brown, J. W. S., Maier, U. G., Schwall, M., Schmitz, L., Wandelt, C. and Feix, G. (1988) The structure and function of zein genes of maize. *Biochemie und Physiologie der Pflanzen,* **183**, 99–106
Brown, J. W. S., Wandelt, C. and Feix, G. (1986) The upstream regions of zein genes: sequence analysis and expression in the unicellular green alga *Acetabularia. European Journal of Cell Biology,* **42**, 161–170

Chourey, P. S. and Zurawski, D. B. (1981) Callus formation from protoplasts of a maize cell culture. *Theoretical Applied Genetics*, **59**, 341–344

Ellis, J. G., Llewellyn, D. J., Walker, J. C., Dennis, E. S. and Peacock, W. J. (1987) The ocs element: a 16 base pair palindrome essential for activity of the octopine synthase enhancer. *EMBO Journal*, **6**, 3203–3208

Forde, B. G., Heyworth, A., Pywell, J. and Kreis, M. (1985) Nucleotide sequence of a B1 hordein gene and the identification of possible upstream regulatory elements in endosperm storage protein genes from barley, wheat and maize. *Nucleic Acids Research*, **13**, 7327–7339

Heidecker, G. and Messing, J. (1986) Structural analysis of plant genes. *Annual Review of Plant Physiology*, **37**, 439–466

Hooft van Huijsduijen, R. A. M., Bollekens, J., Dorn, A., Benoist, C. and Mathis, D. (1987) Properties of a CCAAT box-binding protein. *Nucleic Acids Research*, **15**, 7265–7282

Langridge, P., Brown, J. W. S., Pintor-Toro, J. A. and Feix, G. (1985) Expression of zein genes in *Acetabularia mediterranea*. *European Journal of Cell Biology*, **39**, 257–264

Langridge, P. and Feix, G. (1983) A zein gene of maize is transcribed from two widely separated promoter regions. *Cell*, **34**, 1015–1022

Maier, U.-G., Brown, J. W. S., Schmitz, L. M., Schwall, M., Dietrich, D. and Feix, G. (1988). Mapping of tissue-dependent and independent protein binding sites to the 5' upstream region of a zein gene. *Molecular and General Genetics*, **212**, 241–245

Maier, U.-G., Brown, J. W. S., Toloczyki, C. and Feix, G. (1987) Binding of a nuclear factor to a consensus sequence in the 5' flanking region of zein genes from maize. *EMBO Journal*, **6**, 17–22

Nelson, O. E. and Burr, B. (1973) Biochemical genetics of higher plants. *Annual Review of Plant Physiology*, **24**, 493–518

Ow, D. W., Jacobs, J. D. and Howell, S. H. (1987) Functional regions of the cauliflower mosaic virus 35S RNA promoter determined by use of the firefly luciferase gene as a reporter of promoter activity. *Proceedings of the National Academy of Sciences, USA*, **84**, 4870–4874

Pinto-Toro, J. A., Langridge, P. and Feix, G. (1982) Isolation and characterization of maize genes coding for zein proteins of the 21 000 dalton size class. *Nucleic Acids Research*, **10**, 3845–3860

Quayle, T. J. A., Brown, J. W. S. and Feix, G. (1989) Analysis of distal flanking regions of maize 19-kDa zein genes. *Gene*, **80**, 249–258

Righetti, P. G. and Gianazza, E. (1977) Heterogeneity of storage proteins in maize. *Planta*, **136**, 115–123

Schwall, M. and Feix, G. (1988) Zein promoter activity in transiently transformed protoplasts from maize. *Plant Science*, **56**, 161–166

Spena, A., Viotti, A. and Pirrotta, V. (1983) Two adjacent genomic zein sequences: structure, organization and tissue-specific restriction pattern. *Journal of Molecular Biology*, **169**, 799–811

Wasylyk, B. (1988) Transcription elements and factors of RNA polymerase B promoters of higher eukaryotes. *CRC Critical Reviews in Biochemistry*, **23**, 77–119

Willmitzer, L. (1988) The use of transgenic plants to study plant gene expression. *Trends in Genetics*, **4**, 13–18

15

HORMONAL CONTROL OF WHEAT α-AMYLASE GENES

A. K. HUTTLY* and D. C. BAULCOMBE†

Institute of Plant Science Research Cambridge Laboratory, Maris Lane, Trumpington, Cambridge, UK

Introduction

The plant growth regulator gibberellin (GA) is involved in several aspects of plant development including extension growth, flowering, fruit set and germination (Reid, 1986). It is only in one case, however, the germinating cereal grain and, in particular, events occurring in the aleurone layer, that much progress has been made towards understanding the mechanism of GA action.

Several features of the aleurone layer have led to its use as a system in which to study the involvement of GA. First, it is a simple target tissue that is comprised of a single cell type not undergoing division. Second, it can be easily isolated from any other living tissue at grain maturity and at the same time from the source of the stimulus. Third, the response of the tissue is relatively simple and can be replicated *in vitro* in isolated aleurone layers and protoplasts. It is hoped that by elucidating the mechanism by which GA acts in this simple tissue, the involvement of the growth regulator in other tissues and developmental programmes of the plant that are more complex than the aleurone layer will be better understood.

The aleurone layer is part of the endosperm of the cereal grain. Following formation of the triploid endosperm nucleus during fertilization, several free divisions occur before cell wall formation commences, but by 4 d.p.a. (days post anthesis) a layer of meristematic cells is present lining the endosperm sac. These cells give rise to the starchy endosperm cells by serial tangential cell division and then go on to differentiate into the aleurone layer (Evers, 1970). They are first recognizable as developing aleurone cells about 10 d.p.a. and at this stage they have extremely thin cell walls, large nuclei and many vacuoles. By 14 d.p.a. they have ceased to divide and have begun to form the characteristically thick cell walls of mature aleurone cells and to accumulate storage reserves in the forms of lipid droplets and aleurone grains containing protein and phytin deposits (Morrison, Kuo and O'Brien, 1975). At grain maturity, following desiccation, the aleurone layer is the only living part of the endosperm remaining. At this stage it represents an already highly developed tissue, poised to undergo the final stage of its differentiation as germination starts.

*Present address: Long Ashton Research Station, University of Bristol Department of Agricultural Sciences, Long Ashton, Bristol, UK
†Present address: Sainsbury Laboratory, John Innes Institute, Colney Lane, Norwich, UK

Germination

During germination one of the earliest visible events noted in aleurone cells is the proliferation of rough endoplasmic reticulum. This is accompanied by decreases in the volume of aleurone grains and phytin inclusions and in the numbers of lipid droplets (Tomos and Laidman, 1979; Jones, 1980) as these stored reserves are utilized. Preceding this, however, is the induction of intermediary metabolic pathways such as glycolysis, gluconeogenesis and the oxidative pentose phosphate and TCA cycles, together with increases in lipase activity concurrent with the appearance of active pathways of fatty acid metabolism and net synthesis of phospholipid (Doig *et al.*, 1975a,b). Within hours of rehydration the aleurone cells develop into a highly metabolic tissue which is then capable of synthesizing and secreting the various hydrolytic enzymes that are required for the mobilization of the reserve materials contained within the endosperm. In addition, aleurone cells are net exporters of amino acids and sugars and an important source of mineral ions for the embryo (Tomos and Laidman, 1979). Extensive breakdown of the aleurone cell walls also occurs during germination (Ashford and Gubler, 1984) which, as well as providing extra carbohydrate resources for the embryo, might facilitate the release of hydrolytic enzymes from the aleurone cells.

GA-controlled events in the aleurone layer

The function of GA in germination is thought to be that of messenger from the embryo/scutellum, where the GA is synthesized (Bewley and Black, 1978), to the aleurone layer, thus ensuring that a coordinated programme of germination occurs. The precise role of GA in the events described above has not been resolved. Induction of intermediary metabolism is thought to be independent of GA as most enzymes and organelles are present in the desiccated tissue and simply await rehydration. GA is, however, involved in the activation of lipase and phytase (Gabard and Jones, 1986; Fernandez and Staehelin, 1987) and induction of fatty acid and phospholipid metabolism (Tomos and Laidman, 1979). The most well characterized effects of GA relate to induction of the synthesis of the hydrolytic enzymes destined for the endosperm and in the control of the release of these and other enzymes into the surrounding medium (Table 15.1). Just how GA might influence secretion is not clear, but it is generally thought that the thick aleurone cell walls form an impermeable barrier to the release of the enzymes. Evidence in favour of this suggestion lies with the fact that esterase, peroxidase, acid phosphatase and α-amylase can only be found in the cavities and channels that are formed in the wall system in GA treated tissue (Ashford and Gubler, 1984; Gubler, Ashford and Jacobsen, 1987). GA induces the synthesis of several enzymes capable of aleurone cell wall hydrolysis (Table 15.1), which could explain the role of GA in the process of secretion. However, visible hydrolysis of the walls and the secretion of some aleurone enzymes, such as [1→3]-β-glucanase, precedes GA induced synthesis of at least some of the enzymes considered responsible for the cell wall hydrolysis. A second hypothesis suggests that GA may induce release of enzymes from the cell wall via cation exchange (Tomos and Laidman, 1979), and GA has been shown to influence, albeit indirectly, K^+ efflux from aleurone protoplasts (Bush *et al.*, 1988). Other possibilities include a GA-induced effect on transport across the plasmalemma (Hooley, 1984) and involvement in the synthesis of the intracellular transport system (Vakharia *et al.*, 1987).

Table 15.1 ENZYMES FROM THE ALEURONE LAYER INFLUENCED BY GA

	Secretion	*Synthesis*	*Reference*
Cell wall hydrolysis			
[1→3]-β-glucanase	+		Jones (1971)
[1→3,1→4]-β-glucanase	+	+	Mundy and Fincher (1986), Stuart, Loi and Fincher (1986)
[1→4]-β-xylanase		+	Taiz and Honigaiman (1976)
β-xylopyranosidase		+	Taiz and Honigaiman (1976)
α-arabinofuranosidase		+	Taiz and Honigaiman (1976)
Protein hydrolysis			
aleurain (thiol protease)		+	Rogers, Dean and Heck (1985)
carboxypeptidase	+	+	Hammerton and Ho (1986), Baulcombe, Barker and Jarvis (1987)
Nucleic acid hydrolysis			
endonuclease		+	Brown and Ho (1987)
ribonuclease	+	+	Chrispeels and Varner (1967) Hooley (1982)
Starch hydrolysis			
α-glucosidase		+	Hardie (1975)
limit dextrinase		+	Hardie (1975)
α-amylase	+	+	Filner and Varner (1967)
Others			
acid phosphatase	+	+	Gabard and Jones (1986), Hooley (1984)
esterase		+	Ashford and Jaccobsen (1974)
peroxidase		+	Gubler and Ashford (1983)

Enzymes whose *de novo* synthesis is controlled by GA include those for protein, starch, nucleic acid and phytin degradation and cell wall hydrolysis (*see* Table 15.1). GA has been shown to induce increases in the mRNA levels for most of these proteins (Rogers, Dean and Heck, 1985; Baulcombe, Barker and Jarvis, 1987) and for one enzyme, α-amylase, accumulation of its mRNA is due to the increased transcription of its genes (Jacobsen and Beach, 1985; Zwar and Hooley, 1986). It is probable that GA influences the transcription of all of these protein genes but, as yet, this has not been demonstrated and it remains a possibility that GA could effect mRNA stability and/or post-transcriptional events.

GA is also implicated in the activation of β-amylase. β-Amylase is produced in developing grain in the endosperm and is sequestered in a relatively inactive form bound to gluten bodies during desiccation of the grain. The bound form is released in germination by action of an SH-proteinase (Sopanen and Laurière, 1989); as GA is known to stimulate the synthesis of just such a protease (aleurain; Rogers, Dean and Heck, 1985) the role of GA in this case is probably indirect.

The antagonistic effects of abscisic acid (ABA)

Application of abscisic acid (ABA) in combination with GA to isolated aleurone layers prevents normal GA-stimulated events including synthesis and secretion of the hydrolytic enzymes such as α-amylase (Nolan and Ho, 1988). When applied in isolation, ABA induces the synthesis of a characteristic set of proteins which includes an α-amylase inhibitor (Mundy, 1984), a protein showing homology to one induced in cotton embryos by ABA (Hong, Uknes and Ho, 1988), and a polypeptide induced in other plant tissues by water stress (Chandler *et al.,* 1988). ABA, it is suggested, coordinates the maturation of the grain, including prevention of germination-specific events, accumulation of storage reserves before desiccation occurs (Quatrano, 1987), and preparation of the embryo to withstand desiccation. This last role apparently involves the synthesis of specific polypeptides, whose function is as yet unknown, but which are also produced in response to ABA by other plant tissues when stressed by lack of water (Gomez *et al.,* 1988; Mundy and Chua, 1988). The identification of ABA-induced proteins in mature aleurone layers suggests that ABA plays a similar role in developing aleurone cells. ABA may also have a physiological role in mature aleurone cells in suspending germination during water stress.

α-Amylase

The starch endohydrase, α-amylase, has received the most intensive study of all the enzymes produced by the aleurone layer, probably because it constitutes the bulk of the newly synthesized and secreted enzymes and is so important in endosperm starch breakdown crucial to the brewing industry. It has been shown to be almost entirely synthesized *de novo* by aleurone cells during germination and that this occurs only in the presence of GA (Akazawa and Miyata, 1982).

α-Amylase actually exists as an array of isozymes; upwards of 30 are produced by the aleurone layer of wheat (Ainsworth *et al.,* 1985). This number is due in part to the hexaploid nature of this cereal and accordingly fewer exist in barley. When separated on isoelectric focusing gels the wheat isozymes fall into two groups, a high pI (6.5–7.5) group (α-*Amy1*) and a low pI (4.8–6.2) group (α-*Amy2*) which are antigenically distinct (Baulcombe and Buffard, 1983). Subsequent work using cDNA and genomic clones in wheat and barley has proved that the two groups of isozymes are the products of two distinct multigene families residing on different chromosomes and whose genes have characteristic differences relating to the number of introns and additional coding capacity at the 3' end (Muthukrishran *et al.,* 1984; Lazarus, Baulcombe and Martienssen, 1985; Baulcombe *et al.,* 1987).

Characterization of several cloned wheat genes by cross-hybridization and restriction endonuclease digestion indicated that subgroups also exist in wheat within the two main families (Lazarus, Baulcombe and Martienssen, 1985; Huttly, Martienssen and Baulcombe, 1988). Chromosome location of cloned α-*Amy2* genes demonstrated that these subgroups are not the result of association of the genes with a particular constituent genome of hexaploid wheat; they can, however, be correlated with differential expression.

Expression of α-amylase isozymes

In isolated aleurone layers from Himalayan barley, a small quantity of the low pI isozymes accumulates in the absence of GA; however, after hormone treatment induction of both groups of isozymes occurs within about 8 h (Jacobsen and Chandler, 1987; Nolan, Lin and Ho, 1987). The high pI isozymes rapidly accumulate, becoming the dominant group, but then decline over a period of

12–24 h, while the low pI group increase gradually and only start to decline after 32–40 h. It was also noted that induction of the high pI group requires a higher concentration of GA compared with that capable of inducing the low pI group (Nolan, Lin and Ho, 1987). In germinating wheat grains and half-seeds a similar situation prevails with the exception that the low pI group of isozymes are not present in the tissue in the absence of GA and are first detected 1–2 days after the appearance of the high pI group (Sargent, 1980).

Using cloned α-amylase genes as probes it is possible to show that changes in isozyme levels correlate well with increases in the levels of mRNA specific for each group (Lazarus, Baulcombe and Martienssen, 1985; Jacobsen and Chandler, 1987; Nolan and Ho, 1988). Thus in barley, mRNA specific for the high pI genes rises sharply from 4–8 h but declines after 12–16 h, while that for the low pI genes is present in hydrated aleurone cells and increases gradually for up to 32 h. Similar induction of mRNA was noted in wheat half-grains, although the onset of accumulation was first detected between 12 and 24 h for mRNA of both isozyme groups. Analysis of gene-specific expression of several cloned members from the different barley groups confirmed that within a family there is coordinated control in the timing of mRNA accumulation and effect of GA concentration (Rogers and Milliman, 1984; Jacobsen and Chandler, 1987). As the patterns of mRNA accumulation are similar to that of isozyme accumulation, it led to the suggestion that α-amylase mRNA was probably translated without delay; a result also suggested from translation *in vitro* of isolated α-amylase mRNA (Higgins, Zwar and Jacobsen, 1976). Hence it seemed likely that the main effect of GA on control of α-amylase expression would be at the level of transcription of the genes. Evidence in support of this comes from comparing the rate of transcription of α-amylase genes in nuclei extracted from untreated protoplasts with nuclei from those incubated in the presence of GA (Jacobsen and Beach, 1985; Zwar and Hooley, 1986). In both wild oat and barley, GA significantly stimulated transcription of the α-amylase genes, while simultaneous addition of ABA could suppress this. It is therefore concluded that for α-amylase the major effects of GA and ABA lie at the level of transcription of the genes.

Control of transcription

Our present knowledge of eukaryotic transcription and its control comes mainly from work on mammalian, yeast or *Drosophila* systems (*see* La Thangue and Rigby, 1988), but accumulating evidence suggests that similar mechanisms operate in plants (Ptashne, 1988). Transcription probably involves at least a two-stage process, as eukaryotic genes often exist highly packaged with histones into chromatin. Thus before initiation of transcription can occur the DNA template, it is assumed, must be somehow made available. The mechanisms responsible for this are unknown but have been suggested to involve binding to the nucleoskeleton (Goldman, 1988) and alteration in patterns of methylation (Cedar, 1988). Initiation of transcription is also a complex process involving several different protein factors, including an RNA polymerase, and the DNA template. Identification of such protein factors came first from identification of the specific sequence motifs to which they bind. A basal level of transcription is observed with a minimum or core-promoter which is generally centred on a TATA box motif. The TATA motif

```
                 -480              -460              -440              -420              -400              -360
AMY54      ATTACTCGTCGCAGAAATGAATGTATCTAGAACTAAAATACAACTAGATACATCTATTTCTGCGACGAATAATTCGAAACGGAGGAGTACTTATTTTGCACTGAGTTTACTGCCATGGG
AMY8       TGTTGCTGTACCACATTAACAACAGTATCAGTCTCAGATCTGTTCTGCAAGCAAGTCCATTGGACGGCAGGCGGTGTGCAGTCGTACAGATAC---------C-----TC-----G---A
AMY53      TGAAGGCTTTATTAATTTAAAGCCGGACGTCCCTGACGTCTATGTTCTAAAAAAAGATCAAACAGTGCCAGTTTTAAACGAAAAATTGTCTTCTCTGACAGAAACTATGATGTTAGTCTG
AMY34      TCGATCTGATGTTTCCATGGCCCTTCTGCCAAGCACGCCGCAGTTCTCGGTTCCAGCTTCAGATTTCGTCCGCGTCAGAATATTTACGTGCAAATCAACAAGCCAATCTTGGAACATTGG
B-32b      TCCGCGGTTGGTTTGTGAACCCAAAAGGTATATCCTGCGTAATATTTCTGTTACTGCACCACATTAAGAACAGTTTATATGGTCTGCATTGAAATATCTGCGTGCAAGTGACAGGTGTAC
AMY46                     *CCGCGGCCCAAGTTTGCGCCGCTTTTGGGGTGAAACGGACACCACGTGGACGCGCGGCCGTCTGCGCGTGTCCTTCCCTGGCCCGCCTGTCGGTGGCACAGGG

                 -360              -340              -320              -300              -280              -260
AMY54      AACATGAAAAATCTTGTGCTCCATGTTGTTCAATTCATTTCCTCCCCATC.TTTCTTCGCG.CCCATGCTT..ATTTCCCATTGGAAAATCGGTTA.TCCAAAAAGGCAAGGGCAACGCT
AMY8       --------.....--T-TT-G--CTA---------A--------------T----------C---------AT-------G----------A----.---T------T------------
AMY53      AAAAAACTGTCATCCCCTCACAACTTAAAATGCCAATGAAATCGTTCGCAAGTGTCATC-CCT--CCCT-CCTGGCGAA----C-CC--CTAAGGGG---------.T------------
AMY34      TAAGTCAATCGCCGAGAGGTGTGTTCTGCCTTGAAGGCATGACAATCTTTTGCACGAAT-CTT--------..----TT-T-CA--C---------GC---G.----CA--A-----T---
B-32b      AGATACATTTTGCATCGAGTTCGTTACCACGGAAACATGAAAAATCTTGCAAAATCTTGCAAAATCTTGCACT--G-T-A------------A----C--A------TA------------
B-p155.3                                                                                  *-----------GC--C..----CA--A-----T---
AMY46      CGGGCCTTTTTCTATCCGCCCCCCCTTCCTCCCTGCGGCCGCACCCCCTCCACTCTTCCCTATTCTTTCCCTCGCTGCCGCCGCCGCTGCCATTGCAGCCTTGCAGTTCGGACGCGC

                 -240              -220              -200              -180              -160              -140
AMY54      GGGTTTCGATCCCAACTTGGATGGTGCCATCTTTTTCCATGGAGTTT......GTGCCACCCCGGATTGACTTGACCGTCATCGGCTGCAC.....CTTTTTTCGTAACAGAGTCTGGTA
AMY8       ---------------------------T-------CG------A---......------G------------------------------.....-----C-------G----------
AMY53      -T--------------------A----T-------C-------A---......------G------------------A-----T-T-----.....-----C-------G----------
AMY34      -CT--G-----T-----------CA-.-G---G-------------GC-GTGCCA-----G----TAC-----------A-----AAT-A-T-CCTCT-----C.-A---------TC----
B-32b      ---...T-------G-------A----T-------C-------A---......-----GG---------------.--A-----T-T-----.....-----C------------------
B-p155.3   -C---------T-----------CA-.-----A-------------GC-......------A---A-C------------A---C-A-T-A-T-CA...T----C.-A--------G-C----
AMY46      GCTCGCCTAGCCCCTTCCCCTGCACCGCCAAGCTACCCAATCACGGCCACATGGTTTGCGCACCCGCCGGCCGGATTTGGGGCGGATCCGGTCGGTCTTGAGCTCGCCCGGCTGTGGGCC

                 -120              -100              -80               -60               -40               -20
AMY54      TCCATGCAGTGCCTCCAAGCAACAC.................AC.TCTACGGTACG..TAGCTCGCGTTAAATACTG.....CCATGCCATCCACTGCCTATAAATACCAAGCACGCGGGA
AMY8       -------------------------TCC..............--GG-----.----TA--------------------TTGA..G----G----G----------------G----.-----T-
AMY53      -------------------------TCC..............--GGGG---.......-------T-------------.....-----G----G--------------------------TC-A-
AMY34      C--C----T-----G-----------TCC..............--AGTG.--.--.-C..------T-C-----T---GAGTA----CGAG---C--------------------------TA-T-
B-32b      -------------------------TCC..............--GGGG---.......-------T---------C-.....-TG--G----G---T--------------------TA-A-
B-p155.3   C-------T-----G-----------TCCCTGTCTATCACACCACAGC---.----C..------T-CG----T-CTT...........G--------------------------TA-A-
AMY46      GTCTCGTCTGTTTAGCACGAGCCTTTCC..............--AGTG.--.C---C..------T-C-----T---GAGTA.---CGAG------------------------------T--T-

                    0                20                40                60                80               100
AMY54      CACTT.GTGGCCATCAGTCGATCAGCCAGT.CAGCCAATCATCCATCCATCCGGAGAAGAAGAAGAGTCTACTGTG....CTGTGGGATCGACACTAGCGTAGCGTACGCGATGGGAAG
AMY8       -----CT-A--------------------.----------CT--GT--A----....----------GAC-----....-----A------------------------ATGCGGAAT
AMY53      -----.--AA------...----.-----GT----.-----T------.......-A---------G..-A---AACAG----A-------------...---.----ATGGGGAAG
AMY34      -C---.--A-------AC--...------...............---T-----------C-----G.C--........-T--A---A--...............--CATGGGGAAG
B-32b      -T---.--AA------...----.-----CT.-TG------T------C.......----C------G..-A-C-AACAG----A--------G----...---.----ATGGGGAAG
B-p155.3   -----.--A-------AC------A---..........C-CG---G-G--T-----C-----C-----G.C--........-T--A-C----...............--CATGGGGAAG
AMY46      -G---.--A-------AC-C...------...............---T---G------AC-----G.C--........----A---A--...............--CATGGGGAAG
```

Figure 15.1 Alignment of the upstream sequences of α-*Amy2* genes from wheat and barley. Sequences are displayed relative to α-*Amy2/54*, (---) indicates a direct match. The start of translation and a potential TATA box motif are underlined. The numbering is from the start of transcription of the genes and includes the padding characters (...) inserted into the sequences to produce the best alignments. Sequence data from the two barley genes B-32b and B-p155.3 were obtained from Whittier, Dean and Rogers (1987) and Knox *et al.* (1987).

is found in many genes 25–30 bp upstream of the located start of transcription. The factors binding to this box appear to be responsible for determining the correct start of initiation of transcription, a function also operating in plants (Bruce and Gurley, 1987). Further upstream *cis*-acting elements bind proteins which modulate TATA-box driven transcription (*see* La Thangue and Rigby, 1988). These elements have been somewhat arbitrarily grouped into two classes in *Drosophila* and mammals, based mainly on the position they are found in relation to the TATA box. Upstream promoter elements, of which the CCAAT box is an example, are generally located within 100–200 bp upstream of the start of transcription, while enhancer sequences work in either orientation and are fairly position independent, even functioning if placed 3′ to the gene (Maniatis, Goodbourn and Fischer, 1987). The proteins which bind to these motifs can be developmentally specific or present in a large number of tissues and cells. They can promote or inhibit transcription (Shore and Nasmyth, 1987; Glass *et al.*, 1988) depending on the presence of other protein factors, bind to more than one sequence (Landschulz *et al.*, 1988), and may require activation such as phosphorylation (Cherry *et al.*, 1989) or glycosylation (Jackson and Tjian, 1988). Multiple factors have also been found to bind to the same motif (Schaffner, 1989) and often they are multisubunit complexes (Chodosh *et al.*, 1988). The mechanism by which some of these proteins function relies on the presence of an amphipathic α-helix, and this particular structure can activate transcription in mammals, yeast, insects and plants (Ptashne, 1988).

As a preliminary search among the cloned and sequenced members of wheat and barley α-amylase genes for common sequence motifs which might be responsible for mediating GA control, the 5′ upstream regions were aligned for maximum homology (Figure 15.1). Within the α-*Amy2* family there is considerable sequence similarity for several hundreds of bases upstream of the start of transcription (Huttly, Martienssen and Baulcombe, 1988). Immediately obvious within this region is a highly conserved potential TATA box motif, 25–30 bp upstream of the start of transcription. The remaining sequence similarity ceases at around −300 bp, with two notable exceptions. Additional homology, extending for approximately a further 100 bp upstream is present between the two wheat genes α-*Amy2/54* and α-*Amy2/8*, while the wheat gene α-*Amy2/46* diverges from all the other genes at about −90 bp (*see* Figure 15.1; Huttly, Martienssen and Baulcombe, 1988).

In order to determine whether these conserved regions contain within them sequences important for coordinated expression, it was first necessary to study the pattern of specific expression of the individual cloned genes. In view of the general homology between all cloned genes, S1-nuclease protection analysis was used (Huttly, Martienssen and Baulcombe, 1988). The DNA probes in each case were constructed from the same region, containing the start of transcription, and for three of the wheat genes α-*Amy2/54*, α-*Amy2/8* and α-*Amy2/34* were shown, by using mRNA isolated from aneuploid lines of wheat, to be gene-specific. For the remaining wheat genes α-*Amy2/53* and α-*Amy2/46* this was not the case and S1-nuclease fragments were still protected when the aneuploid line of wheat was nullisomic for the chromosome containing the gene in question. This suggests the existence of uncloned genes in wheat which are closely related to α-*Amy2/53* and α-*Amy2/46*.

Expression of the genes was examined in several tissues of the plant: leaf, root, developing grain and germinating grain (Huttly, Martienssen and Baulcombe, 1988). None of the cloned genes was expressed in root or leaf tissue but mRNA coding for the two genes α-*Amy2/54* and α-*Amy2/8* was present in both developing and germinating grain; the remaining genes were germination specific. The presence of α-amylase isozymes in developing grain both in wheat and other cereals has been known for some time; they are thought to be located in the pericarp and to be present between 10–20 d.p.a. (MacGregor, 1983). The pI of these pericarp isozymes was shown to be identical to some α-*Amy2* (low pI isozymes) expressed during germination (Gale and Ainsworth, 1984), suggesting that some genes could be expressed in both tissues. The analysis described above confirmed this and identified two such genes. By using mRNA isolated from the *Rht* (GA-insensitive) lines of wheat it was possible to show that, in developing grain, expression of the same gene is not dependent on GA (Huttly, Martienssen and Baulcombe, 1988). Since the same start of transcription operates in both tissues this indicates that the promoters of these two genes must be complex.

Analysis of the wheat gene α-*Amy2/46* was difficult, since protection of S1-nuclease fragments was weak and variable, thus suggesting that α-*Amy2/46* might be an inactive gene. Since α-*Amy2/46* also lacks part of the upstream homology shared by all the other genes, while being an otherwise intact gene (D. Baulcombe, unpublished data) it seems possible that those sequences might be important for expression in aleurone tissue. Similarly, the extended region of homology found between α-*Amy2/54* and α-*Amy2/8* correlates with expression of these two genes in the pericarp of developing grain.

Cis-acting elements that have been defined in other eukaryotes are in the region of 8–12 bp long, whereas the homology shared by α-*Amy2* genes greatly exceeds this, thus suggesting that functionally significant sequences may be buried within the general similarity between members of the α-*Amy2* multigene family. The sequence alignments were thus extended to include α-*Amy1* genes and also genes for a carboxypeptidase and cathepsin B-like protein (G. Murphy, personal communication) both of which are induced by GA in germinating grain. Within the α-*Amy1* families of wheat and barley there are also considerable regions of upstream homology for up to 600–700 bp (Martienssen and Baulcombe, 1989), but attempted alignment of α-*Amy1* and α-*Amy2* genes and the other GA-responsive genes showed that apart from homology surrounding potential TATA box motifs there are no regions of similarity shared by every gene. This could be for a number of reasons: it may be that different *trans*-acting factors and thus *cis*-acting sequences operate within the different sets of genes, or alternatively, since some degree of heterogeneity can be tolerated within sequence motifs (Levine and Hoey, 1988), they may not be obvious by eye or to conventional computer comparison programs. In order to proceed further it was clear that a method of determining the functional significance of a particular DNA sequence was required.

Although some wheat seed genes have been successfully transferred to tobacco by *Agrobacterium* transformation and have been demonstrated to be correctly regulated in the new host (Colot *et al.*, 1987), this plant has no tissue analogous to the aleurone layer of cereals. It thus seemed inappropriate to study α-amylase expression in transgenic tobacco. It has recently become possible to transform some cereals (de La Pena, Lörz and Schell, 1987; Yang *et al.*, 1988) but this procedure is both slow and at the time was not available. The method therefore chosen as a functional assay system was that of transient expression in cereal protoplasts. The upstream region of the α-amylase gene α-*Amy2/54* was successfully used to drive the reporter gene β-glucuronidase (*GUS*; Jefferson, 1987) following transformation into protoplasts isolated from oat aleurone layers (Huttly and Baulcombe, 1989).

Transient expression in aleurone protoplasts

Aleurone protoplasts from wild oats and barley are still responsive to GA following isolation and also neither re-form cell walls nor divide (Hooley, 1982; Jacobsen, Zwar and Chandler, 1985). During culture of 1–5 days and in the presence of GA the protoplasts utilize their stored reserves, becoming highly vacuolated and, after a lag phase dependent on the species, start to produce α-amylase. In the absence of GA the protoplasts still begin to hydrolyse the aleurone grains and lipid droplets but more slowly than in the presence of GA and they synthesize and secrete little or no α-amylase. These protoplasts appeared to be the ideal system in which to study GA-regulated expression if they could be transformed with DNA and if they retained their responsiveness towards GA. The obvious choice of host cell for analysis of a wheat gene would have been wheat aleurone protoplasts, but these could not be isolated in sufficient numbers to be transformed, therefore a cultivated oat was used (Huttly and Baulcombe, 1989).

Following PEG-mediated uptake of DNA, cultivated oat aleurone protoplasts can be shown to respond to GA in a manner that is similar to wild oat protoplasts and intact aleurone cells (Hooley, 1982), while transient expression from the wheat α-*Amy2/54* promoter, in the construct pα2GT, is specifically controlled by GA in parallel with expression of the endogenous oat α-amylase. Expression of *GUS* when driven by this promoter is hence dependent on the presence of GA in the incubation medium, in a dose responsive manner (Figure 15.2) and cannot be detected until after 2 days following transformation and application of GA (Figure 15.3). GA-stimulated expression is also abolished by simultaneous addition of ABA with GA (Figure 15.4). *GUS* expression directed by the cauliflower mosaic virus (CaMV) 35S promoter, on the other hand, is not affected by the presence or absence of GA or ABA (Figure 15.4) and is detectable less than 24 h after transformation (Figure 15.3). The effect of GA on expression is thus promoter specific and not the result of a general stimulation of transcription and translation in the aleurone protoplasts. As the transformed DNA is not likely to be chromosomal it suggests that a major part at least of the GA-responsive regulation is not affected by chromosomal structure or DNA methylation.

Two other promoter:*GUS* fusions introduced into the protoplasts were derived from upstream sequences of the endosperm storage proteins HMW and LMW glutenins (Colot *et al.,* 1987). These were included in the analysis to determine if the protoplasts retained any measure of tissue specificity. HMW and LMW glutenin subunits are co-regulated and are known to accumulate only in the protein bodies of the starchy endosperm cells during grain maturation (Miflin, Burgess and Shewry, 1981). Thus, they are thought to be not normally expressed in aleurone cells of developing or germinating grain, although this has not been specifically shown for HMW glutenin (Mundy *et al.,* 1986). In transient assays in aleurone protoplasts, however, differential expression of the two glutenin promoters is observed. The HMW glutenin promoter is highly active but not responsive to GA while the LMW glutenin is not expressed at all (Huttly and Baulcombe, 1989). Expression of the HMW glutenin gene in the mature aleurone protoplasts could be the result of the protoplasts producing an aberrant *trans*-acting factor or ceasing to produce an inhibitory one, implying that they have lost or changed their state of differentiation. The LMW and HMW glutenin genes, however, were always thought to be co-regulated, and it seems unlikely that their expression would operate through a separate set of *trans*-acting factors. An alternative explanation,

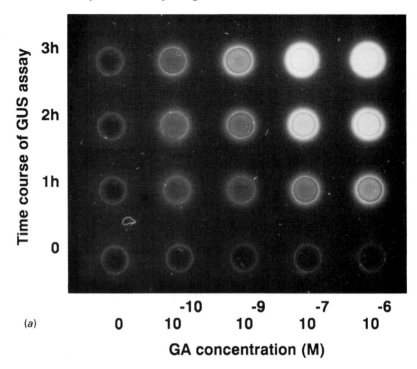

(a)

GA concentration (M)

GA concentration (M)	GUS Activity pmoles MU/min/ 10^4 protoplasts	α-Amylase Activity arbitary units
0	4.4	0
10^{-10}	18	1.0
10^{-9}	58	6.0
10^{-8}	117	11.0
10^{-7}	136	12.6
10^{-6}	128	11.8

(b)

Figure 15.2   The effect of increasing GA$_3$ concentration on transient expression from the α-*Amy2/54* promoter in the plasmid pα2GT. Aleurone protoplasts were transformed with 20 μg of pα2GT then divided into six. Protoplasts were incubated in differing concentrations of GA$_3$ as indicated and *GUS* assays were performed after 5 days of culture. (a) Visualization of the fluorescent product of the *GUS* assay. Protoplasts were lysed according to Huttly and Baulcombe (1989) and the fluorogenic substrate 4-methylumbelliferyl-β-D-glucuronide added to the extract. An aliquot was taken from the reaction at time 0 and at hourly intervals up to 3 h incubation. Na$_2$CO$_3$ was added to stop the reaction and increase fluorescence from the product, 4-methylumbelliferone (MU) (b) *GUS* activity expressed as pmoles MU/min/10^4 protoplasts was calculated from the above *GUS* assays. Endogenous α-amylase activity secreted into the incubation media by the transformed protoplasts was determined using alphachrome substrate and is expressed relative to units of barley malt α-amylase

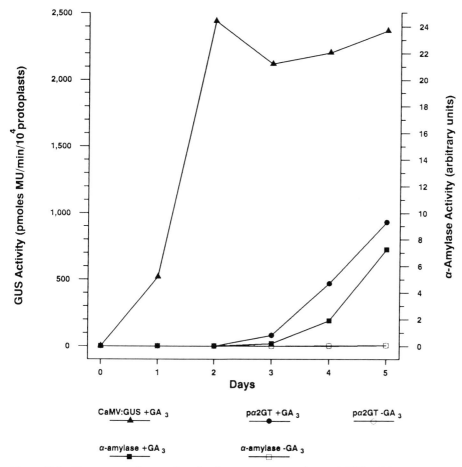

Figure 15.3 Time course of expression of endogenous oat α-amylase and *GUS* activity from aleurone protoplasts transformed by pα2GT or CaMV:GUS. The production of *GUS* by transformed protoplasts was determined over a 1–5 day incubation period. Those protoplasts transformed with pα2GT were divided and incubated in the presence or absence of 10^{-6}M GA$_3$. *GUS* activity is expressed as pmoles MU/min/10^4 protoplasts. Endogenous oat α-amylase produced by the transformed protoplasts is also indicated and expressed relative to units of barley malt α-amylase

therefore, is that this particular HMW glutenin promoter might contain sequences capable of being bound by *trans*-acting factors present legitimately in mature aleurone cells and that expression *in vivo* is regulated by chromosome structure or methylation.

Definition of important regions of the α-amylase promoter

As a start to the delineation of those regions required for correct GA-regulated control within the *α-Amy2/54* promoter, a 5' deletion series was constructed from the initial *α-Amy2:GUS* fusion, pα2GT. The resultant plasmids were then analysed

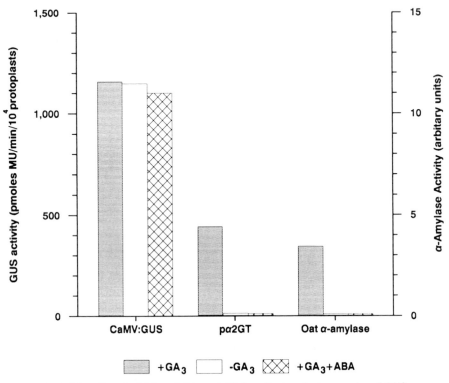

Figure 15.4   The effect of addition of GA$_3$ and ABA on the transient expression of *GUS* directed by CaMV:GUS and pα2GT. Following transformation, aliquots of protoplast were incubated for 5 days in the absence of GA$_3$, the presence of 10^{-6} M GA$_3$ or 10^{-6} M GA$_3$ with 2.5×10^{-5} M ABA. Production of endogenous oat α-amylase by transformed protoplasts is also indicated

in the transient expression system. A decline in the level of expression was apparent as the upstream regions were removed sequentially and two constructs with 300–400 bp of promoter sequence remaining had 50% of the level of expression of the initial construct (Figure 15.5). However, two further small deletions inside of this point reduced expression to just 9% (pα2GT.36), then to a basal level similar to uninduced expression (pα2GT.27). The promoter in pα2GT.14 has been deleted to just upstream of the point at which the various wheat and barley α-*Amy2* genes diverge (*see* Figure 15.1). The deletions in pα2GT.36 and pα2GT.27 extend into the conserved region; since GA-dependent expression from these constructs was greatly reduced the analysis confirms that the conserved regions of α-*Amy2* genes contain *cis*-acting sequences important for GA-stimulated expression. Further analysis of the promoter in transient expression experiments is now being carried out to locate these regions more precisely. Recently, the GA-induced binding of a protein to the upstream regions of a rice α-*Amy2*-type gene has been reported (Ou-Lee, Turgeon and Wu, 1988). The 3′ boundary of the region protected by this factor was determined; dissimilarity between the wheat and rice genes precludes accurate alignment, but this point is within the region identified above as essential for promoter function.

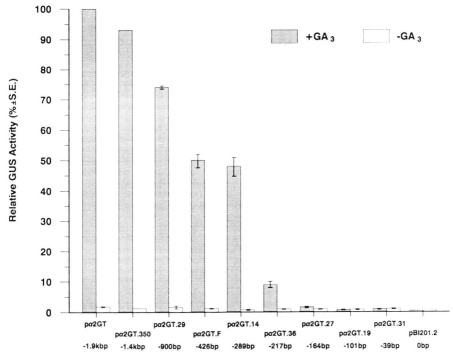

Figure 15.5 Expression of *GUS* from plasmid constructs containing deletions of the α-*Amy2/54* promoter. Expression was calculated relative to that obtained from pα2GT which contains 1.9 kbp of upstream promoter sequence. Each construct was transformed a minimum of three times with the exception of pα2GT.350; error bars indicate the standard error of the means of the replicate experiments. The end point position of each deletion is indicated, pBI202.2 is a control plasmid which contains the *GUS* gene and nopaline synthase terminator only and no promoter sequences

Expression from pα2GT.14 could still be abolished by simultaneous addition of ABA and GA to the protoplast incubation medium. ABA was shown to prevent GA-induced increases in transcription of α-amylase genes by Zwar and Hooley (1986). The mechanism by which ABA might operate is unknown, but if separate *trans*-acting factors are involved then they must bind to similar regions of the α-amylase promoter as do those affected by GA.

Control of transcription by GA

Gibberellins are superficially similar to the steroid hormones of animals: the biosynthetic pathway for both groups of compounds are common up to the C_{15} level in the isoprenoid pathway and there are structural and physical similarities including lipid solubility. The steroid hormones operate by interacting with receptor proteins (Evans, 1988); as a hormone–protein complex they are then able to bind to DNA and influence transcription. In the case of the hormone oestradiol, binding of the hormone to the receptor induces the formation of receptor dimers which are essential for tight binding of the complex to the palindromic sequence motif found in the upstream region of genes regulated by the hormone (Kumar and Chambon, 1988). Such a pattern of events is perhaps the most direct model for GA to increase transcription of α-amylase and other GA-regulated genes. A GA receptor has proved elusive, although recently some evidence suggests such a protein might reside on the plasmalemma (R. Hooley, personal communication). Unless the receptor can subsequently detach from the membrane it would be necessary to include further steps between perception of GA by the cell by binding to the receptor and control of α-amylase expression.

There is a characteristic lag phase between addition of GA and production of α-amylase and other GA-stimulated hydrolases, the duration of which differs between plants (Bewley and Black, 1978). In the past this has been suggested to indicate that GA might not directly influence the synthesis of α-amylase but, rather, operate through effects on synthesis of the endomembrane system, since α-amylase, as a secreted protein, is dependent of the endoplasmic reticulum (ER) and Golgi apparatus for its synthesis, processing and secretion (Walter, Gilmore and Blobel, 1984; Gubler, Ashford and Jacobsen, 1986). Initiation of endomembrane proliferation is now not thought to be dependent on GA, but GA has been shown to influence turnover of phosphatidyl-choline (Vakharia *et al.*, 1987). The possible explanation of this turnover is that the GA is either involved in redirecting pre-ER into ER or that this represents an early event in the signal transduction pathway as has been proposed in animals. The lag phase is evident when the α-*Amy2/54* promoter, in the plasmid pα2GT, directs expression in oat aleurone protoplasts (*see* Figure 15.3), despite this construct lacking the α-amylase signal sequence and therefore not being dependent on the presence of the endomembrane system to complete translation of the mRNA. Since the CaMV 35S promoter was capable of directing *GUS* expression less than 24 h after transformation, the lag phase is consistent with a period in which specific *trans*-acting factors are produced or modified, as a part of a signal transduction pathway.

The work on the rice α-*Amy2* gene *trans*-acting factor suggest that the protein is only present in nuclei of GA-treated aleurone cells (Ou-Lee, Turgeon and Wu, 1988), hence indicating that GA specifies production of an enhancer-binding protein, or its modification, and does not operate through modification of an already present repressor. Work is currently under way to make cDNA clones encoding this *trans*-acting factor and from there to determine other steps in the signal transduction pathway operating in this tissue.

Acknowledgements

The authors thank Shell UK for continued support for AKH. We would also like to thank Andy Phillips for help in the preparation of this manuscript.

References

Ainsworth, C. C., Doherty, P., Edwards, K. G. K., Martienssen, R. A. and Gale, M. D. (1985) Allelic variation at α-amylase loci in hexaploid wheat. *Theoretical and Applied Genetics,* **70**, 400–406

Akazawa, T. and Miyata, S. (1982) Biosynthesis and secretion of α-amylase and other hydrolases in germinating cereal seeds. *Essays in Biochemistry,* **18**, 41–78

Ashford, A. E. and Gubler, F. (1984) Mobilization of polysaccharide reserves from endosperm. In *Seed Physiology 2*, (ed. D. R. Murray), Academic Press, Australia, pp. 117–161

Ashford, A. E. and Jacobsen, J. V. (1974) Cytochemical localization of phosphatase in barley aleurone cells: the pathway of gibberellic-acid-induced enzyme release. *Planta,* **120**, 81–105

Baulcombe, D. C., Barker, R. F. and Jarvis, M. G. (1987) A gibberellin responsive wheat gene has homology to yeast carboxypeptidase Y. *Journal of Biological Chemistry,* **262**, 13726–13735

Baulcombe, D. C. and Buffard, D. (1983) Gibberellic-acid-regulated expression of α-amylase and six other genes in wheat aleurone layers. *Planta,* **157**, 493–501

Baulcombe, D. C., Huttly, A. K., Martienssen, R. A., Barker, R. F. and Jarvis, M. G. (1987) A novel wheat α-amylase gene (α-*Amy3*). *Molecular and General Genetics,* **209**, 33–40

Bewley, J. D. and Black, M. (1978) *Physiology and Biochemistry of Seeds vol. 1.* Springer-Verlag, Berlin, Heidelberg, New York

Brown, P. H. and Ho, T.-H.D. (1987) Biochemical propedrties and hormonal regulation of barley nuclease. *European Journal of Biochemistry,* **168**, 357–364

Bruce, W. B. and Gurley, W. B. (1987) Functional domains of a T-DNA promoter active in crown gall tumors. *Molecular Cell Biology,* **7**, 59–67

Bush, D. S., Hedrich, R., Schroeder, J. I. and Jones, R. L. (1988) Channel-mediated K^+ flux in barley aleurone protoplasts. *Planta,* **176**, 368–377

Cedar, H. (1988) DNA methylation and gene activity. *Cell,* **53**, 3–4

Chandler, P. M., Walker-Simmons, M., King, R. W., Crouch, M. and Close, T. J. (1988) *Journal of Biology,* suppl. 12C, 143, (abstract L-033)

Cherry, J. R., Johnson, T. R., Dollard, C., Shuster, J. R. and Denis, C. L. (1989) Cyclic AMP-dependent protein kinase phosphorylates and inactivates the yeast transcriptional activator ADR1. *Cell,* **65**, 409–419

Chodosh, L. A., Olesen, J., Hahn, S., Baldwin, A. S., Guarente, L. and Sharp, P. A. (1988) A yeast and a human CCAAT-binding protein have heterologous subunits that are functionally interchangeable. *Cell,* **53**, 25–35

Chrispeels, M. J. and Varner, J. E. (1967) Gibberellic acid-enhanced synthesis and release of α-amylase and ribonuclease by isolated barley aleurone layers. *Plant Physiology,* **42**, 398–406

Colot, V., Robert, L. S., Kavanagh, T. A., Bevan, M. W. and Thompson, R. D. (1987) Localization of sequences in wheat endosperm protein genes which confer tissue-specific expression in tobacco. *EMBO Journal,* **6**, 3559–3564

de La Pena, A., Lörz, H. and Schell, J. (1987) Transgenic rye plants obtained by injecting DNA into young floral tillers. *Nature,* **325**, 274–276

Doig, R. I., Colborne, A. J., Morris, G. and Laidman, D. L. (1975a) The induction of glyoxysmal enzyme activities in the aleurone cells of germinating wheat. *Journal of Experimental Botany,* **26**, 387–398

Doig, R. I., Colborne, A. J., Morris, G. and Laidman, D. L. (1975b) Enzymes of intermediary metabolism in the aleurone cells of germinating wheat. *Journal of Experimental Botany,* **26**, 399–410

Evans, R. M. (1988) The steroid and thyroid receptor superfamily. *Science,* **240**, 889–895

Evers, A. D. (1970) Development of the endosperm of wheat. *Annals of Botany,* **34**, 547–555

Fernandez, D. E and Staehelin, L. A. (1987) Effect of gibberellic acid on lipid degradation in barley aleurone layers. In *Molecular Biology of Plant Growth Control*, Alan R. Liss, Inc., New York, pp. 323–334

Filner, P. and Varner, J. E. (1967) A simple and unequivocal test for *de novo* synthesis of enzymes: density labelling of barley α-amylase with H_2O^{18}. *Proceedings of the National Academy of Sciences, USA*, **58**, 1520–1526

Gabard, K. A. and Jones, R. L. (1986) Localization of phytase and acid phosphtase isozymes in aleurone layers of barley. *Physiology of Plants*, **67**, 182–192

Gale, M. D. and Ainsworth, C. C. (1984) The relationship between α-amylase species found in developing and germinating wheat grain. *Biochemistry and Genetics*, **22**, 1031–1036

Glass, C. K., Hilloway, J. M., Devary, O. V. and Rosenfeld, M. G. (1988) The thyroid hormone receptor binds with opposite transcriptional effects to a common sequence motif in thyroid hormone and estrogen response elements. *Cell*, **54**, 313–323

Goldman, M. A. (1988) The chromatin domain as a unit of gene regulation. *BioEssays*, **9**, 50–55

Gomez, J., Sanchez-Martinez, D., Stiefel, V., Rigan, J. and Puigdomenech Pages, M. (1988). A gene induced by the plant hormone abscissic acid in response to water stress encodes a glycine-rich protein. *Nature*, **334**, 262–264

Gubler, F. and Ashford, A. E. (1983) Changes in peroxidase in isolated aleurone layers in response to gibberellic acid. *Australian Journal of Plant Physiology*, **10**, 87–97

Gubler, F., Ashford, A. E. and Jacobsen, J. V. (1986) Involvement of the Golgi apparatus in the secretion of α-amylase from gibberellin-treated barley aleurone cells. *Planta*, **168**, 447–452

Gubler, F., Ashford, A. E. and Jacobsen, J. V. (1987) The rlease of α-amylase through gibberellin-treated barley aleurone cell walls. *Planta*, **172**, 155–161

Hammerton, R. W. and Ho, T.-H.D. (1986) Hormonal regulation of the development of protease and carboxypeptidase activities in barley aleurone layers. *Plant Physiology*, **80**, 692–697

Hardie, D. G. (1975) Control of carbohydrase formation by gibberellic acid in barley endosperm. *Phytochemistry*, **14**, 1719–1722

Higgins, T. J. V., Zwar, J. A. and Jacobsen, J. V. (1976) Gibberellic acid enhances the level of translatable mRNA for α-amylase in barley aleurone layers. *Nature*, **260**, 166–169

Hong, B., Uknes, S. J. and Ho, T.-H.D. (1988) Cloning and characterization of a cDNA encoding an mRNA rapidly-induced by ABA in barley aleurone layers. *Plant Molecular Biology*, **11**, 495–506

Hooley, R. (1982) Protoplasts isolated from aleurone layers of wild oat (*Avena fatua* L.) exhibit the classic response to gibberellic acid. *Planta*, **154**, 29–40

Hooley, R. (1984) Gibberellic acid controls specific acid-phosphatase isozymes in aleurone cells and protoplasts of *Avena fatua* L. *Planta*, **16**, 355–360

Huttly, A. K. and Baulcombe, D. C. (1989) A wheat α-*Amy 2* promoter is regulated by gibberellin in transformed oat aleurone protoplasts. *EMBO Journal*, **8**, 1907–1913

Huttly, A. K., Martienssen, R. A. and Baulcombe, D. C. (1988) Sequence heterogeneity and differential expression of the α-*Amy2* gene family in wheat. *Molecular General Genetics*, **214**, 232–240

Jackson, S. P. and Tjian, R. (1988) O-Glycosylation of eukaryotic transcription factors: implications for mechanisms of transcriptional regulation. *Cell*, **55**, 125–133

Jacobsen, J. V. and Beach, L. R. (1985) Control of transcription of α-amylase and rRNA genes in barley aleurone protoplasts by gibberellin and abscisic acid. *Nature*, **316**, 275–277

Jacobsen, J. V. and Chandler, P. M. (1987) Gibberellic and abscisic acid in germinating cereals. In *Plant Hormones and their Role in Plant Growth and Development*, (ed. P. J. Davis), Martinus Hijhoff Pub., Dordrecht

Jacobsen, J. V., Zwar, J. A. and Chandler, P. M. (1985) Gibberellic-acid-responsive protoplasts from mature aleurone of Himalaya barley. *Planta*, **163**, 430–439

Jefferson, R. A. (1987) Assaying chimeric genes in plants: the GUS gene fusion system. *Plant Molecular Biology Reporter*, **5**, 387–405

Jones, R. L. (1971) Gibberellic acid-enhanced release of β-1,3-glucanase from barley aleurone cells. *Plant Physiology*, **47**, 412–416

Jones, R. L. (1980) Quantitative and qualitative changes in the endoplasmic reticulum of barley aleurone layers. *Planta*, **150**, 70–81

Knox, C. A. P., Sonthayanon, B., Chandra, G. R. and Muthukrishnan, S. (1987) Structure and organization of two divergent α-amylase genes from barley. *Plant Molecular Biology*, **9**, 3–17

Kumar, V. and Chambon, P. (1988) The estrogen receptor binds tightly to its responsive elements as a ligand-induced homodimer. *Cell*, **55**, 145–156

La Thangue, N. B. and Rigby, P. W. J. (1988) *Trans*-acting protein factors and the regulation of eukaryotic transcription. In *Transcription and Splicing*, (eds B. D. Hames and D. M. Glover), IRL Press, Oxford, pp. 1–42

Landschulz, W. H., Johnson, P. F., Adashi, E. Y., Graves, B. J. and McKnight, S. L. (1988) Isolation of a recombinant copy of the gene encoding C/EBP. *Genes and Development*, **2**, 786–800

Lazarus, C. M., Baulcombe, D. C. and Martienssen, R. A. (1985) α-Amylase genes of wheat are two multigene families which are differentially expressed. *Plant Molecular Biology*, **5**, 13–24

Levine, M. and Hoey, T. (1988) Homeobox proteins as sequence-specific transcription factors. *Cell*, **55**, 537–540

MacGregor, A. W. (1983) Cereal α-amylases: synthesis and action pattern. In *Seed Proteins*, (eds J. Daussant, J. Mosse and J. Vaughan), Academic Press, pp. 1–33

Maniatis, T., Goodbourne, S. and Fischer, J. A. (1987) Regulation of inducible and tissue-specific gene expression. *Science*, **236**, 1237–1245

Martienssen, R. A. and Baulcombe, D. C. (1989) An unusual wheat insertion sequence (WIS1) lies upstream of an α-amylase gene in hexaploid wheat, and carries a 'minisatellite' array. *Molecular General Genetics*, **217**, 401–410

Miflin, B., Burgess, S. and Shewry, P. (1981) The development of protein bodies in the storage tissues of seeds: subcellular separations of homogenates of barley, maize, and wheat endosperms and of pea cotyledons. *Journal of Experimental Botany*, **32**, 199–219

Morrison, I. N., Kuo, J. and O'Brien, T. P. (1975) Histochemistry and fine structure of wheat aleurone cells. *Planta*, **123**, 105–116

Mundy, J. (1984) Hormonal regulation of α-amylase inhibitor synthesis in germinating barley. *Carlsberg Research Communication*, **49**, 439–444

Mundy, J. and Chua, N.-H. (1988) Abscisic acid and water stress induce the expression of a novel rice gene. *EMBO Journal*, **7**, 2279–2286

Mundy, J. and Fincher, G. B. (1986) Effects of gibberellic acid and abscisic acid on levels of translatable mRNA [1→3,1→]-β-D-glucanase in barley aleurone. *FEBS Letters*, **198**, 349–352

Mundy, J., Heigaard, J., Hansen, A., Hallgren, L., Jorgensen, K. G. and Munck, L. (1986) Differential synthesis *in vitro* of barley aleurone and starchy endosperm proteins. *Plant Physiology,* **81**, 630–636

Muthukrishrarn, S., Gill, B. S., Swegle, M. and Chandra, G. (1984) Structural genes for α-amylases are located on barley chromosomes 1 and 6. *Journal of Biological Chemistry,* **259**, 13637–13639

Nolan, R. C. and Ho, T.-H.D. (1988) Hormonal regulation of gene expression in barley aleurone layers. *Planta,* **174**, 551–560

Nolan, R. C., Lin, L.-S. and Ho, T.-H.D. (1987) The effect of abscisic acid on the differential expression of α-amylase isozymes in barley aleurone layers. *Plant Molecular Biology,* **8**, 13–22

Ou-Lee, T.-M., Turgeon, R. and Wu, R. (1988) Interaction of a gibberellin-induced factor with the upstream region of an α-amylase gene in rice aleurone tissue. *Proceedings of the National Academy of Sciences, USA,* **85**, 6366–6369

Ptashne, M. (1988) How eukaryotic transcriptional activators work. *Nature,* **335**, 683–689

Quatrano, R. S. (1987) The role of hormones during seed development. In *Plant Hormones and their Role in Plant Growth and Development,* (ed. P. J. Davis), Martinus Nijhoff Pub., Dordrecht, pp. 494–514

Reid, J. B. (1986) Gibberellin mutants. In *Plant Gene Research, A Genetic Approach to Plant Biochemistry,* (eds A. D. Blanstein and P. J. King), Springer-Verlag, pp. 1–34

Rogers, J. C., Dean, D. and Heck, G. R. (1985) Aleurain: a barley thiol protease closely related to mammalian cathepsin H. *Proceedings of the National Academy of Sciences, USA,* **82**, 6512–6516

Rogers, J. C. and Milliman, C. (1984) Coordinate increase in major transcripts from the high pI α-amylase multigene family in barley aleurone cells stimulated with gibberellic acid. *Journal of Biological Chemistry,* **259**, 12234–12240

Sargent, J. G. (1980) α-Amylase isozymes and starch degradation. *Cereal Research Communication,* **8**, 77–86

Schaffner, W. (1989) How do different transcription factors binding the same DNA sequence sort out their jobs? *Trends in Genetics,* **5**, 37–39

Shore, D. and Nasmyth, D. (1987) Purification and cloning of a DNA binding protein from yeast that binds to both silencer and activator elements. *Cell,* **51**, 721–732

Sopanen, T. and Laurière, C. (1989) Release and activity of bound β-amylase in germinating barley grain. *Plant Physiology,* **89**, 244–249

Stuart, I. M., Loi, L. and Fincher, G. B. (1986) Development of [1→3,1→4]-β-D-glucan endohydrolase isozymes in isolated scutella and aleurone layers of barley. *Plant Physiology,* **80**, 310–314

Taiz, L. and Honigaiman, W. A. (1976) Production of cell wall hydrolysing enzymes by barley aleurone layers in response to gibberellic acid. *Plant Physiology,* **58**, 380–386

Tomos, A. D. and Laidman, D. L. (1979) The control of mobilization and metabolism in the aleurone tissue during germination. In *Recent Advances in the Biochemistry of Cereals,* (eds D. L. Laidman and R. G. Wyn Jones), Academic Press, pp. 119–146

Vakharia, D. N., Brearley, C. A., Wilkinson, M. C., Galliard, T. and Laidman, D. L. (1987) Gibberellin modulation of phosphatidyl-choline turnover in wheat aleurone tissue. *Planta,* **172**, 502–507

Walter, P., Gilmore, R. and Blobel, G. (1984) Protein translocation across the endoplasmic reticulum. *Cell,* **38**, 5–8

Whittier, R. F., Dean, D. A. and Rogers, J. C. (1987) Nucleotide sequence analysis of α-amylase and thiol protease genes that are hormonally regulated in barley aleurone cells. *Nucleic Acids Research,* **15**, 2515–2535

Yang, H., Zhang, H. M., Davey, M. R., Mulligan, B. J. and Cocking, E. C. (1988) Production of kanamycin resistant rice tissues following DNA uptake into protoplasts. *Plant Cell Reports,* **7**, 421–425

Zwar, J. A. and Hooley, R. (1986) Hormonal regulation of α-amylase gene transcription in wild oat (*Avena fatua* L.) aleurone protoplasts. *Plant Physiology,* **80**, 459–463

16

EXPRESSION OF GENES FOR PHOTOSYNTHETIC ELECTRON TRANSFER COMPONENTS IN TRANSGENIC PLANTS

JOHN C. GRAY, DAVID I. LAST, PAUL DUPREE,
BARBARA J. NEWMAN and ROSALIND E. SLATTER
Botany School, University of Cambridge, Cambridge, UK

Introduction

The ability to introduce genes into plants presents unprecedented opportunities for studying many facets of plant biology. Rapid advances in our understanding of the regulation of plant gene expression have been made by studying the behaviour of genes and gene constructs in transgenic plants (Schell, 1988; Willmitzer, 1988; Benfey and Chua, 1989). In addition, transgenic plants offer many exciting possibilities for improving our understanding of basic biochemical processes. For example, the ability to manipulate the amount of an individual enzyme or protein, by the use of sense or antisense gene constructs (Rodermel, Abbott and Bogorad, 1988; Smith *et al.*, 1988; van der Krol *et al.*, 1988), should allow the quantitative estimation of the importance of that enzyme or protein for a particular function in the plant. In particular, transgenic plants appear to offer the possibility of determining flux control coefficients (Kacser and Burns, 1973) for individual steps in a biochemical pathway, thus providing quantitative measures of the contribution of a particular enzyme to the overall regulation of the pathway. Transgenic plants may also provide useful tools for studying the importance of a particular polypeptide in the assembly of multisubunit complexes or in other protein–protein interactions. For example, it may be possible to unravel the pathway of assembly of a multisubunit complex by altering separately the expression of genes for each individual subunit of the complex, and examining the amount of the complex and of any precursor complexes that accumulate.

Transgenic plants thus appear to offer excellent prospects for understanding the assembly and function of complex biochemical pathways such as photosynthetic electron transfer. The photosynthetic electron transfer chain consists of three intrinsic multisubunit complexes, photosystem II, the cytochrome *bf* complex and photosystem I, located in the chloroplast thylakoid membrane and three extrinsic proteins, plastocyanin, ferredoxin and ferredoxin-NADP$^+$ reductase (FNR). This system is reponsible for the transfer of electrons from water to NADP$^+$, producing oxygen and a source of reducing power and generating a transmembrane proton gradient which is used to drive ATP synthesis. The genes for components of this electron transfer chain are distributed between the nuclear and chloroplast genomes, with each of the intrinsic membrane complexes containing both nuclear and chloroplast gene products (Gray, 1987). In contrast, all of the extrinsic proteins

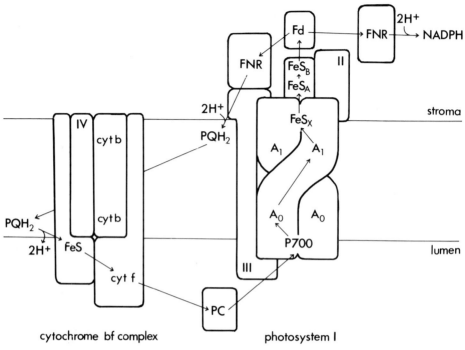

Figure 16.1   A scheme for the involvement of plastocyanin (PC) and ferredoxin-NADP$^+$ reductase (FNR) in cyclic and non-cyclic photosynthetic electron transfer. The membrane-bound form of FNR is shown in association with subunit III of photosystem I, with which it may operate as a ferredoxin-plastoquinone reductase (E. C. Davies and D. S. Bendall, unpublished data)

are encoded by nuclear genes. We are interested in the synthesis and assembly of functional electron transfer chains in chloroplasts of higher plants and are using transgenic plants to examine the regulation of expression of nuclear genes, to study the pathways of assembly of the membrane complexes and to obtain qualitative and quantitative estimates of the role of the nuclear-encoded proteins in the regulation of photosynthetic electron transfer. In this chapter we describe some of our recent work on the expression of pea nuclear genes for plastocyanin and FNR in transgenic tobacco plants. The involvement of plastocyanin and FNR in electron flow in the vicinity of photosystem I is shown in Figure 16.1. The initial aim was to determine if fully functional pea proteins can be synthesized and correctly localized in transgenic tobacco plants.

Plastocyanin

Plastocyanin is a 10 kDa copper protein located in the lumen of the thylakoid membrane system where it transfers electrons from cytochrome *f* to the primary donor P700 of photosystem I (*see* Figure 16.1). Plastocyanin is confined to chloroplast-containing tissues of plants, and its synthesis appears to be under the control of phytochrome. Haslett and Cammack (1974) showed that plastocyanin was present in only small amounts in etiolated bean leaves, but its accumulation was stimulated by red light and this stimulation was reversed by subsequent illumination with far-red light. Takabe, Takabe and Akazawa (1986) found, using Western blotting, that the amount of the plastocyanin polypeptide (per unit fresh weight) increased about 30-fold during a 72 h greening period in pea, wheat and barley, and Nielsen and Gausing (1987) showed that plastocyanin mRNA was 5–20-fold higher in green barley leaves compared with etiolated leaves. In contrast, the amount of plastocyanin active in electron transfer (per unit fresh weight) in barley seedlings has been found to remain constant during greening (Plesnicar and Bendall, 1972), perhaps suggesting a post-translational regulation of the amount of the plastocyanin holoprotein. In *Chlamydomonas reinhardtii* plastocyanin accumulation is controlled at a post-transcriptional level by the availability of copper (Merchant and Bogorad, 1986).

Plastocyanin is encoded in the nucleus and synthesized on cytoplasmic ribosomes as a precursor with an *N*-terminal presequence which is proteolytically removed during import of the protein into chloroplasts (Grossman *et al.*, 1982). A *Silene pratensis* cDNA clone encoding the complete plastocyanin precursor has been characterized (Smeekens *et al.*, 1985) and the deduced amino acid sequence suggests that the presequence is composed of a chloroplast import domain and a thylakoid transfer domain. Following import of the precursor into chloroplasts the amino-terminal domain of the presequence is removed by a stromal protease (Smeekens *et al.*, 1986) and the resulting intermediate form is then imported into the thylakoid lumen where it is processed to the mature size by a second protease (Hageman *et al.*, 1986; Kirwin, Elderfield and Robinson, 1987). A similar two-domain structure of the plastocyanin presequence is suggested by the nucleotide sequences of the plastocyanin genes from spinach (Rother *et al.*, 1986), barley (Nielsen and Gausing, 1987) and *Arabidopsis thaliana* (Vorst *et al.*, 1988).

A cDNA clone for pea plastocyanin was identified in a pea leaf cDNA library in the *Eco*K selection vector M13K8.2 (Waye *et al.*, 1985) by hybridization with a mixed oligonucleotide probe (5'CCNGCRTTRTTYTTRAA3') deduced from the determined sequence Phe-Lys-Asn-Asn-Ala-Gly of pea plastocyanin (Boulter *et al.*, 1979). This clone was used to identify another five cDNA clones by hybridization. The sequences of all the clones were identical in the regions in which they overlapped, suggesting that they arose from a single type of mRNA. Hybridization of a cDNA clone to a Southern blot of pea genomic DNA digested with *Eco*RI, *Hin*dIII and *Bam*HI indicated the presence of a single-copy gene residing on a 3.5 kbp *Eco*RI fragment, a 4.7 kbp *Hin*dIII fragment and a 16 kbp *Bam*HI fragment. The 16 kbp *Bam*HI fragment was isolated from a λEMBL3 library of *Bam*HI-digested pea nuclear DNA and characterized by restriction mapping (Figure 16.2) and partially by nucleotide sequencing (Last and Gray, 1989).

The pea plastocyanin gene encodes a protein of 168 amino acid residues, of which 69 residues constitute the presequence and 99 residues make up the mature polypeptide (Last and Gray, 1989). A comparison of plastocyanin presequences is shown in Figure 16.3. There is considerable variation in the central region of the peptide, suggesting that it may be relatively unimportant structurally and may merely act as a link between the more conserved and functionally more important *N*-terminal and *C*-terminal sequences. The hydrophobic *C*-terminal region of the presequence is the most highly conserved, perhaps suggesting constraints imposed by the thylakoid import machinery.

To examine the expression of the pea plastocyanin gene in a foreign plant, the

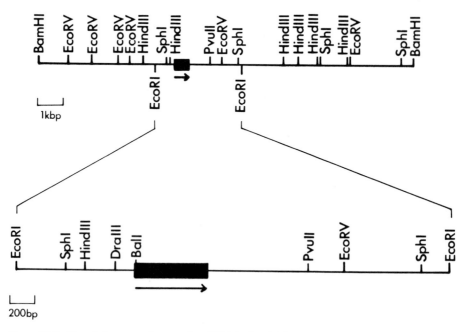

Figure 16.2   Restriction map of pea nuclear DNA in the vicinity of the plastocyanin gene. The upper map is of the 16 kbp *Bam*HI fragment isolated from a pea genomic library in λEMBL3. The lower map shows the 3.5 kbp *Eco*RI fragment which was transferred to tobacco

3.5 kbp *Eco*RI fragment, containing the coding region with approximately 1 kbp upstream sequence and 2 kbp of downstream sequence, was inserted into the binary vector pBin19 (Bevan, 1984) and mobilized to *Agrobacterium tumefaciens* LBA4404 by triparental mating. The gene was transferred to tobacco by *Agrobacterium*-mediated leaf disc transformation (Horsch *et al.*, 1985). Primary transformed plants were examined for the presence of the pea plastocyanin gene by Southern hybridization and for the presence of pea plastocyanin by non-denaturing polyacrylamide gel electrophoresis of leaf extracts. The results obtained with five plants selected at random are shown in Figure 16.4. Three plants (A1, A3 and A12) showed a single hybridizing band corresponding to the 3.5 kbp pea DNA fragment, and electrophoresis of leaf extracts demonstrated the presence of pea plastocyanin, migrating just ahead of tobacco plastocyanin. The plants B4 and B6 did not contain an unrearranged pea plastocyanin gene and did not contain any pea plastocyanin. However, there appeared to be no correlation between the amount of pea plastocyanin and the copy number of the introduced pea plastocyanin gene in primary transgenic plants. Plant A1 contained less pea plastocyanin than plants A3 and A12, although plants A3 and A12 each contained a single copy of the pea plastocyanin gene per nucleus, whereas plant A1 contained two copies per nucleus. Quantitative data on the amounts of pea plastocyanin in several other transgenic tobacco plants were obtained by scanning densitometry of the stained bands on non-denaturing gels, related to absolute amounts of purified pea and tobacco plastocyanins. The mass ratios of pea plastocyanin to tobacco plastocyanin in some transgenic plants with unrearranged pea genes were in the range 0.23–1.84. The

```
ps  MATVTSTT-VAIPSFSGLKTNAATKVSAMAKIPTSTSQS---PRLCVRASL--KDIGVALVATAASAVLASNALA V
    :::::::  :::::  ::  :::::   .    .  ::  :  ::::::::: :::: :.::: :::::: :
sp  MATVTSSAAVAIPSFAGLKASSTTRAATVKVA--V--AT---PRMSIKASL--KDVGVVVAATAAAGILA-NAMA A
    :::::::::::::  ::::::  ::: .    .      .  :::::::: :::::::::::::::::::: :
so  MATVASSAAVAVPSFTGLKASGSIKPTTAKIIPT-TTA-V--PRLSVKASL--KNVGAAVVATAAAGLLAGNAMA V
    :::  :::::::::  ::: :  :: ::: :  ::  :  :  :::::::: :::::::::::::::::::: 
at  MAAITSAT-VTIPSFTGLKLAVSSKPKTLSTISRSSSATRAPPKLALKSSL--KDFGVIAVATAASIVLAGNAMA M
    :   ::  :: ::::  :::  ::   .    .   :::   ::::: :
hv  MAAL-SSAAVSVPSFA-----AATPMRS----SRSS------RMVVRASLGKKAASAAVAMAAGAMLLGGSAMA Q
```

Figure 16.3 Comparison of the presequences of the plastocyanin precursor from pea (ps), *Silene pratensis* (sp), spinach (so), *Arabidopsis thaliana* (at) and barley (hv). Sequences are compared with the species above; identical residues are denoted with a colon (:), and conservative changes are marked with a stop (.). Gaps introduced into the sequences to aid alignment are marked with a dash (-). The cleavage site for the thylakoidal protease is shown by an arrow

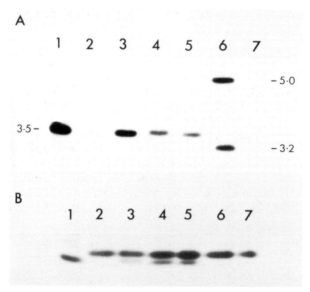

Figure 16.4 Expression of the pea plastocyanin gene in transgenic tobacco plants.
Individual transformed plants were analysed by: (A) Southern blotting of *Eco*RI-digested
nuclear DNA and hybridization with a pea plastocyanin cDNA probe. (B) PAGE blue
staining of leaf proteins separated by non-denaturing polyacrylamide gel electrophoresis.
Track 1, pea; 2, wild-type tobacco; 3–7, transgenic plants A1, A3, A12, B4 and B6
respectively

amount of pea plastocyanin in leaves of these transgenic plants was in the range
2–10 mg/g fresh weight, although there was considerable variation depending on
the age and growth conditions of the plants.

The effect of gene copy number on the amount of pea plastocyanin was examined
by quantitative analysis of ten kanamycin-resistant plants obtained by self-
pollination of plant A3 which contained a single pea plastocyanin gene per nucleus.
Four plants were homozygous for the introduced DNA and contained two copies of
the pea plastocyanin gene, whereas the remainder were hemizygous and contained
a single pea plastocyanin gene. Clear differences between the ratios of pea and
tobacco plastocyanins in extracts of leaves at a similar developmental stage (ninth
leaf up from the base of the stem) from homozygous and hemizygous plants were
observed. The mean value of this ratio for hemizygous plants was 0.54 ± 0.05,
whereas for homozygous plants it was 0.85 ± 0.07. Doubling the plastocyanin
gene-copy number in the homozygous plants therefore gave an almost twofold
increase in the amount of pea plastocyanin, but had no effect on the amount of
tobacco plastocyanin. The amounts of tobacco plastocyanin in hemizygous and
homozygous plants were not significantly different when expressed either relative
to the total amount of protein or on a chlorophyll basis, indicating that the
expression of the introduced pea genes did not affect the expression of the
endogenous tobacco genes. Analysis of these plants indicated that expression of
increased numbers of plastocyanin genes is not limited by the amounts of factors
necessary for transcriptional or post-transcriptional events. The synthesis of *Silene
pratensis* plastocyanin in transgenic tomato plants has also been reported not to
affect the expression of the endogenous plastocyanin genes (de Boer *et al.*, 1988).

CHARACTERIZATION OF PEA PLASTOCYANIN FROM TRANSGENIC TOBACCO PLANTS

Pea plastocyanin was localized to the thylakoid lumen of chloroplasts from transgenic tobacco plants, by acetone extraction of washed thylakoids prepared from purified intact chloroplasts. The extracted plastocyanin had the same electrophoretic mobility in non-denaturing polyacrylamide gels as authentic pea plastocyanin. de Boer *et al.* (1988) have also shown that *Silene* plastocyanin is correctly targeted to the thylakoid lumen in transgenic tomato plants. To confirm that the pea plastocyanin had been correctly processed, the protein was purified from leaves of transgenic tobacco plants. During elution from DEAE-cellulose with 100 mM sodium phosphate (pH 6.9) the plastocyanin was resolved into two distinct blue bands. Electrophoresis under non-denaturing conditions indicated that the first band was pea plastocyanin and the second band was tobacco

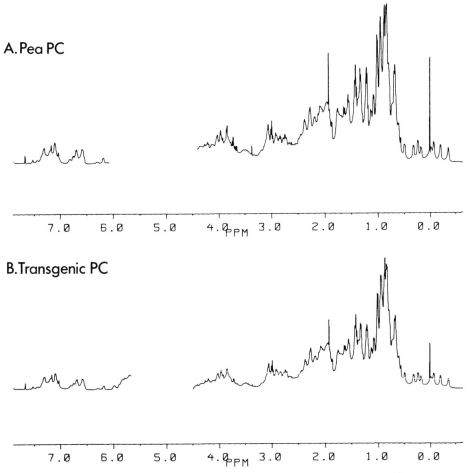

Figure 16.5 ¹H NMR spectra of reduced pea plastocyanin. (A) Natural pea plastocyanin purified from pea leaves. (B) Pea plastocyanin purified from kanamycin-resistant progeny of self-pollinated transgenic tobacco plant A3

plastocyanin. The identities of the proteins were confirmed by gas-phase protein sequencing; the *N*-terminal sequence of plastocyanin from the first peak was found to be Val-Glu-Val-Leu-Leu-Gly-Ala-Ser-Asp-Gly, whereas the sequence for the second peak was found to be Ile-Glu-Val-Leu-Leu-Gly-Ser-Asp-Asp-Gly. These sequences are identical to the published *N*-terminal sequences of pea and tobacco plastocyanins, respectively (Boulter *et al.*, 1979). This indicated that the *N*-terminal presequence of the pea plastocyanin precursor had been correctly removed in transgenic tobacco plants, and the isolation of a blue protein indicated that copper had been incorporated into the protein structure.

To determine whether the pea plastocyanin was correctly folded in transgenic tobacco plants, the purified pea plastocyanin was compared, by ^{1}H NMR spectroscopy, to natural pea plastocyanin purified from pea leaves (Figure 16.5). This indicated that pea plastocyanin from transgenic tobacco plants was structurally very similar, and most probably identical, to natural pea plastocyanin. The spectra of the two pea plastocyanin preparations are much more similar to each other than they are to the ^{1}H NMR spectrum of french bean plastocyanin (King and Wright, 1986), which shows 81% sequence identity with pea plastocyanin (Milne, Wells and Ambler, 1974). French bean plastocyanin has a completely different set of chemical shifts in the high-field region indicating that the ^{1}H NMR spectra are highly sensitive to slight variations in the structure of the protein.

Pea plastocyanin purified from transgenic tobacco plants has been found to be active in electron transfer *in vitro* by measuring its rate of reduction in the presence of digitonin-treated pea chloroplasts as a source of the cytochrome *bf* complex (D. S. Bendall, personal communication). Second-order rate constants of approximately 10^8 M^{-1} s^{-1} for both authentic pea plastocyanin and pea plastocyanin from transgenic plants indicate that the proteins were essentially identical. This demonstration that pea plastocyanin synthesized by transgenic tobacco is indistinguishable from natural pea plastocyanin and that it can easily be separated from tobacco plastocyanin suggests the use of tobacco as an expression system for the production of altered plastocyanins created by mutagenesis *in vitro*.

REGULATION OF EXPRESSION OF THE PEA PLASTOCYANIN GENE IN TRANSGENIC TOBACCO PLANTS

The pea plastocyanin gene was expressed in the same light-regulated and organ-specific manner in transgenic tobacco plants as in pea plants, as shown by Northern hybridization and Western blotting. Northern hybridization of RNA extracted from leaves of dark-grown and greening peas, and from dark-adapted and greening transgenic tobacco plants, identified a single band of 900 bases hybridizing to the plastocyanin cDNA probe. Hybridization with the RNA from greening peas was greater than with RNA from dark-grown peas, indicating that the accumulation of mRNA for pea plastocyanin was stimulated by light. A 900 b RNA was also detected in greening transgenic tobacco plants, but not in dark-adapted plants, under hybridization conditions that failed to detect transcripts of the endogenous tobacco plastocyanin genes. This indicated that the pea plastocyanin gene was light-regulated in transgenic tobacco plants. Light-stimulated increases in plastocyanin mRNA have previously been demonstrated with barley leaves (Barkardottir *et al.*, 1987; Nielsen and Gausing, 1987).

The pea plastocyanin gene was shown to be expressed in leaves, but not roots, of pea plants and the transgenic tobacco plants A1 and A3, by probing Western blots of leaf and root extracts with an antiserum to pea plastocyanin. Plastocyanin is found only in chloroplast-containing tissues of higher plants, and it has been shown previously that plastocyanin mRNA is not present in coleoptiles and roots of barley (Nielsen and Gausing, 1987).

Comparison of the 5' upstream sequences of the pea, spinach and *Arabidopsis thaliana* plastocyanin genes has not revealed any similar conserved sequences that might be involved in regulation of gene expression. The sequence CTTTATCAT located just upstream of the TATA box in the pea gene is similar to the consensus sequence CCTTATCAT recognized by Grob and Stüber (1987) and suggested to function as a light-response element involved in phytochrome regulation. A sequence similar to the Box II sequence of the pea *rbc*S genes, which binds the protein factor GT-1 (Green *et al.*, 1988), is present in the reverse orientation in the upstream region of the pea plastocyanin gene. This sequence, TTCCACAC, is located 137 bp upstream from the translation start site. A functional analysis of the sequences responsible for light-regulated and organ-specific expression is currently being carried out.

Ferredoxin-NADP$^+$ reductase

Ferredoxin-NADP$^+$ reductase (FNR) is a 34 kDa flavoprotein located in the stromal compartment of chloroplasts; about half of the enzyme can be released from chloroplasts by simple osmotic shock, whereas the remainder remains more tightly bound to the thylakoid membrane. FNR is probably attached to the membrane by a specific binding protein (Vallejos, Ceccarelli and Chan, 1984; Matthijs, Coughland and Hind, 1986). FNR catalyses the final step of the linear photosynthetic electron transfer chain by mediating the passage of electrons from ferredoxin to NADP$^+$, and has also been implicated in cyclic electron flow around photosystem I (*see* Figure 16.1). FNR is therefore situated at a branch-point in electron flow, and may play a key role in regulating the relative amounts of cyclic and non-cyclic electron flow to meet the changing demands of the plant for ATP and reducing power. The functional significance of the stromal and membrane-bound forms of the enzyme is not clear, although the membrane-bound form shows light-mediated enhancement of its catalytic activity (Carrillo, Lucero and Vallejos, 1981). The physical differences between stromal and membrane-bound forms are also not clear. The two forms are antigenically indistinguishable (Süss, 1979), but both are heterogeneous in size. In spinach, the membrane-bound form can be resolved into two size classes of 33 kDa and 35 kDa, and up to eight species of the stromal enzyme have been separated by isoelectric focusing (Gozzer *et al.*, 1977; Ellefson and Krogmann, 1979; Hasumi, Nagata and Nakamura, 1983). The expression of a single gene for FNR in transgenic plants may help to resolve some of these uncertainties. It should be possible to determine if a single FNR gene can give rise to multiple forms of FNR and also if the stromal and membrane-bound forms are the product of a single gene.

FNR is encoded in nuclear DNA and is synthesized in the cytoplasm as a higher molecular weight precursor (Grossman *et al.*, 1982; Tittgen *et al.*, 1986). The precursor is imported post-translationally into the chloroplast where it is processed to the mature size (Grossman *et al.*, 1982; Tittgen *et al.*, 1986). A partially purified preparation of the stromal processing protease was reported to cleave the FNR precursor from wheat and barley to the mature size (Robinson and Ellis, 1984). The FNR precursor has been reported to be able to associate with FAD to produce a catalytically active protein (Carrillo, 1985), but it is not known if this is physiologically relevant. Synthesis of FNR is light-regulated (Haslett and Cammack, 1976; Sluiters-Scholten, Moll and Stegwee, 1977), and appears to be under phytochrome control. The accumulation of FNR activity in bean leaves was stimulated by red light, and the stimulation was prevented by subsequent illumination with far-red light (Haslett and Cammack, 1976).

cDNA clones for pea FNR were identified in a pea leaf cDNA library in λgt11 (Gantt and Key, 1986) by immunochemical screening with rabbit antibodies to the stromal FNR of pea chloroplasts (Newman and Gray, 1988). A full-length clone was characterized by nucleotide sequencing and shown to encode a polypeptide of 360 amino acid residues. Comparison of the deduced amino acid sequence with the determined *N*-terminal sequence of the mature FNR protein suggested that the protein was synthesized initially with a presequence of 52 amino-acid residues. However, it is possible that the presequence is shorter than 52 amino acid residues and the cleaved protein undergoes further processing at the N-terminus. Heterogeneity of *N*-terminal sequences has been obtained for spinach and pea FNR (Hasumi, Nagata and Nakamura, 1983; Newman and Gray, 1988), and Karplus, Walsh and Herriot (1984) have established that the *N*-terminal residue of spinach FNR is pyroglutamate, which is presumably formed from a glutamine residue in the protein. A comparison of the presequences from pea, spinach (Jansen *et al.*, 1988) and *Mesembryanthemum crystallinum* (Michalowski, Schmitt and Bohnert, 1989) is shown in Figure 16.6. Michalowski, Schmitt and Bohnert

```
ps   MAAAVTAAVSLPYSNSTSLPIRTSIVA-PERLVFKKVSL--NNVSISGRVGTIRAQVT
     ::::::::::: . :: : ::: :    :.. : :: :     :::. :.::::: ..
mc   MAAAVTAAVSFPSTKSTPLSTRTSSVITHEKINFNKVPLYYRNVSVGGKVGTIRAVAS
     :  ::::::::::::.: :: :.:::. .:: . :::::::::::. :: : ::
so   MTTAVTAAVSFPSTKTTSLSARSSSVISPDKISYKKVPLYYRNVSATGKMGPIRAQIA
```

Figure 16.6 Comparison of the presequences of the FNR precursor from pea (ps), *Mesembryanthemum crystallinum* (mc) and spinach (so). Sequences are compared with the species above; identical residues are denoted with a colon (:), and conservative changes are marked with a stop (.). Gaps introduced into the sequences to aid alignment are marked with a dash (-). Proposed cleavage sites are marked with arrows

(1989) have suggested that cleavage occurs between Thr-52 and Ile-53 in *Mesembryanthemum crystallinum* on the basis of the similarity between the upstream sequence Ser-Val-Gly-Gly-Lys-Val-Gly and the sequence Ser-Gly-Gly-Val-Arg-Gln-Gly which precedes the cleavage site of the presequence for the small subunit of ribulose biphosphate carboxylase. The mature FNR proteins contain a high proportion of identical amino-acid residues; the mature pea protein is 90% identical to spinach (Karplus, Walsh and Herriot, 1984), 87% to *Mesembryanthemum crystallinum* (Michalowski, Schmitt and Bohnert, 1989) and 57% to the cyanobacterium *Spirulina* sp. (Yao *et al.*, 1984).

Hybridization of a cDNA clone to Southern blots of pea genomic DNA suggested the presence of two copies of the FNR gene in the pea haploid genome (Newman and Gray, 1988). Two copies of the FNR gene have also been suggested for spinach (Jansen *et al.*, 1988) and *Mesembryanthemum crystallinum* (Michalowski, Schmitt and Bohnert, 1989). Two different *Bam*HI fragments showing hybridization to a pea FNR cDNA clone have been isolated from a pea genomic library in λEMBL3. Preliminary sequence analysis of a 3.7 kbp *Bam*HI fragment has identified two small exons separated by an intron of 118 bp and flanked by a non-coding sequence. It is not yet clear if other small exons are present in the fragment. The coding sequence shows several changes from the full-length cDNA indicating that this region of DNA does not encode the previously characterized FNR precursor. A second pea genomic clone contained a 13 kbp *Bam*HI fragment which encodes the FNR precursor. Nucleotide sequence analysis of a 4.9 kbp *Bam*HI-*Hin*dIII fragment indicated that the coding region was divided into nine exons separated by eight introns (Figure 16.7). The sequence of the exons was identical to that of the cDNA clone characterized previously (Newman and Gray, 1988). The 4.9 kbp *Bam*HI-*Hin*dIII fragment contains approximately 1 kbp of upstream sequence and 1 kbp of downstream sequence.

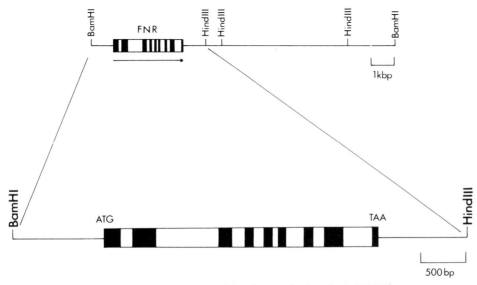

Figure 16.7 Maps of a pea genomic clone containing the gene for ferredoxin-NADP$^+$ reductase. Coding regions are shown as filled boxes

The 4.9 kbp *Bam*HI-*Hin*dIII fragment containing the pea FNR gene was transferred into tobacco using *Agrobacterium*-mediated transformation as described above for the pea plastocyanin gene. Transformed plants were examined for the presence of pea FNR activity in leaf extracts. Leaf extracts were subjected to non-denaturing polyacrylamide gel electrophoresis and stained for diaphorase activity with nitro-blue tetrazolium and NADPH. FNR displays a high level of diaphorase activity (Avron and Jagendorf, 1956). Wild-type tobacco plants gave two equally staining bands, whereas pea plants gave a single band of lower mobility (Figure 16.8). Extracts of some transgenic plants produced all three bands, indicating that the pea FNR gene was being expressed (Figure 16.8). The presence of active pea FNR in these transgenic tobacco plants suggests that the prosthetic group FAD was correctly associated with the FNR polypeptide. The identical electrophoretic mobility of pea FNR in pea and transgenic tobacco plants suggests that the FNR precursor was correctly processed in transgenic tobacco plants. These

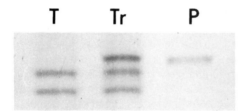

T   Tr   P

Figure 16.8   Expression of the pea FNR gene in a transgenic tobacco plant.
NADPH-nitro-blue tetrazolium reductase activity was detected after non-denaturing
polyacrylamide gel electrophoresis of leaf extracts from wild-type tobacco (T), an
individual transgenic tobacco plant (Tr) and pea (P)

plants offer promising material for investigating the basis for the stromal and
membrane-bound forms in chloroplasts, and for studying the effect of increased
amounts of FNR on the partition of electron flow between the cyclic and non-cyclic
pathways.

Conclusions

These preliminary studies have established that pea nuclear genes for plastocyanin
and FNR are expressed in transgenic tobacco plants to produce correctly processed
functional proteins. This indicates that the differences in the primary structures of
the pea and endogenous tobacco proteins do not markedly affect the processing
and accumulation of the pea proteins. The proteins and enzymes involved in
transmembrane transfer, proteolytic processing and prosthetic group addition are
able to function even with proteins that may be considerably different in primary
structure, and presumably in conformation, from the endogenous proteins. It is
likely, of course, that highly variable regions of the proteins are not directly
involved as recognition sites for processing or the addition of prosthetic groups.

Detailed biochemical analysis of the effects of increased amounts of plastocyanin
and FNR in the transgenic plants will be carried out in the future, but inspection of
the mature plants indicates that outwardly they are indistinguishable from
wild-type tobacco plants. This perhaps is not unexpected with the relatively low
increases in the amounts of plastocyanin (up to 200% more than endogenous) and
FNR (up to 50% more), although it should be possible to increase the amounts of
the proteins further by crossing individual transgenic plants. These hybrid plants
may indicate if higher levels of gene-dosage interfere with gene expression.
Antisense constructs of the plastocyanin and FNR genes may enable the amounts
of these proteins to be decreased in transgenic plants, and this should provide
information on the quantitative importance of the proteins in photosynthetic
electron transfer.

Acknowledgements

We are grateful to T. Kavanagh, S. Mason, M. Bevan for introducing us to plant
transformation, to R. Wilson and A. Loynes for help with tissue culture, to E.
Laue, J. Barna and D. Bendall for NMR and electron transfer measurements and

to L. Packman for protein sequence analysis. This work was supported by grants from the Science and Engineering Research Council and the Gatsby Charitable Foundation.

References

Avron, M. and Jagendorf, A. T. (1956) A TPNH diaphorase from chloroplasts. *Archives of Biochemistry and Biophysics*, **65**, 475–490

Barkardottir, R. B., Jensen, B. F., Kreiberg, J. D., Nielsen, P. S. and Gausing, K. (1987) Expression of selected nuclear genes during leaf development in barley. *Developmental Genetics*, **8**, 495–511

Benfey, P. N. and Chua, N.-H. (1989) Regulated genes in transgenic plants. *Science*, **244**, 174–181

Bevan, M. (1984) Binary *Agrobacterium* vectors for plant transformation. *Nucleic Acids Research*, **12**, 8711–8721

Boulter, D., Peacock, D., Guise, A., Gleaves, J. T. and Estabrook, G. (1979) Relationships between the partial amino acid sequences of plastocyanin from members of ten families of flowering plants. *Phytochemistry*, **18**, 603–608

Carrillo, N. (1985) Biosynthesis of ferredoxin-NADP$^+$ oxidoreductase. Evidence for the formation of a functional preholoenzyme in the cytoplasmic compartment. *European Journal of Biochemistry*, **150**, 469–474

Carrillo, N., Lucero, H. A. and Vallejos, R. H. (1981) Light modulation of chloroplast membrane-bound ferredoxin-NADP oxidoreductase. *Journal of Biological Chemistry*, **256**, 1058–1059

de Boer, D., Cremers, F., Teerstra, R., Smits, L., Hille, J., Smeekens, S. *et al.* (1988) *In vivo* import of plastocyanin and a fusion protein into developmentally different plastids of transgenic plants. *EMBO Journal*, **7**, 2631–2635

Ellefson, W. L. and Krogmann, D. W. (1979) Studies of the multiple forms of ferredoxin-NADP oxidoreductase from spinach. *Archives of Biochemistry and Biophysics*, **194**, 593–599

Gantt, J. S. and Key, J. L. (1986) Isolation of nuclear encoded plastid ribosomal protein cDNAs. *Molecular and General Genetics*, **202**, 186–193

Gozzer, C., Zanetti, G., Galliano, M., Sacchi, G. A., Minchiotti, L. and Curti, B. (1977) Molecular heterogeneity of ferredoxin-NADP reductase from spinach leaves. *Biochimica et Biophysica Acta*, **485**, 278–290

Gray, J. C. (1987) Genetics and synthesis of chloroplast membrane proteins. In *Photosynthesis*, (ed. J. Amesz), Elsevier, Amsterdam, pp. 319–342

Green, P. J., Yong, M.-H., Cuozzo, M., Kano-Murakami, Y., Silverstein, P. and Chua, N.-H. (1988) Binding site requirements for pea nuclear protein factor GT-1 correlates with sequences required for light-dependent transcriptional activation of the *rbcS-3A* gene. *EMBO Journal*, **7**, 4035–4044

Grob, U. and Stüber, K. (1987) Discrimination of phytochrome dependent light inducible from non-light inducible plant genes. Prediction of a common light-responsive element (LRE) in phytochrome dependent light inducible plant genes. *Nucleic Acids Research*, **15**, 9957–9973

Grossman, A. R., Bartlett, S. G., Schmidt, G. W., Mullet, J. E. and Chua, N.-H. (1982) Optimal conditions for post-translational uptake of proteins by isolated chloroplasts. *Journal of Biological Chemistry*, **257**, 1558–1563

Hageman, J., Robinson, C., Smeekens, S. and Weisbeek, P. (1986) A thylakoid processing protease is required for complete maturation of the lumen protein plastocyanin. *Nature,* **324**, 567–569

Haslett, B. G. and Cammack, R. (1974) The development of plastocyanin in greening bean leaves. *Biochemical Journal,* **144**, 567–572

Haslett, B. G. and Cammack, R. (1976) Changes in the activity of ferredoxin-NADP reductase during the greening of bean leaves. *New Phytologist,* **76**, 219–226

Hasumi, H., Nagata, E. and Nakamura, S. (1983) Molecular heterogeneity of ferredoxin-NADP reductase from spinach leaves. *Biochemical and Biophysical Research Communications,* **110**, 280–286

Horsch, R. B., Fry, J. E., Hoffmann, N., Eichholtz, D., Rogers, S. G. and Fraley, R. T. (1985) A simple and general method of transferring genes into plants. *Science,* **227**, 1229–1231

Jansen, T., Reiländer, H., Steppuhn, J. and Herrmann, R. G. (1988) Analysis of cDNA clones encoding the entire precursor-polypeptide for ferredoxin:NADP$^+$ oxidoreductase from spinach. *Current Genetics,* **13**, 517–522

Kacser, H. and Burns, J. A. (1973) The control of flux. *Symposia of the Society for Experimental Biology,* **27**, 65–104

Karplus, P. A., Walsh, K. A. and Herriot, J. R. (1984) Amino acid sequence of spinach ferredoxin-NADP oxidoreductase. *Biochemistry,* **23**, 6576–6583

King, G. C. and Wright, P. E. (1986) Proton NMR studies of plastocyanin/ assignment of aromatic and methyl group resonances from two-dimensional spectra. *Biochemistry,* **25**, 2364–2374

Kirwin, P. M., Elderfield, P. D. and Robinson, C. (1987) Transport of proteins into chloroplasts. Partial purification of a thylakoidal processing peptidase involved in plastocyanin biogenesis. *Journal of Biological Chemistry,* **262**, 16386–16390

Last, D. I. and Gray, J. C. (1989) Plastocyanin is encoded by a single-copy gene in the pea haploid genome. *Plant Molecular Biology,* **12**, 655–666

Matthijs, H. C. P., Coughlan, S. J. and Hind, G. (1986) Removal of ferredoxin-NADP oxidoreductase from thylakoid membranes, rebinding to depleted membranes, and identification of binding sites. *Journal of Biological Chemistry,* **261**, 12154–12158

Merchant, S. and Bogorad, L. (1986) Regulation by copper of the expression of plastocyanin and cytochrome *c*-552 in *Chlamydomonas reinhardti. Molecular and Cellular Biology,* **6**, 462–469

Michalowski, C. B., Schmitt, J. M. and Bohnert, H. J. (1989) Expression during salt stress and nucleotide sequence of cDNA for ferredoxin-NADP$^+$ oxidoreductase from *Mesembryanthemum crystallinum. Plant Physiology,* **89**, 817–822

Milne, P. R., Wells, J. R. E. and Ambler, R. P. (1974) The amino acid sequence of plastocyanin from french bean (*Phaseolus vulgaris*). *Biochemical Journal,* **143**, 691–701

Newman, B. J. and Gray, J. C. (1988) Characterisation of a full-length cDNA clone for pea ferredoxin-NADP$^+$ reductase. *Plant Molecular Biology,* **10**, 511–520

Nielsen, P. S. and Gausing, K. (1987) The precursor of barley plastocyanin. Sequence of cDNA clones and gene expression in different tissues. *FEBS Letters,* **225**, 159–162

Plesnicar, M. and Bendall, D. S. (1973) The photochemical activities and electron carriers of developing barley leaves. *Biochemical Journal,* **136**, 803–812

Robinson, C. and Ellis, R. J. (1984) Transport of proteins into chloroplasts. Partial purification of a chloroplast protease involved in the processing of imported precursor polypeptides. *European Journal of Biochemistry,* **142**, 337–342

Rodermel, S. R., Abbott, M. S. and Bogorad, L. (1988) Nuclear-organelle interactions: nuclear antisense gene inhibits ribulose biphosphate carboxylase enzyme levels in transformed tobacco plants. *Cell,* **55**, 673–681

Rother, C., Jansen, T., Tyagi, A., Tittgen, J. and Herrmann, R. G. (1986) Plastocyanin is encoded by an uninterrupted nuclear gene in spinach. *Current Genetics,* **11**, 171–176

Schell, J. S. (1988) Transgenic plants as tools to study the molecular organization of plant genes. *Science,* **237**, 1176–1183

Sluiters-Scholten, C. M. T., Moll, W. A. W. and Stegwee, D. (1977) Ferredoxin and ferredoxin-NADP oxidoreductase in leaves of *Phaseolus vulgaris. Planta,* **133**, 289–294

Smeekens, S., Bauerle, C., Hageman, J., Keegstra, K. and Weisbeek, P. (1986) The role of the transit peptide in the routing of precursors toward different chloroplast compartments. *Cell,* **46**, 365–375

Smeekens, S., De Groot, M., Van Binsbergen, J. and Weisbeek, P. (1985) Sequence of the precursor of the chloroplast thylakoid lumen protein, plastocyanin, *Nature,* **317**, 456–458

Smith, C. J. S., Watson, C. F., Ray, J., Bird, C. R., Morris, P. C., Schuch, W. and Grierson, D. (1988) Antisense RNA inhibition of polygalacturonase gene expression in transgenic tomatoes. *Nature,* **334**, 724–726

Süss, K.-H. (1979) Isolation and partial characterisation of membrane-bound ferredoxin-NADP reductase from chloroplasts. *FEBS Letters,* **101**, 305–310

Takabe, T., Takabe, T. and Akazawa, A. T. (1986) Biosynthesis of P700-chlorophyll *a* protein, plastocyanin, and cytochrome b_6-f complex. *Plant Physiology,* **81**, 60–66

Tittgen, J., Hermans, J., Steppuhn, J., Jansen, T., Jansson, C., Andersson, B. *et al.* (1986) Isolation of cDNA clones for fourteen nuclear-encoded thylakoid membrane proteins. *Molecular and General Genetics,* **204**, 258–265

Vallejos, R. H., Ceccarelli, E. and Chan, R. (1984) Evidence for the existence of thylakoid intrinsic protein that binds ferredoxin-NADP oxidoreductase. *Journal of Biological Chemistry,* **259**, 8048–8051

van der Krol, A. R., Lenting, P. E., Veenstra, J., van der Meer, I. M., Koes, R. E., Gerats, A. G. M. *et al.* (1988) An anti-sense chalcone synthase gene in transgenic plants inhibits flower pigmentation. *Nature,* **333**, 866–869

Vorst, O., Oosterhoff-Teerstra, R., Vankan, P., Smeekens, S. and Weisbeek, P. (1988) Plastocyanin of *Arabidopsis thaliana*: isolation and characterisation of the gene and chloroplast import of the precursor protein. *Gene,* **65**, 59–69

Waye, M. M. Y., Verhoeyen, M. E., Jones, P. T. and Winter, G. (1985) EcoK selection vectors for shotgun cloning into M13 and deletion mutagenesis. *Nucleic Acids Research,* **13**, 8561–8571

Willmitzer, L. (1988) The use of transgenic plants to study plant gene expression. *Trends in Genetics,* **4**, 13–18

Yao, Y., Tamura, T., Wada, K., Matsubara, H. and Kodo, K. (1984) *Spirulina* ferredoxin-NADP reductase – the complete amino acid sequence. *Journal of Biochemistry (Tokyo),* **95**, 1513–1516

17

CLONING AND MOLECULAR ANALYSIS OF ANTHOCYANIN GENES IN *ZEA MAYS*

UDO WIENAND, JAVIER PAZ-ARES and HEINZ SAEDLER

Max-Planck Institut für Züchtungsforschung, Cologne, W. Germany

Introduction

Anthocyanins are pigments belonging to the group of flavonoids. They are present in various plant tissues, predominantly in flowers and fruits. In *Zea mays* (maize), the synthesis of anthocyanins has been studied in great detail using transposable element (TE)-induced mutations of genes involved in the pathway. Genetical investigation and molecular analysis of these genes have made the anthocyanin pathway one of the best characterized pathways in plants. Since many genes are involved in the synthesis of anthocyanins, this pathway represents a unique example for the study of coordinated gene expression and gene regulation.

Isolation of maize anthocyanin genes using transposable elements

The activity of transposable elements can be detected if a TE integrates into a gene, thereby inducing a mutation. If a gene involved in pigmentation is affected by this mutation, a recognizable change in phenotype is produced. Transposition into such a gene very often blocks its function which usually causes a colourless phenotype. However, due to the instability of most TE-induced mutations, gene function is restored upon excision of the element, which somatically can be seen as sectors or spots. In maize, about a dozen transposable elements have been analysed genetically, most of them belonging to the so-called two element systems (for reviews see Fedoroff, 1983; Nevers, Shepherd and Saedler, 1985; Döring and Starlinger, 1986; Peterson, 1986). Such systems consist of an autonomous and a non-autonomous element. The autonomous element can actively transpose, whereas the non-autonomous element can only be activated in *trans* (by the autonomous element). The interaction between autonomous and non-autonomous elements is specific. Only those elements belonging to the same two component system can interact. The functions of TEs in maize have been mainly investigated by their interaction with genes involved in pigmentation.

The genetical analysis of anthocyanin biosynthesis has shown that a number of genes are involved in the pathway. About a dozen loci have been found which affect pigmentation (Coe, 1957; Styles and Ceska, 1977; Dooner, 1983). Some genes code for enzymes necessary for the stepwise synthesis of anthocyanins and have been called structural genes. Structural genes of the pathway that have been investigated are *c2,a1,a2,bz1,bz2* and *pr* (Figure 17.1). In contrast to this, various regulatory genes have been characterized which either act in a tissue specific manner or influence the degree of pigmentation. Such genes are *c1,pl,r1,vp1,clf,b* and *in*. *c1*, for example, regulates the expression of pigments only in various tissues of the maize kernel, whereas the genetically homologous *pl* locus acts as a regulatory gene in the other tissues of maize. In the case of *c1*, the regulatory nature of this gene has been concluded from the analysis of the expression of structural genes like *c2*, *a1* and *bz1* in lines with different *c1* alleles. From this analysis it became clear that, in the absence of *c1* function, *c2*, *a1* and *bz1* are not expressed (Dooner, 1983, Cone, Burr and Burr, 1986)

Recently TE-induced mutations have been used for gene cloning using the so-called transposon tagging strategy (for review *see* Wienand and Saedler, 1987). This technique uses TE-induced mutations where a genetically defined element interacts with the gene of interest. If this element is known molecularly, it can be used as a specific probe to isolate the gene of interest. Several maize elements have been cloned and so far three of them have been used for gene tagging. These are the *Ac* element (*activator;* Fedoroff, Wessler and Shure, 1983), the *En/Spm* element (*enhancer* equivalent to *suppressor mutator;* Pereira *et al.*, 1985) and the *Mu1* element (*mutator;* Barker *et al.*, 1984). The probability of cloning a locus from a TE-induced mutant is higher using a mutant where an autonomous element is present at that locus. This is because the structure of such an element is predictable in terms of size and restriction cuts. In addition, the copy number of autonomous elements (like *Ac* or *En/Spm*) compared with non-autonomous elements is much lower (usually less than five copies/genome) in most maize lines. A genomic library constructed from a mutant where an autonomous element is present at the gene of interest is therefore screened only for clones containing a full size autonomous element. An important prerequisite for the usage of TEs in gene cloning is that no cross-hybridization between elements belonging to different systems so far has been observed.

In maize the *Ac*, *En1* and *Mu1* elements have been used for the isolation of the structural genes *c2,a1,bz1*, and bz2 (Fedoroff, Furtek and Nelson, 1984; O'Reilly *et al.*, 1985; Wienand *et al.*, 1986; Theres, Scheele and Starlinger, 1987), as well as for the isolation of the regulatory genes *c1,r1,p* and *vp1* (Cone, Burr and Burr, 1986; Lechelt, Laird and Starlinger, 1986; Paz-Ares *et al.*, 1986; Peterson and Schwarz, 1986; Dellaporta *et al.*, 1987; McCarty, *et al.*, 1988). The fact that eight anthocyanin genes have been cloned using transposon tagging, four of which have regulatory functions, demonstrates that transposon tagging is a powerful cloning strategy.

Cloning and analysis of the structural genes *a1* and *c2*

The reasons for cloning the structural genes *a1* and *c2* were to analyse the structure and function of these genes as well as to investigate a series of transposable element-induced mutations of both genes. Also *a1* and *c2* could serve as target genes to study the interaction between structural and regulatory genes of the pathway.

Both genes have been cloned using variegated mutant alleles where the *En1* element is present at either *a1* or *c2* (O'Reilly *et al.*, 1985; Wienand *et al.*, 1986).

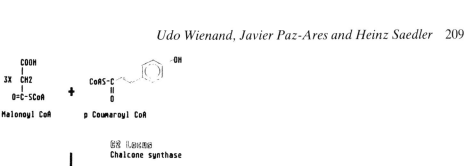

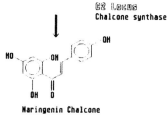

C2 Locus
Chalcone synthase

Naringenin Chalcone

Chalcone isomerase

Flavanone

Hydroxylase

Dihydroquercitin

A1 Locus
Dihdroquercitin
reductase

Leucocyanidin

A2 Locus   **Bz1 Locus**
               UFGT

Cyanidin   Cyanidin 3-glucoside

Figure 17.1   Flow diagram of anthocyanin biosynthesis in maize aleurone. Structural genes involved in the pathway as known so far are *c2,a1,a2,bz1* and *bz2*. The enzymatic steps of genes *c2*, *a1* and *bz1* are indicated. Regulatory genes necessary for anthocyanin expression in aleurone are *c1,r1,c1f1* and *vp1*

Genomic lambda libraries were constructed from these mutants and screened for clones containing a full size *En1* element (8.3 kb; Pereira *et al.*, 1985). In both cases such clones could be isolated and were further analysed. Sequences adjacent to the element were isolated and used as probes in Southern experiments using DNA isolated from wild type, mutant and revertant alleles of either *a1* or *c2*. The different restriction pattern among the different alleles observed in these experiments indicated that the original clones isolated indeed contained *a1* and *c2* specific sequences.

The analysis of the structure of both genes was done comparing cDNA and genomic sequences. In the case of *a1* this comparison revealed three short introns and four exons (219, 179, 194 and 834 bp) (Schwarz-Sommer *et al.*, 1987). The amino acid sequence derived from the *a1* specific cDNA encodes a 40 kD protein. This protein shares homology with the protein product of the *pallida* locus in *Antirrhinum majus*. This indicates that both proteins might have identical functions. Recently this was confirmed by Rohde *et al.* (1986) and Reddy *et al.* (1987) showed that the *a1* gene product is a dihydroflavanone reductase.

The *c2* locus was analysed by sequencing genomic as well as cDNA clones. The comparison of both sequences showed that the *c2* locus contains two exons, 427 bp and 1013 bp in length (Ulla Niesbach-Klösgen, personal communication). The *c2* encoded protein is 44 kD in size. The sequence of this protein shares 60–80% homology to chalcone synthase proteins isolated from other species. This confirmed the assumption (Dooner, 1983) that the *c2* locus encodes chalcone synthase.

Analysis of the regulatory *c1* locus

The analysis of genes which have a regulatory function is of great interest in terms of the structure and function of these genes. The products and functions of regulatory genes have not been identified so far. Furthermore, the expression of these genes is probably rather low which additionally complicates the cloning of these genes. Gene tagging so far has been the only possible method to isolate regulatory genes. Since in maize a number of regulatory genes involved in pigmentation have been identified genetically, this provides a unique model system to study at the molecular level basic features of gene regulation.

One such regulatory gene is the *c1* locus, which regulates in a tissue-specific manner the synthesis of anthocyanins. The *c1* locus only affects pigmentation in the scutellum and aleurone tissue of the maize kernel. A number of interesting *c1* mutants have been isolated. Two of these mutants (*c1-I* and *c1-p*) might be important for the understanding of *c1* function. The *c1-I* allele is a dominant inhibitor of *c1* function. This allele, if crossed to the wild-type *c1* allele decreases pigmentation almost completely. The *c1-p* mutant represents a conditionally coloured mutant producing anthocyanin during germination under the influence of light. At least three further regulatory genes have to be active beside *c1* in order to achieve pigment formation in the maize kernel. These genes are *r1*, *vp1* and *c1f1*.

The isolation of the *c1* locus was also achieved using the gene tagging strategy, by cloning it from a variegated *c1* mutant (Cone, Burr and Burr, 1986; Paz-Ares *et al.*, 1986). This mutation of *c1* was caused by the insertion of an *En* element. Cloning of the *c1* locus was performed similarly to that of the *a1* and *c2* loci. From a library constructed from the DNA of this mutant a clone could be isolated containing the full size *En* element. Using sequences flanking the *En* element it was possible to isolate the wild-type and several mutant alleles of *cl*. The analysis of a series of transposable element-induced *cl* mutants showed that the locus was about 3 kb in size.

Using *cl* specific probes in Northern experiments revealed several transcripts in wild-type polyA+ RNA isolated from 30-day-old kernels. cDNA cloning and sequence analysis led to a definition of the structure of *cl* (Paz-Ares *et al.*, 1987). The gene has three exons (150 bp, 129 bp and 720 bp) and two small introns (88 bp and 144 bp). From the cDNA analysis the putative sequence of the *cl* encoded protein could be determined. The protein has 273 amino acids and a molecular weight of approximately 29 kD. The protein contains two major domains, a basic one at the amino-terminus (amino acids 1–120) and an acidic one at the carboxy-terminus (amino acids 230–273) (Figure 17.2). The basic domain of this

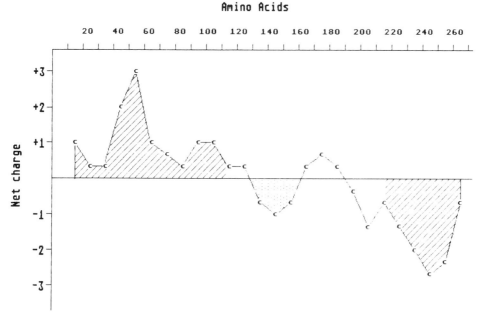

Figure 17.2   Charge distribution in the putative *cl* protein. The average of net charges over successive 30 amino acids is shown, measured at 10 amino-acid intervals. The basic domain at the amino-terminus extends until amino acid 120. The major acidic domain of the protein is located between amino acids 230 and 270

protein shows a 40% homology to the products of *myb* proto-oncogenes from animals and humans (Katzen, Kornberg and Bishop, 1985; Majello, Kenyon and Dalla-Favera, 1986). *myb* proto-oncogene products are known to be located in the nucleus (Boyle *et al.*, 1984). The basic domain of these proteins shows DNA binding properties (Klempnauer and Sippel, 1987). Recently it was shown for the viral *myb* proto-oncogene protein that the interaction of this protein with DNA is sequence specific (Biedenkamp *et al.*, 1988). The homology of the *cl* encoded protein with the *myb* proto-oncogene proteins suggests that the *cl* protein also might be a DNA binding protein. Since the analysis of various *c1* mutants showed that *c1* affects transcription of several structural genes involved in the anthocyanin pathway (*a1, c2, bz1*; Cone, Burr and Burr, 1986) the *c1* protein might function as a transcriptional activator. The structural similarity of the *c1* gene product to transcription activators in yeast supports this idea. The analysis of the yeast transcriptional activators GAL4 and GCN4, showed that the products of these genes are DNA binding proteins which contain a basic domain (for DNA binding) and an acidic domain for the activation of transcription (Hope and Struhl, 1986; Ma and Ptashne, 1987).

Further support for *c1* being a transcriptional activator comes from the analysis of the dominant inhibitory allele *c1-I*. The major difference between this allele and the *c1* wild-type allele is a 3.5 kb insert present in the third exon of the *c1-I* allele. The analysis of a *c1-I* specific cDNA showed that an 8 bp insertion causes a frame-shift mutation, which subsequently leads to a shortened protein, 252 amino acids in length. This putative *c1-I* encoded protein lacks the acidic domain present in the wild-type protein. This loss of acidity at the carboxy-terminus of the *c1-I* protein could be the reason for the inhibitory effect of the *c1-I* allele.

For analysis of the function of the *c1* protein the *c1* cDNA was expressed in *Escherichia coli* and the *c1* protein partially purified. Filter binding assays using various DNA fragments of the structural *a1* gene showed that the *c1* encoded protein is a DNA binding protein. Binding of the *c1* protein to *a1* has been found at the 5' (promoter) part as well as at the 3' part of the gene. Further experiments to identify sequence specific binding of the *c1* protein are in progress.

References

Barker, R. F., Thompson, D. V., Talbot, D. R., Swanson, J. and Bennetzen, J. L. (1984) Nucleotide sequence of the maize transposable element Mu1. *Nucleic Acids Research*, **12**, 5955

Biedenkamp, H., Borgmeyer, U., Sippel, A. E. and Klempnauer, K. H. (1988) Viral *myb* oncogene encodes a sequence specific DNA-binding activity. *Nature*, **335**, 835

Boyle, W. J., Lambert, M. A., Lipsick, J. S. and Baluda, M. A. (1984) Avian myeloblastosis E26 virus oncogene products are nuclear proteins. *Proceedings of the National Academy of Sciences, USA*, **81**, 4265

Coe, E. H. Jr (1957) Anthocyanin synthesis in maize. A gene sequence construction. *American Nature*, **91**, 381

Cone, K. C., Burr, F. A. and Burr, B. (1986) Molecular analysis of the maize regulatory locus c1. *Proceedings of the National Academy of Sciences, USA*, **83**, 9631

Dellaporta, S. L., Greenblatt, I., Kermicle, J. L., Hicks, J. B. and Wessler, S. R. (1987) Molecular cloning of the maize R-nj allele by transposon tagging with Ac. In *Chromosome Structure and Function, 18th Stadler Genetics Symposium*, (eds J. P. Gustafson and R. Appels), Plenum Press, New York and London, p. 263

Dooner, H. K. (1983) Coordinate genetic regulation of flavonoid biosynthetic enzymes in maize. *Molecular and General Genetics*, **189**, 136

Döring, H. P. and Starlinger, P. (1986) Molecular genetics of transposable elements in plants. *Annual Review of Genetics*, **20**, 175

Fedoroff, N. (1983) Controlling elements in maize. In *Mobile Genetic Elements*, (ed. J. Shapiro). Academic Press, New York

Fedoroff, N., Furtek, D. and Nelson, O. E. (1984) Cloning of the bronze locus in maize by a simple and generalizable procedure using the transposable element Ac. *Proceedings of the National Academy of Sciences, USA*, **81**, 3825

Fedoroff, N., Wessler, S. and Shure, M. (1983) Isolation of the transposable maize controlling elements Ac and Ds. *Cell*, **35**, 235

Hope, I. A. and Struhl, K. (1986) Functional dissection of a eucaryotic transcriptional activator protein, GCN4 of yeast. *Cell*, **46**, 885

Katzen, A. L., Kornberg, T. B. and Bishop, J. M. (1985) Isolation of the proto-oncogene *c-myb* from *D. melanogaster*. *Cell*, **41**, 449

Klempnauer, K. H. and Sippel, A. E. (1987) The highly conserved amino-terminal region of the protein encoded by the v-myb oncogene functions as a DNA-binding domain. *EMBO Journal*, **6**, 2719

Lechelt, C., Laird, A. and Starlinger, P. (1986) Cloning of DNA from the P locus. *Maize Genetics Cooperation News Letters*, **60**, 40

Ma, J. and Ptashne, M. (1987) Deletion analysis of GAL4 defines two transcriptional activating segments. *Cell*, **48**, 847

McCarty, D. R., Carson, C. B., Stinard, P. S. and Robertson, D. S. (1989) Molecular analysis of viviparous-1: an abscisic acid-insensitive mutant of maize. *The Plant Cell*, **1**, 523

Majello, B., Kenyon, L. C. and Dalla-Favera, R. D. (1986) Human *c-myb* proto-oncogene: nucleotide sequence of cDNA and organization of the genomic locus. *Proceedings of the National Academy of Sciences, USA*, **83**, 9636

Nevers, P., Shepherd, N. and Saedler, H. (1985) Plant transposable elements. *Advances in Botanical Research*, **12**, 102

O'Reilly, C., Shepherd, N. S., Pereira, A., Schwarz-Sommer, Zs., Bertram, I., Robertson, D. S. et al. (1985) Molecular cloning of the a1 locus of *Zea mays* using the transposable element En and Mu. *EMBO Journal*, **4**, 877

Paz-Ares, J., Ghosal, D., Wienand, U., Peterson, P. A. and Saedler, H. (1987) the regulatory c1 locus of *Zea mays* encodes a protein with homology to *myb* proto-oncogene products and with structural similarities to transcriptional activators. *EMBO Journal*, **6**, 3553

Paz-Ares, J., Wienand, U., Peterson, P. A. and Saedler, H. (1986) Molecular cloning of the c locus of *Zea mays*: a locus regulating the anthocyanin pathway. *EMBO Journal*, **5**, 829

Pereira, A., Schwarz-Sommer, Zs., Gierl, A., Bertram, I., Peterson, P. A. and Saedler, H. (1985) Genetic and molecular analysis of the Enhancer (En) transposable element system in *Zea mays*. *EMBO Journal*, **4**, 17

Peterson, P. A. (1986) Mobile elements in maize. *Plant Breeding Reviews*, **4**, 81

Peterson, T. and Schwarz, D. (1986) Isolation of a candidate clone of the maize P locus. *Maize Genetics Cooperation News Letters*, **60**, 36

Reddy, A. R., Britsch, L., Salamini, F., Saedler, H. and Rohde, W. (1987) The *A1* (anthocyanin-1) locus in *Zea mays* encodes dihydroquercitin reductase. *Plant Science*, **52**, 7

Rohde, W., Barzen, E., Marocco, A., Schwarz-Sommer, Zs., Saedler, H. and Salamini, F. (1986) Isolation of genes that could serve as traps for transposable elements in *Hordeum vulgare*. *Barley Genetics*, **V**, 533

Schwarz-Sommer, Zs., Shepherd, N., Tacke, E., Gierl, A., Rohde, W., Leclercq, L. et al. (1987) Influence of transposable elements on the structure and function of the A1 gene of *Zea mays*. *EMBO Journal,* **6**, 287

Styles, D. E. and Ceska, O. (1977) The genetic control of flavanoid synthesis in maize. *Canadian Journal of Genetics and Cytology,* **19**, 289

Theres, K., Scheele, T. and Starlinger, P. (1987) Cloning of the *bz2* locus of *Zea mays* using the transposable element Ds as a gene tag. *Molecular and General Genetics,* **209**, 193

Wienand, U. and Saedler, H. (1987) Plant transposable elements: unique structures for gene tagging and gene cloning. In *Plant DNA Infectious Agents,* (eds Th. Hohn and J. Schell), Springer Verlag, Wein, New York, pp. 205

Wienand, U., Weydemann, U., Niesbach-Klösgen, U., Peterson, P. A. and Saedler, H. (1986) Molecular cloning of the *c2* locus of *Zea mays*, the gene coding for chalcone synthease. *Molecular and General Genetics,* **203**, 202

18

TRANSGENIC PETUNIA PLANTS DIFFERING IN THE EXPRESSION OF MAIZE A1 GENE

FELICITAS LINN, HEINZ SAEDLER
Max-Planck Institut für Züchtungsforschung, Cologne, W. Germany

IRIS HEIDMANN and PETER MEYER
Max-Delbrück-Laboratorium in der MPG, Cologne, W. Germany

Introduction

The pathway of anthocyanin synthesis in *Petunia hybrida* is one of the best analysed systems genetically and biochemically (De Vlaming *et al.*, 1984). The major components contributing to flower pigmentation are flavonoids. In general, flower coloration is determined by the conversion of the three dihydroflavonols, dihydrokaempferol, dihydroquercetin, and dihydromyricetin into pelargonidin, cyanidin and delphinidin derivatives (Figure 18.1).

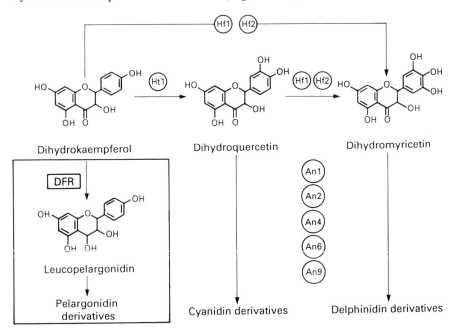

Figure 18.1 The introduction of the A1-gene from *Zea mays* which encodes DFR leads to the production of leucopelargonidin in *Petunia hybrida* which is processed into brick-red pelargonidin-glycosides. Ht1 stands for flavonoid 3'-hydroxylase and Hf1 and Hf2 for flavonoid 3',5'-hydroxylase 1 and 2 respectively

215

In petunia, the enzyme dihydroflavonol 4-reductase (DFR) has a high affinity for dihydromyricetin and dihydroquercetin but does not convert dihydrokaempferol into pelargonidin derivatives (Forkman and Ruhnau, 1987). Therefore only blue delphinidin derivatives and red cyanidin derivatives are known in petunia varieties and no pelargonidin pigments are synthesized, which would give rise to a brick-red flower coloration. DFR enzymes from other species exhibit a broader substrate specificity than the petunia DFR. Maize DFR, which is encoded by the A1-locus, converts dihydroquercetin into leucopelargonidin (Reddy *et al.*, 1987), but it also accepts dihydrokaempferol as a substrate to convert it into leucopelargonidin. In petunia, leucopelargonidin can be further processed into pelargonidin pigments as has been shown by feeding experiments with leucopelargonidin (Gert Forkman, unpublished data). The cDNA of *Zea mays* DFR, encoded by the A1 gene (Schwarz-Sommer *et al.*, 1987), was cloned between the viral 35S promoter and its corresponding terminator. This construct was introduced by direct gene transfer into protoplasts of the petunia line RL01. Due to an enzymatic block in its anthocyanin pathway the RL01-line accumulates dihydrokaempferol (Stotz *et al.*, 1985) which in transgenic plants, that express the A1 cDNA, is metabolized to leucopelargonidin and subseqeuntly to pelargonidin-3-glucoside giving rise to brick-red coloration of the flowers (Meyer *et al.*, 1987). In this chapter we report on the stability of the pelargonidin phenotype in genetic crosses and the molecular analysis of the transgenic plants where a high variance of A1-expression is observed in different transformants.

Results

CROSSES OF TRANSGENIC A1 PLANTS WITH COMMERCIAL VARIETIES

When a brick-red transformant with small flowers was crossed to a commercial line of large flowers and intense cyanidin pigmentation, the F_1 plants displayed a cyanidin-red phenotype with only a minor proportion of pelargonidin. The F_2 generation from a selfed F_1 plant featured three different types of pigmentation: a cyanidin type, a pelargonidin type, and a mixed type with a major proportion of cyanidin as already observed in the F_1 plants. These different colours seen at variable intensities in large as well as in small flowers demonstrating usefulness of this material for commercial breeders.

TRANSGENIC PLANTS DISPLAY STRONG DIFFERENCES IN A1-EXPRESSION AND PELARGONIDIN PRODUCTION

Among different transformants three phenotypes were observed, which either showed no flower pigmentation, pigmentation only in some floral cells, or coloration of the whole flower. These three phenotypes termed 'white', 'variegated', and 'red' were analysed at the molecular level with respect to the number of integrated copies of the maize DFR cDNA and the methylation of a *Hpa*II site within the cauliflower mosaic virus (CaMV) 35S promoter region.

Correlation of A1 expression and copy number

Seventy five per cent (6 out of 8) of the red plants contained one copy of the A1 construct, while 80% (4 out of 5) of the variegated plants and 88% (7 out of 8) of the white transformants had at least two of the A1-genes integrated in the genome.

Correlation of the methylation of a HpaII *site in the 35S promoter and the number of integrated copies*

Four 'one-copy-integrants' which were unmethylated at a *Hpa*II-site within the 35S promoter displayed a red phenotype, while three other plants which also contained one integrated A1-copy were methylated at this site. This group of three plants exhibited one of each phenotype, i.e. one white, one variegated, and one red plant. This indicates that methylation of the *Hpa*II-site is not in all cases equivalent to an inactive promoter, and that it is probably a neighbouring region whose methylation status is important for gene activity. If a larger part of the promoter is undermethylated, this would also affect the *Hpa*II site, which explains why the four transformants which contain a non-methylated *Hpa*II-site exhibited a red phenotype. Among the transformants with more than one integrated copy, there was only one white plant with an unmethylated *Hpa*II-site within the 35S promoter, but 10 other plants were methylated at all integrated promoter regions. Five of these plants were white, four were variegated, and only one plant was red (Table 18.1).

Although the methylation studies were performed with leaf material, which does not necessarily represent the situation in flower tissue, the data strongly suggest

Table 18.1 CORRELATION BETWEEN THE METHYLATION PATTERN AT A *Hpa*II-SITE OF THE 35S PROMOTER AND THE NUMBER OF INTEGRATED COPIES

Transgenic plants with one integrated A1 copy (A1- and nptII*-gene are linked)*

A1-methylation pattern in leaf material	Flower pigmentation		
	Red	Variegated	White
Unmethylated	4	0	0
Methylated	1	1	1

Transgenic plants with more than one integrated A1 copy (A1- and nptII*-gene are not necessarily linked)*

A1-methylation pattern in leaf material	Flower pigmentation		
	Red	Variegated	White
Unmethylated	0	0	1
Methylated	1	4	5

that undermethylation of the 35S promoter is preferentially found in one-copy-integrants where it results in continuous expression of the A1 gene. There was only one exception where a white plant with multiple integrated copies contained an unmethylated 35S promoter. However, in this case it remains unclear whether this promoter is linked to an intact A1 copy.

VARIABILITY OF THE A1 EXPRESSION

A variegated plant, RL01-16, containing several copies of the A1 gene, developed branches with fully coloured flowers (named 16^+) next to branches which did not show any A1 expression (named 16^-). A branch of each type was cut and rooted to produce a red 16^+ plant and a white 16^- plant. Southern blot analysis showed that both plants contained an identical pattern of integrated A1 copies. When plant 16^- was selfed, the F_1 progeny was homogeneously white flowering. Surprisingly, however, the progeny of the cross $16^- \times 16^+$ were also homogeneously white flowering, while selfing of plant 16^+ and the cross $16^+ \times 16^-$ approximately resulted in a 1:1 ratio of white and red progeny. The data indicate that the A1 copy which had been activated in plant 16^+ is only expressed in the next generation if plant 16^+ is the female partner of the cross, whereas when the gene is transmitted via pollen, it is inactivated. Southern blot analysis of the white and red progeny showed no particular A1 copy present in red but not in white plants. Therefore it seems to be the methylation status of the egg cell that determines whether or not the progeny will show expression of the A1 gene.

Discussion

Our data show that transgenic plants which contain one copy of the A1 construct preferentially express the A1 gene, and that the expression is correlated with non-methylation of a *C*-residue at a *Hpa*II-site within the 35S promoter. If multiple copies of the A1 construct are integrated, continuous expression of the gene is only observed in one out of 10 cases and also the *Hpa*II-sites of the integrated promoters are found to be methylated in nine of 10 plants.

The results suggest that in transformants which have only one integrated A1 copy in their genome, the A1 gene is linked to the selectable marker, i.e. the *npt*II-gene. In multiple copy-integrants the intact A1 construct is not necessarily located next to the active *npt*II gene, since deletion of parts of the introduced plasmid is a phenomenon commonly observed in direct gene transfer experiments. The expressed *npt*II gene can be linked to a truncated A1 copy and vice versa. If we assume that genes which integrate into the genome are predominantly methylated and therefore transcriptionally inactive, an A1 copy would not be expressed in most cases. However, if it is linked to a transcriptionally active *npt*II gene, it will be located in a selected genomic region which would allow proteins to bind to the promoter region and thus prevent methylation at this site.

Acknowledgement

This work is supported by the Bundesministerium für Forschung und Technologie BCT 0365 2-2/3.

References

De Vlaming, P., Geraths, A. G. M., Wiering, H., Wijsman, H. J. W., Cornu, A., Farey, E. *et al.* (1984) *Petunia hybrida*: a short description of the action of 91 genes, their origin and their map location. *Plant Molecular Biology Reporter,* **2**, 21–42

Forkmann, G. and Ruhnau, B. (1987) Distinct substrate specificity of dihydroflavonol 4-reductase from flowers of *Petunia hybrida. Zeitschrift für Naturforschung,* **42c**, 1146–1148

Meyer, P., Heidmann, I., Forkmann, G. and Saedler, H. (1987) A new petunia flower colour generated by transformation of a mutant with a maize gene. *Nature,* **330**, 677–678

Reddy, A. R., Britsch, L., Salamini, F., Saedler, H. and Rohde, W. (1987) The A1 (anthocyan-1) locus in *Zea mays* encodes dihydroquercetin reductase. *Plant Science,* **52**, 7–12

Schwarz-Sommer, Z., Shepherd, N., Tacke, E., Gierl, A., Rohde, W., Leclercq, L. *et al.* (1987) Influence of transposable elements on the structure and function of the A1 gene of *Zea mays. EMBO Journal,* **2**, 287–294

Stotz, G., De Vlaming, P., Wiering, H., Schramm, A. W. and Forkmann, G. (1985) Genetic and biochemical studies on flavoid 3'-hydroxylation in flowers of *Petunia hybrida. Theoretical and Applied Genetics,* **70**, 300–305

19

MODULATION OF PLANT GENE EXPRESSION

WOLFGANG SCHUCH, MARY KNIGHT, ALISON BIRD,
JAQUELINE GRIMA-PETTENATI
*ICI Seeds, Plant Biotechnology Section, Jealott's Hill Research Station, Bracknell,
Berkshire, UK*

and *ALAIN BOUDET*
Université Paul Sabatier, Centre de Physiologie Végétale, Toulouse, France

Introduction

The use of genetic engineering techniques has enabled us to isolate and characterize plant genes, as well as to study their function in transgenic plants. Recently, methods have been developed for the control *in vivo* of plant gene expression using antisense RNA (for review *see* van der Krol, Mol and Stuitje, 1988). We have applied this approach to the generation of transgenic plants in which an enzyme specific to lignin biosynthesis, cinnamyl alcohol dehydrogenase (CAD), has been down-regulated.

Lignin is a complex polymer found in cell walls of all vascular plants. It is an essential component of cell walls of differentiated specialized plant cells, in particular the xylem tissue. Lignins play an important role in the conducting function of the xylem by reducing the permeability of the cell wall to water. Lignification of the xylem and sclerenchyma cells contributes to the mechanical strength and rigidity of the plants. Lignin is also deposited in plant cell walls after wounding, leading to the production of an effective barrier against progression of plant pests. In these instances, lignin (or lignin-like substances) are laid down in cells which do not normally lignify, e.g. epidermis or parenchyma cells.

Lignin is composed of three different kinds of monomeric substrates: coumaryl, coniferyl and sinapyl alcohols (Sarkanen and Ludwig, 1971). The ratios of these different substrates varies between lignins of different plant species: gymnosperm lignin is rich in coniferyl alcohol; angiosperm lignin also contains considerable amounts of sinapyl and coumaryl alcohols. The cross-linking of the different monomers to each other, and the degree of secondary modification can also vary in lignin.

There exists considerable natural variation in the lignin content of plants depending on the environmental conditions and developmental stage of the growth of the plant (Grand, Boudet and Ranjeva, 1982). This study has recently been extended to an investigation of different lines of maize (Larroque, 1989). As much as 50% difference in lignin contents was found between different maize genotypes. There is, however, no major phenotypic indication of this reduced lignin content.

221

There are also a number of naturally occurring mutants in which enzymes of the lignin biosynthetic pathway have been affected. These are the brown midrib mutants of maize and sorghum (Kuc and Nelson, 1964; Bucholtz *et al.*, 1980). In these plants CAD activity has been reduced by as much as 50%, but the plants show a normal phenotype apart from the brown midrib.

The biosynthesis of lignin precursors which is part of the phenylpropanoid pathway is well understood (Robertson, 1986). The biosynthetic pathway of lignin is shown in Figure 19.1. Only steps 7, 8 and 9 in the diagram are specific to lignin biosynthesis. The others are shared for the production of a number of other phenolic compounds.

The enzymes specific to the lignin pathway are: cinnamoyl-coA reductase (CCR) (step 7), cinnamoyl alcohol dehydrogenase (CAD) (step 8) and peroxidase (step 9). These enzymes have been extensively studied at the biochemical level. CAD from poplar has been purified to homogeneity (Sarni, Grand and Boudet, 1984), and antibodies have been raised to the protein. These have been used to identify CAD cDNA clones (Walter *et al.*, 1988; Knight *et al.*, unpublished data). The availability of a bean CAD cDNA makes it now possible to attempt to modulate the expression of CAD using antisense RNA, and to study the effect of altered CAD expression on plant development and function.

Results and discussion

ISOLATION OF A BEAN CAD cDNA CLONE

We have isolated a bean CAD cDNA from a cDNA library constructed from RNA extracted from bean tissue culture cells treated with fungal elicitor (Edwards *et al.*, 1985). This clone is referred to as MG10. We have compared this cDNA to cDNA clones isolated by Grand (Grand, personal communication) (Figure 19.2). Comparison of clones 4A and 4B and sequence data published subsequently by Walter *et al.* (1988) indicates that MG10 contains a near full-length bean CAD cDNA.

ISOLATION AND PARTIAL CHARACTERIZATION OF A TOBACCO CAD GENE

We have used the Pst1 insert of this clone to probe a tobacco genomic library. We have isolated 54 putative CAD genomic clones, 20 of which have been plaque purified. DNA has been prepared from 11 of these, all of which have been analysed by restriction enzyme analysis. This indicated that all clones contained the same CAD genomic fragment. A restriction map of a representative clone, #15 is shown in Figure 19.3. The two *Eco*R1 fragments (1.2 kb and 8.7 kb) which hybridize to MG10, have been identified by Southern hybridization to genomic DNA (data not shown). The orientation of the CAD gene in this cloned fragment has been determined using the cDNA clones 4B and 4A as probes. Figure 19.3 shows the orientation of the CAD gene in the genomic clone. We have chosen to characterize further the 1.2 kb *Eco*R1 fragment by sequence analysis. This has revealed strong nucleotide and amino acid sequence homology in this region: we have identified two exons of 147 and 78 bases respectively. Amino-acid sequence homology in these exons is 85% and 88% respectively. These exons are separated by an intron

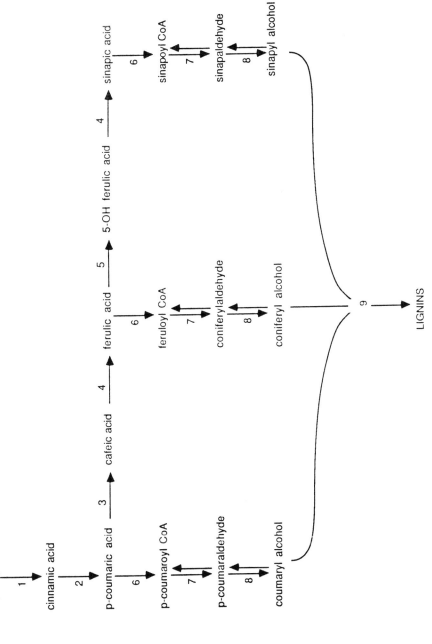

Figure 19.1 Biosynthesis of lignins. Reactions 1–6 are shared with the synthesis of other phenolic compounds. Only reactions 7–9 are specific for lignin biosynthesis. Numbers represent the following enzymes: 1: Phenylalanine ammonia lyase; 2: cinnamate 4-hydroxylase; 3: *p*-coumarate 3-hydroxylase; 4: catechol *O*-methylase; 5: ferulate 5-hydroxylase; 6: hydroxycinnamate:CoA ligase; 7: cinnamoyl:CoA reductase; 8: cinnamoyl alcohol dehydrogenase (CAD); 9: peroxidase

BEAN CAD cDNA CLONES

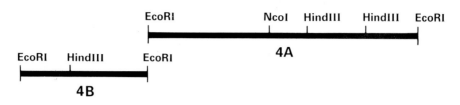

Figure 19.2   Structure of MG10, a bean CAD cDNA. The clone is compared with two clones, 4A and 4B, isolated by Grand (personal communication)

TOBACCO CAD GENE

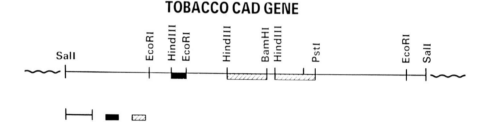

Figure 19.3   Structure of a genomic clone encoding CAD from tobacco. Boxed regions represent regions of the clone hybridizing with either clone 4B (5′ probe) or 4A (3′ probe)

of 102 bases. This fragment also contains additional intron sequences covering 923 bases.

CONSTRUCTION OF CAD ANTISENSE VECTORS

We have used part of the bean cDNA and the tobacco gene in the construction of CAD antisense vectors. The fragments chosen in the first round experiments are shown in Figure 19.4. We have used a 1166 bp *Nco*I fragment in the construction of the bean CAD antisense vector. A 600 bp *Kpn*I fragment which includes the exons mentioned above (225 bases), a total of 325 bp intron sequence and 50 bp derived from the cloning vector was used in the construction of the tobacco gene CAD antisense vector. We decided to use the cauliflower mosaic virus (CaMV) promoter to drive the synthesis of antisense RNA. The rationale behind this decision was as follows: the CaMV promoter shows preferred expression in phloem tissue, however, the location of the major lignification is found in the xylem tissue. We have identified those cells in which secondary lignification takes place by using fusions between the promoter of phenylalanine ammonia lyase (PAL) (Cramer *et al.*, 1989) and the GUS marker gene (Jefferson, Kavanagh and Bevan, 1987). PAL is the first enzyme of the phenylpropanoid pathway leading to the synthesis of

SEQUENCE COMPARISON OF BEAN CAD cDNA AND PART OF TOBACCO CAD GENE

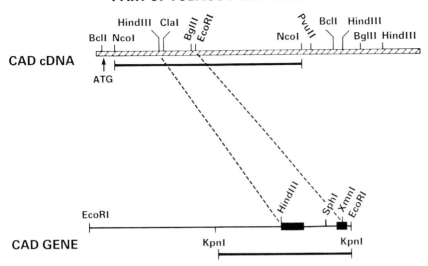

Figure 19.4 Relatedness of bean and tobacco CAD clones and fragments used in the construction of antisense vectors. Fragments underlined were used in the construction of antisense RNA vectors

lignins. We argued that strong inhibition of CAD, as seen in our previous antisense RNA experiments (Smith *et al.*, 1988; Grierson *et al.*, Chapter 10), may lead to significant disruption of normal plant growth and development, and may even lead to plant death due to the collapse of the xylem. We therefore designed the experiments in such a way as to obtain a non-lethal reduction of CAD activity, rather than aim for lethal inhibition. The structure of the antisense constructs is shown in Figure 19.5. These vectors are based on a derivative of the binary plant transformation vector, Bin19, pJR1 (Smith *et al.*, 1988).

TOBACCO CAD VECTORS

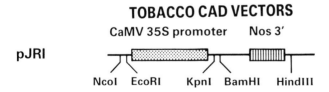

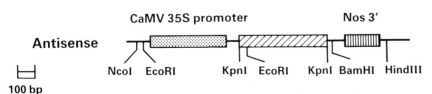

Figure 19.5 Structure of tobacco CAD antisense vectors. DNA fragments indicated in Figure 19.4 were inserted into the plant expression vector pJR1

TRANSFORMATION OF TOBACCO

Tobacco leaf disc transformations have been performed using standard protocols. Plants were rooted twice in the presence of kanamycin. A total of 36 plants transformed with the bean antisense vector and 17 plants transformed with the tobacco antisense vector have been generated.

ANALYSIS OF CAD ACTIVITY IN CONTROL AND TRANSFORMED PLANTS

Control and transformed plants were grown in tissue culture for the same period of time (12 weeks). Plants were then harvested, and the CAD enzyme activity determined both in leaf and stem material. Three leaf samples and two stem samples were assayed separately. The mean of the measurements for each plant was calculated and plotted. Plants which showed at least a 40% reduction in their mean CAD activity in both the leaf and stem samples were selected for further analysis. Results for this analysis of bean antisense construct containing plants are shown in Figure 19.6. The variation of CAD enzyme data obtained is also shown in this figure. A total of four bean antisense and two tobacco antisense plants were taken for further analysis. The presence of the antisense gene in these transformed plants was demonstrated by polymerase chain reaction amplification with oligonucleotides derived from the CaMV 35S promoter and the *nos* 3' region (data not shown).

PHENOTYPES OF TRANSFORMED PLANTS

Plants showing the greatest reduction in CAD activity were kept in tissue culture and were potted and grown to maturity in the growth room. All antisense plants showed a reduction in plant height compared with control plants. BCADAS1 #3 showed the greatest phenotypic change. It was severely stunted even in comparison with the other antisense plants. It also had malformed leaves in which part of the leaf blade was crinkled. However, at flowering, this plant had achieved the same height as the other antisense plants due to an increased growth during late developmental stages. All antisense plants flowered, and set seed apart from BCADAS1 #3. Seeds pods of #3 were 0.5–0.6 cm long and shrivelled compared with 1.5 cm long pods of the other antisense plants and controls. When BCADAS1 #3 was pollinated with pollen from untransformed tobacco plants, seed set was observed indicating that the plant is male sterile.

ANALYSIS OF LIGNINS IN CONTROL PLANTS AND BCADAS1 #3

Stem sections were taken from control and BCADAS1 #3 plants. Sections were stained using phloroglucinol. Control plants showed extensive secondary lignification of the xylem vessels (Figure 19.7a). BCADAS1 #3 showed reduced lignification of the xylem vessels (Figure 19.7b). This was observed both for the degree of lignin deposition as well as the extent of cells lignified.

Similar results have also been observed for the other transformed plants showing reduced CAD enzyme activity. Further experiments are in progress which aim to

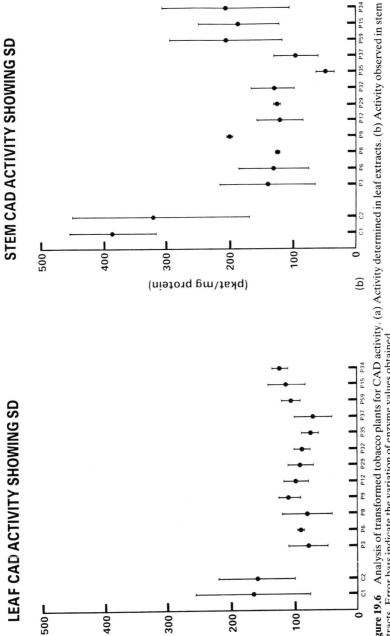

Figure 19.6 Analysis of transformed tobacco plants for CAD activity. (a) Activity determined in leaf extracts. (b) Activity observed in stem extracts. Error bars indicate the variation of enzyme values obtained

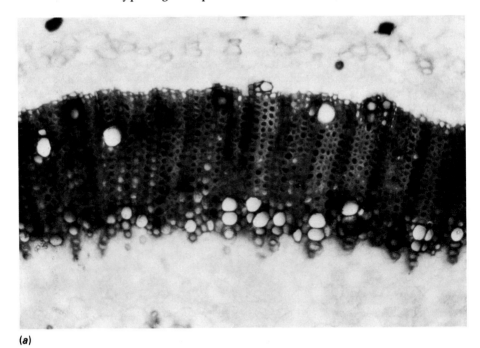

(a)

(b)

Figure 19.7   Phloroglucinol staining of lignins in control and antisense plants. (a) Section through stem of a control plant. (b) Section through stem of BCADAS1 #3

characterize the biochemical and developmental changes manifest in these transformed plants.

We have used the antisense RNA approach to modulate the expression of the CAD gene which plays an important role in the pathway leading to lignin deposition in plant cell walls. We have used a heterologous DNA sequence derived from bean in the construction of antisense RNA vectors. We have also used a genomic DNA fragment from tobacco consisting of intron and exon sequences for the construction of additional antisense RNA vectors. Plants transformed with both vectors have been analysed for a reduction in CAD activity. Out of a total of 53 transformed plants analysed, 16 were selected as showing reduced CAD enzyme activity both in leaf and in stem tissue. There were nine plants containing bean antisense constructs and seven containing tobacco constructs. Apart from one plant, containing the bean antisense construct, the phenotype of these plants differed only in an apparent reduction in height from control plants. Plant structure, habit, flowering and fertility were normal in these plants.

One of these plants which contains the bean CAD construct has been analysed for lignin deposition. We have observed a reduction in the xylem vessels in this plant. We have also seen reduced lignin deposition in those vessels which were lignified. The interpretation of these early experiments is that the introduction of the antisense RNA construct has led to this phenotype. Experiments are now in progress to test for the altered phenotype in progeny plants generated by backcrossing it to an untransformed control plant.

Plants containing both the bean and the tobacco antisense constructs have shown reduced CAD activity. If these observations can be substantiated in the following generations we have demonstrated that heterologous sequences as well as genomic sequences can be successfully used to modulate gene expression using antisense RNA.

Acknowledgements

Part of the work has been carried out with the support of an EEC training fellowship in Biomolecular Engineering. We thank Mr Frederick Plewniak and Dr Mark Olive for the construction of the expression vectors. Work with transgenic plants was carried under MAFF licence PHF 1039.

References

Bucholtz, D. L., Cantrell, R. P., Axtell, P. J. and Lechtenberg, V. L. (1980) *Journal of Agricultural and Food Chemistry,* **28**, 1239–1241

Cramer, C. L., Edwards, K. J. E., Dron, M., Liang, S., Dildine, S. L., Bolwell, G. P. *et al.* (1989) *Plant Molecular Biology,* **12**, 367–384

Edwards, K. J. E., Cramer, C. L., Bolwell, G. P., Dixon, R. A., Schuch, W. and Lamb, C. J. (1985) *Proceedings of the National Academy of Sciences, USA,* **82**, 6731–6735

Grand, C., Boudet, A. M. and Ranjeva, R. (1982) *Holzforschung,* **36**, 217–223

Jefferson, R. A., Kavanagh, T. A. and Bevan, M. W. (1987) *EMBO Journal,* **6**, 3901–3907

Kuc, J. and Nelson, O. E. (1964) *Archives of Biochemistry and Biophysics,* **105,** 103–113

Larroque, C. (1989) *Thesis,* L'Institute National Polytechnique de Toulouse

Robertson, B. (1986) *Physiology and Molecular Plant Physiology,* **28,** 137–148

Sarkanen, K. V. and Ludwig, C. H. (1971) In *Lignin: Occurrence, Formation and Reactions,* (eds. K. V. Sarkanen and C. H. Ludwig). Wiley Interscience, New York, London, Sydney, Toronto, pp. 1–40

Sarni, F., Grand, C. and Boudet, A. M. (1984) *European Journal of Biochemistry,* **139,** 259–265

Smith, C. J. S., Watson, C. F., Ray, J., Bird, C. R., Morris, P. C., Schuch, W. *et al.* (1988) *Nature,* **334,** 724–726

van der Krol, A. R., Mol, J. N. M. and Stuitje, A. R. (1988) *Gene,* **72,** 45–50

Walter, H. M., Grima-Pettenati, J., Grand, C., Boudet, A. M. and Lamb, C. J. (1988) *Proceedings of the National Academy of Sciences, USA,* **85,** 5546–5550

20

REGENERABLE SUSPENSION AND PROTOPLAST CULTURES OF BARLEY AND STABLE TRANSFORMATION VIA DNA UPTAKE INTO PROTOPLASTS

PAUL A. LAZZERI and HORST LÖRZ

Max-Planck Institut für Züchtungsforschung, Cologne, W. Germany

Introduction

As the major group of crop species, the cereals are obvious targets for the application of the emerging techniques of genetic manipulation. However, as cereals are not readily transformed by *Agrobacterium*, alternative transformation methods must be sought (Lazzeri and Lörz, 1988). One such method is that of direct gene transfer to protoplasts (Paszkowski *et al.*, 1984). This technique is advantageous in that protoplasts from different species can be transformed using similar methods, but to obtain transgenic plants it must be possible to regenerate from protoplast-derived cultures. Historically this has proved very difficult in cereals (Jones, 1985), but in the last few years plants have been regenerated from protoplasts of a number of cereal and grass species (Vasil, 1987; Lazzeri and Lörz, 1988). These developments make direct gene transfer feasible for cereal transformation, and the first reports of transgenic plants produced by this method are now appearing (Horn *et al.*, 1988; Rhodes *et al.*, 1988; Toriyama *et al.*, 1988; Zhang and Wu, 1988).

In this chapter we described our progress in the development of a protoplast-mediated transformation and regeneration system for barley.

Initiation of embryogenic suspensions and regeneration

In barley it is routinely possible to initiate embryonic callus cultures and to regenerate plants from them (Lührs and Lörz, 1987). Embryogenic callus cultures are, however, typically compact and slow-growing and do not proliferate on transfer to liquid medium. To overcome this difficulty we have adopted the strategy of recurrent culture on solid medium, with selection for a particular callus phenotype which is both embryogenic and able to proliferate in liquid medium.

Scutella from immature embryos are cultured on L2 medium (Lazzeri, Lührs and Lörz, 1989) containing 2,4-D and with maltose as the carbon source. The resulting calli are subcultured at 4–5 week intervals, with selection for a fast-growing, granular callus type. This callus disperses readily when transferred to liquid medium L1 (Lazzeri, Lührs and Lörz, 1989) and starts to grow directly without a lag or 'adaptation' phase. For the first weeks, suspension cultures are maintained at

231

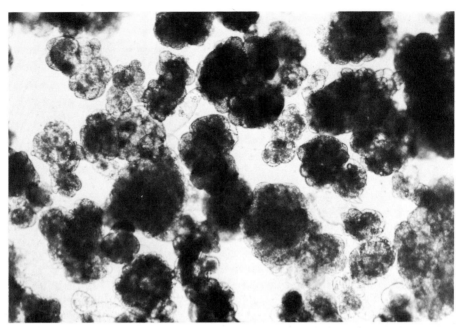

Figure 20.1   Cytoplasmic cell aggregates of an embryogenic barley suspension (cv. Dissa)

a high callus:medium ratio, with approximately 12 day culture intervals. After this time cultures are gradually brought to a lower density and are subcultured at shorter intervals, and they become more homogeneous, until they are composed mainly of cytoplasmic cell aggregates 100–1000 μm in diameter (Figure 20.1). At this stage suspensions are tested for their morphogenetic capacity, by plating on to solid medium.

By these methods, to date some 20 independent cell lines of cv. Dissa and four of cv. Igri have been established (Table 20.1). On plating on to solid media approximately half of these lines have proved embryogenic, and from six lines of cv. Dissa green plantlets have now been transferred to the greenhouse.

Protoplast culture and regeneration

Suspension cultures which are seen to be embryogenic when plated on to solid medium are used for protoplast isolation. Using standard techniques yields of 0.5 to

Table 20.1 SUSPENSION INITIATION AND REGENERATION

Cultivar	No. of lines initiated	No. of lines producing embryos or shoots on solid medium	No. of lines producing plantlets	No. of plantlets produced	
				Green	Albino
Dissa	20	9/20	6/9	20	9
Igri	4	2/4	0	–	–

Table 20.2 CULTURE OF PROTOPLASTS FROM EMBRYOGENIC
SUSPENSIONS

Cultivar/line no.	Average plating efficiency	Culture development
Dissa		
DL 3, 5	No division	–
DL 6, 12, 13	<0.05%, <0.05%, <0.05%	Soft callus
DL 1	<0.05%	Compact callus
DL 2, 4, 11	<0.45%, 0.63%, 0.39%	Embryogenic callus
DL 14, 17	Not determined, 0.22%	Embryos, plantlets
Igri		
IL 1	0.18%	Compact callus

3×10^6 protoplasts/g cells are usually obtained, depending on the cell line and growth conditions. Protoplasts are cultured in agarose-solidified medium based on that used for the parent suspensions, at densities between 2.5 and 5×10^5/ml. The carbohydrate source supplied has been found to have an important effect on protoplast division; maltose is generally superior to glucose, in some cases giving more than 10-fold increases in plating efficiency. Protoplasts from most cell lines appear to be capable of division and microcallus formation; of 12 lines tested 10 yielded dividing protoplasts and five lines had plating efficiencies higher than 0.15% (Table 20.2). Protoplast microcalli from a number of lines have been cultured under the conditions used for regeneration from suspensions. The morphogenetic response of protoplast-derived callus generally reflects that of the parent suspension: embryogenic cultures have been recovered from protoplasts from five cv. Dissa lines and green plantlets have been regenerated from two of these lines (Table 20.2, Figure 20.2).

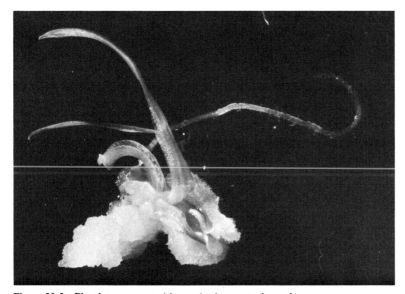

Figure 20.2 Plantlet regenerated from a barley protoplast culture

Protoplast transformation

For the development of a transformation protocol, the protoplast source used was a long-established suspension of cv. Dissa (Lührs and Lörz, 1988). While this line is no longer morphogenic, it provides good yields of homogeneous protoplasts which have plating efficiencies up to 5%. These protoplasts have been transformed using modifications of three different PEG-induced DNA uptake protocols (Krens *et al.*, 1982; Hein, Przewozny and Schieder, 1983; Negrutiu *et al.*, 1987). In all experiments the transforming DNA used was the plasmid pRT99GUS (Töpfer, Schell and Steinbiss, 1988), which contains the *npt*II gene as a selectable marker and the *GUS* gene as a scorable marker, both under control of the cauliflower mosaic virus (CaMV) 35S promoter.

After transformation, protoplasts were embedded in agarose medium and were cultured for 10–15 days before selection for transformed colonies. The selection agent used was G418-sulphate (geneticin), at concentrations of 20–50 mg/l. Initial selection was performed on agarose segments in liquid medium (Figure 20.3) and resistant colonies were then transferred to solid medium, also containing G418 (Figure 20.4). Callus clones showing sustained growth after three passages of selection on solid medium were sampled for Npt enzyme activity.

Figure 20.3   Selection for transformed colonies after direct DNA uptake into protoplasts; segments of agarose containing microcalli cultured in liquid medium. Left: control treatment without selection. Right: transformed colonies growing in the presence of 25 mg/l G418

G418-resistant callus clones were recovered with all three transformation protocols, at frequences varying between 1.6 and 13 in 10^5; approximately 40 resistant clones being recovered from 32×10^6 treated protoplasts, with a mean plating efficiency of 2.5%. Randomly-selected clones tested for Npt activity all gave positive results (Figure 20.5), showing that the G418-selection system

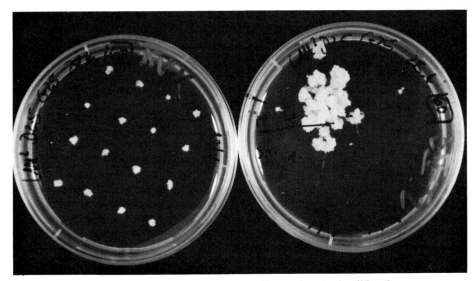

Figure 20.4   Growth of transformed callus on solid medium under selection (25 mg/l
G418). Left: control, non-transformed calli. Right: single transformed clone

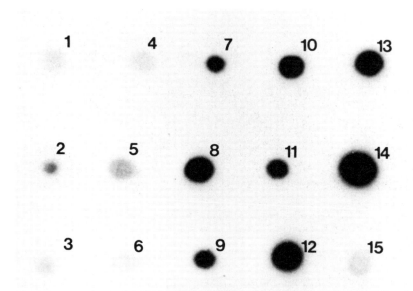

Figure 20.5   Npt dot-blot test of transformed barley callus (70 h exposure). Spots 1, 2 and
3: control, transformed and control tobacco tissue respectively. Spots 4, 5 and 6: control
barley callus. Spots 7–13: transformed barley callus clones. Spots 14 and 15: transformed
and control *Vigna* callus respectively

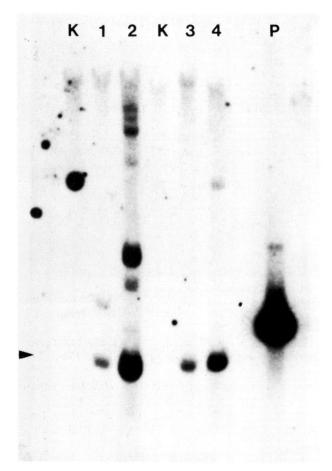

Figure 20.6   Southern blot analysis of transformed barley callus: DNA isolated from
callus restricted with *Pst*I and hybridized to a probe from the *npt*II gene. Lanes K: control,
non-transformed, protoplast-derived barley callus. Lanes 1, 2, 3 and 4: transformed callus
clones showing hybridization to the *npt* probe, with bands at 0.8 kb (arrow); the expected
size of the *npt* fragment. Lane P: plasmid DNA containing the *npt*II gene

functions efficiently. However, when resistant calli were tested for glucuronidase
(GUS) activity using the fluorimetric assay (Jefferson, 1987), a few clones gave
clear positive results while the majority showed weak GUS activity a few-fold
higher than the control callus. When assayed by the histochemical GUS test
(Jefferson, 1987) the 'positive' lines showed blue coloration in around 15–25% of
the tissue, whereas the other lines showed small and localized areas of blue
coloration, often only tens of cells in size.

Southern analysis of DNA from resistant calli confirmed the presence of the *npt*
gene in all lines tested (Figure 20.6). As is expected for cells transformed via direct
gene transfer several lines appeared to contain multiple copies of the gene and
there was evidence of plasmid concatamerization or rearrangement. When DNA
isolated from resistant callus clones was probed with the *GUS* gene, positive

hybridization was again seen in all cases, including clones in which no *GUS* activity could be detected.

Conclusions

The experiments described here show that it is possible to establish reproducibly barley cell suspensions, and that these suspensions can retain the ability to form somatic embryos from which plants may be regenerated. Such morphogenic suspensions provide a source of protoplasts from which plantlets may also be regenerated. Additionally, barley protoplasts have been stably transformed via PEG-mediated direct DNA uptake.

Together, these results form the basis for the production of transgenic barley plants, and current experiments focus on improving the regeneration and gene transfer processes in order to develop a usable transformation system.

References

Hein, T., Przewozny, T. and Schieder, O. (1983) Culture and selection of somatic hybrids using an auxotrophic cell line. *Theoretical and Applied Genetics*, **64**, 119–122

Horn, M. E., Shillito, R. D., Conger, B. V. and Harms, C. T. (1988) Transgenic plants of Orchardgrass (*Dactylis glomerata* L.) from protoplasts. *Plant Cell Reports*, **7**, 469–472

Jefferson, R. A. (1987) Assaying chimeric genes in plants: the *GUS* gene fusion system. *Plant Molecular Biology Reporter*, **5**, 387–405

Jones, M. G. K. (1985) Cereal protoplasts. In *Cereal Tissue and Cell Culture*, (eds S. W. J. Bright and M. G. K. Jones). Martinus Hijhoff/W. Junk Publ., Dordrecht, pp. 204–230

Krens, F. A., Molendijk, L., Wullems, G. J. and Schilperoort, R. A. (1982) *In vitro* transformation of plant protoplasts with Ti-plasmid DNA. *Nature*, **296**, 72–74

Lazzeri, P. A. and Lörz, H. (1988) *In vitro* genetic manipulation of cereals and grasses. *Advances in Cell Culture*, **6**, 291–325

Lazzeri, P. A., Lührs, R. and Lörz, H. (1989) Stable transformation of barley via PEG-mediated direct DNA uptake into protoplasts. (Submitted)

Lührs, R. and Lörz, H. (1987) Plant regeneration *in vitro* from embryogenic cultures of spring- and winter-type barley (*Hordeum vulgare* L.) varieties. *Theoretical and Applied Genetics*, **75**, 16–25

Lührs, R. and Lörz, H. (1988) Initiation of morphogenic cell-suspension and protoplast cultures of barley. *Planta*, **175**, 71–81

Negrutiu, I., Shillito, R., Potrykus, I., Biasini, G. and Sala, F. (1987) Hybrid genes in the analysis of transformation conditions. I. Setting up a simple method for direct gene transfer in plant protoplasts. *Plant Molecular Biology*, **8**, 363–373

Paszkowski, J., Shillito, R. D., Saul, M. W., Mandak, V., Hohn, B. and Potrykus, I. (1984) Direct gene transfer to plants. *EMBO Journal*, **3**, 2717–2722

Rhodes, C. A., Pierce, D. A., Mettler, I. J., Mascarenhas, D. and Detmer, J. J. (1988) Genetically transformed maize plants from protoplasts. *Science*, **240**, 204–207

Töpfer, R., Schell, J. and Steinbiss, H.-H. (1988) Versatile cloning vectors for transient gene expression and direct gene transfer in plant cells. *Nucleic Acids Research,* **16,** 8725

Toriyama, K., Arimoto, Y., Uchimiya, H. and Hinata, K. (1988) Transgenic rice plants after direct gene transfer into protoplasts. *Bio/Technology,* **6,** 1072–1074

Vasil, I. K. (1987) Developing cell and tissue culture systems for the improvement of cereal and grass crops. *Journal of Plant Physiology,* **128,** 193–218

Zhang, W. and Wu, R. (1988) Efficient regeneration of transgenic plants from rice protoplasts and correctly regulated expression of the foreign gene in the plants. *Theoretical and Applied Genetics,* **76,** 835–840

21

REGENERATION AND TRANSFORMATION OF APPLE AND STRAWBERRY USING DISARMED Ti-BINARY VECTORS

DAVID J. JAMES, ANDREW J. PASSEY and DEREK J. BARBARA
Institute of Horticultural Research, East Malling, Maidstone, Kent, UK

Introduction

Advances in molecular biology in recent years have permitted modifications to the naturally occurring process of agrobacteria-mediated gene transfer such that the so-called 'disarmed' vectors or plasmids can now function in place of wild-type plasmids normally present in *Agrobacterium* (Bevan, Flavell and Chilton, 1983; Fraley *et al.*, 1983; Bevan, 1984). In these, the genes encoding growth regulator autonomy have been deleted to avoid interference with the regeneration of normal plants, and have been replaced with dominant scorable and selectable markers. The use of such disarmed vectors has meant that it is possible to transfer foreign genes into plant cells and tissues and subsequently to regenerate viable, fertile normal-looking plants, usually through the intervention of tissue culture techniques. (For a detailed account of these procedures *see* the review by Fraley, Rogers and Horsch, 1986.) For the purposes of this chapter it is sufficient to note that there are two types of disarmed vector available for use in transformation systems: 'co-integrate' and 'binary'. The distinction between these two types and their relative merits have also been extensively reviewed (e.g. Klee, Horsch and Rogers, 1987).

Many plant species, including a range of crop plants, can now be transformed using disarmed vectors in *Agrobacterium* species and the transgenic plants subjected to physiological, biochemical, molecular and genetic studies. Until recently no fruit crops were among these, but recent work in our laboratory has successfully produced transgenic apple (James *et al.*, 1989) and strawberry plants (James *et al.*, unpublished data) using the disarmed T-binary vector pBIN6 (Bevan, 1984). This is a prototype binary vector containing both a scorable marker gene *nos*, encoding the enzyme nopaline synthase (Nos), and a selectable marker, *nptII*, a chimaeric gene composed of the regulating sequences of a nopaline synthase gene and the coding sequences of a bacterial neomycin phosphotransferase gene. Expression of this gene confers resistance to the antibiotic kanamycin and permits the selection of kanamycin-resistant transformed cells. This plasmid has been used to transform single varieties of apple and strawberry using complex explants such as leaf discs and petioles initially obtained from micropropagating cultures. Transgenic plants have been taken through to the greenhouse stage of growth and, in the case of strawberry, inheritance studies have been performed on the segregation of the *nos* gene in the R_1 generation.

The fundamental aim of the work was to investigate ways of making the regeneration of transgenic fruit plants, using agrobacteria-mediated gene transfer, a routine and reproducible procedure.

Regeneration

The importance of developing good micropropagation and regeneration systems for introducing foreign genes into plant species is paramount and particular problems relating to fruit plants in this area have already been reviewed (James, 1987). Although micropropagation procedures have been available for both apple and stawberry for some years (Boxus, 1974; Jones, 1976) only recently have regeneration protocols for complex explants been described (*see* review, James, 1987). A panoply of environmental and tissue-culture derived variables are now known to affect the regeneration process in fruit plants (James, Passey and Rugini, 1988; Welander, 1988; Fasolo, Zimmerman and Fordham, 1989).

THE IMPORTANCE OF THE SOURCE MATERIAL

It is well appreciated that perhaps most important of all is choice of species and cultivar. This is true of practically all plant species and was acknowledged from an early stage in our work. After some extensive early screening we chose one cultivar of apple, Greensleeves, and one cultivar of strawberry, Rapella, as subjects for the insertion of foreign genes using pBIN6 as the vector. In both species it was possible to define conditions that reproducibly gave levels of shoot regeneration approaching 100% on a per explant basis. Not surprisingly different conditions were required for the two species, a notable difference being the inhibitory effect of light on regeneration in apple (James, Passey and Rugini, 1988) and for its requirement in strawberry (James *et al.*, unpublished data). Both species however benefited from the presence of the antibiotic cefotaxime in the regeneration media. Cefotaxime is a cephalosporin now widely used in transformation studies to control growth of surviving *Agrobacteria* after infection. Although effective in both apple and strawberry its effects were markedly concentration dependent. Stimulatory effects of cefotaxime on regeneration were first reported for wheat embryos (Mathias and Boyd, 1986), but its effects seem to be applicable to a wide range of fruit species as well, including pear (Predieri *et al.*, 1989).

The choice of cultivar was also extremely important since one set of conditions that routinely gave regeneration approaching 100% in our chosen apple cultivar Greensleeves gave no regeneration in the rootstock M.9 (James, Passey and Rugini, 1988). However, changing the concentration of benzyldenine or changing to another cytokinin, thidiazuron, could, in association with cefotaxime, yield more than 30% regeneration. This unfortunately only emphasizes the empirical nature of this work and the need to develop individual protocols for individual subjects within the same species. It does however also reflect a need to understand better the physiology and biochemistry of the source material or at least to standardize its physiological state as much as possible before using it in transformation experiments.

In a recent experiment in which leaf discs were excised from rooted shoots *in vitro* grown for periods of 5–8 weeks, the 'age' of the leaf directly influenced the

Table 21.1 EFFECTS OF LEAF AGE FROM TIME OF ROOTING *IN VITRO* ON
SUBSEQUENT ABILITY OF EXCISED LEAF DISCS TO REGENERATE ADVENTITIOUS
SHOOTS. FIFTY DISCS PER TREATMENT

Treatment	Age of leaf (weeks)	Total no. shoots	Mean no. (s.e.) shoots per disc	Discs regenerated (%)
A	8	65	1.97 (0.56)	66
B	7	210	4.57 (0.47)	92
C	6	247	5.37 (0.47)	92
D	5	390	7.96 (0.46)	98

	Anovar: detailed comparisons	
Treatments	Variance ratio	Significance of difference
A v. B	129.4	$P = 0.001$
B v. C	14.8	$P = 0.001$
C v. D	159.1	$P = 0.001$

Anovar: analysis of variance

regeneration efficiency; older leaves producing significantly fewer regeneration events than younger ones (Table 21.1).

Transformation and selection

The techniques used for transformation were those defined by Horsch *et al.* (1985). Using this technique we have examined a number of variables that may affect shoot regeneration after transformation and include the effects of the plant phenolic acetosyringone on bacterial virulence (Stachel *et al.,* 1985), the effect of the length of the co-cultivation period, the type and quality of agar or agarose, the choice of plant growth regulators during regeneration and the length of the selection period on kanamycin. Only the last of these will be dealt with since space will not permit a detailed description of the effects of the other parameters. They will instead be the subject of future publications.

To assess the effect of these variables on transformation the proportion of explants that callused after extended exposure to kanamycin is taken as a measure of the efficiency of transformation. When a scorable marker is available, as was the case with Nos in pBIN6, it is possible to screen large numbers of calli for the presence of enzyme activity in tissue extracts using a simple paper electrophoresis method (Otten and Schilperoort, 1978). When shoots arose from calli these could also be screened at an early stage since samples of tissue as small as 10 mg would suffice. Any samples showing evidence of Nos activity could then be micropropagated to give sufficient shoot numbers (minimum of 30) for rooting bioassays in the presence of kanamycin and for molecular analyses by Southern blotting (Maniatis, Fritsch and Sambrook, 1982; Reed and Mann, 1985). The latter was carried out using a 7.5 kbp *Bgl* II fragment from pBIN1, a precursor of pBIN6, which spans the T-DNA region of pBIN6 (James *et al.,* 1989).

STRATEGIES FOR KANAMYCIN SELECTION

After infection, explant tissues were exposed to concentrations of 50 µg/ml kanamycin for various periods to select transformed cells. The exposure time was varied in order to effect a compromise between producing only transformed callus and having to screen large numbers of 'escapes', i.e. putative transformants that resulted from too brief an exposure to kanamycin after infection. We tried four strategies:

1. infection without selection
2. limited selection (1–22 days)
3. continuous selection (at least 60 days)
4. delayed selection (1–10 days).

In all strategies cultures were kept on regeneration media for a minimum of 60 days before recording.

Efficiency of regeneration

Infection without selection
The process of infecting apple or strawberry explants with virulent bacteria without kanamycin selection had no effect on regeneration efficiency compared with uninfected controls. Consequently this meant that initially there were many shoots to screen for Nos activity. Using this method proved to be inefficient since no shoots of either species ever gave a positive result despite examining many samples (Table 21.2).

Table 21.2 EFFICIENCY OF TRANSFORMATION OF APPLE LEAF DISCS USING THE BINARY VECTOR *pBIN6* IN LBA4404

Selction period (days)	Discs regenerating (%)	Mean no. shoots/ disc	Nos assays		No. discs
			Tested	+ve	
0	87	6.1	50	0	100
1–3	49	1.6	130	0	200
4–7	11	1.0	45	3	400
10–22	1	0.5	1	0	200
60+	2	0.75	4	3	200

Limited selection
Apple pBIN6   Periods of exposure to kanamycin as short as 24 h can reduce by half the proportion of leaf discs undergoing regeneration (Table 21.2). In addition, the mean number of shoots per explant decreased by nearly 80%. Nevertheless, it has been possible to select transgenic shoots after as little as 4 days selection on kanamycin (James *et al.*, 1989; Table 21.2). However, the time taken to regenerate is not greatly reduced compared with those exposed to the antibiotic for longer periods and the average time taken for a transgenic shoot to differentiate after a

4–7 day exposure to kanamycin was about 3 months. Calculated on a per leaf disc basis this represents an efficiency of 0.75% (3/400). Calculated on the basis of 'number of transgenic shoots per number of expected shoots without kanamycin selection' the efficiency is much lower, i.e. 0.13% (3/2400). For selection periods greater than 10 days the proportion of discs regenerating shoots was as low as that for 60 or more days indicating that 10 days was sufficient to kill all untransformed cells.

The transgenic nature of the Nos⁺ shoots was subsequently confirmed by the ability of DNA extracts to hybridize to the probe (James *et al.*, 1989).

Strawberry pBIN6   Only preliminary data are available from limited selection experiments with strawberry, but the situation is similar to that in apple in that, as judged by the presence of Nos activity and rooting bioassays in the presence of kanamycin, transgenic plants can be regenerated from leaf discs or petioles after exposure periods as short as 6 days, at an efficiency of 2–4% on a per explant basis. Longer exposure times of 12 or 18 days increased the efficiency to 4–16%. However, confirmation of these results requires evidence from Southern blotting.

Continuous selection
Apple pBIN6   Selection periods greater than 60 days produced transgenic shoots as evidenced from Nos assays, rooting bioassays and Southern blotting (James *et*

(a)   (b)

Figure 21.1   (a) pBIN6 transformed apple (cv. Greensleeves) plant grown for 6 months under greenhouse conditions and (b) micropropagated control (×1/7)

al., 1989). All three of the transgenic clones took slightly longer (about 5 months) to regenerate than those produced after limited exposures. Transgenic plants had no obvious phenotypic defects compared with control plants either at the *in vitro* stage or after transplanting to soil (Figure 21.1).

Strawberry pBIN6   Two clones were positively identified as transgenic plants by biochemical, rooting and molecular testing. The efficiency of production from transformed calli was 0.1–0.2% on a per explant basis. This represented the combined data from several different experiments over a considerable period of time and was based on the proportion of kanamycin-resistant calli undergoing shoot regeneration from a population of more than 1500. More recent data (which are not included in the calculations) suggest that much higher efficiencies of the order of 14% can be achieved. This assumes that the Nos assays used to calculate these figures are a true reflection of transformation. In all our work we have always found complete correlation between Nos assays, Southern blots and rooting bioassays. However, the test for nopaline is not without its faults and both false positives and false negatives may be obtained (Hepburn *et al.*, 1983). In addition, a positive response in culture indicating expression of the gene can be followed by a negative one when the test is carried out on plants transplanted to soil, indicating a termination of expression under normal growth conditions (Horsch *et al.*, 1985). We found, however, that Nos activity was detectable in all the plant parts tested once the plants had been transferred to soil, i.e. stem, root, leaf, petiole, and runners. After flowering it was also detectable in sepals, petals, ovules, stigmas and receptable. After fruit set it was detectable in the receptacle (the 'berry') at all stages of ripening (James *et al.*, unpublished data). Vegetative (Figure 21.2) and floral characteristics were not apparently altered after growth under glasshouse conditions.

One of the original two confirmed transgenic strawberry plants was used in inheritance studies to examine segregation of the *nos* gene after selfing.

Inheritance studies   Seeds from selfed R_0 plants of one of the transgenic clones were germinated aseptically, micropropagated and then tested for Nos or used in a

Figure 21.2   pBIN6 transformed strawberry (cv. Rapella) plant grown for 6 months under greenhouse conditions (right) and micropropagated control (left). Plants were transplanted to soil at the same time and runnered simultaneously ($\times 1/3$)

callus bioassay to assess kanamycin resistance (Kan^R) or sensitivity (Kan^S). In this, leaf pieces from 30 of the surviving seedlings from the R_1 progeny were excised and grown on callus-inducing media in the presence or absence of kanamycin and growth measured. In all seedlings there was complete correlation between Nos assays and callus bioassays, two-thirds of the seedlings being Nos^+/Kan^R and the other third being Nos^-/Kan^S. Where a larger number of seedlings was used (80) the *nos* gene segregated 3:1 Nos^+/Nos^-, the expected Mendelian ratio for a single dominant gene.

Delayed selection
Delaying selection by 7 days gave no shoot regeneration from more than 100 leaf discs, but extending this to 10 days permitted 8% of the explants eventually to regenerate a total of 26 shoots. However, these remained 'bleached' after transfer to the light on a micropropagation medium, indicating they were all 'escapes' and that the extra 3 days of non-selection had permitted the initiation of untransformed shoot primordia.

Discussion

The availability of good micropropagation and regeneration systems for any crop species makes possible the production of transgenic plants as our work with apple and strawberry has shown. These successes need to be qualified however and it will be apparent that the efficiency of transformation, particularly in apple, as expressed on a per explant basis was low even for cultivars that are very responsive to regeneration under non-selecting conditions. It should be emphasized however that our calculations of efficiency are based on averages from many different experiments where we were not able to define optimal conditions. As the effects of more variables become defined it is more than likely that the efficiency will increase.

When using micropropagation systems to provide explant material for regeneration and transformation experiments leaf material is often used (Horsch *et al.*, 1985). In the present work with apple and strawberry (and presumably with other species too) the investigator is faced with the choice of using leaves from shoot cultures undergoing axillary shoot multiplication or from shoots rooted *in vitro*. These two types of leaf are distinct morphologically, physiologically and biochemically. Even within the category of leaves excised from rooted plantlets there is still the decision to be made of how old the leaves should be before excision. Immediately after rooting *in vitro*, apple leaves begin to expand and increase their surface area several-fold over a period of 4–6 weeks. These changes in cell division and cell expansion after rooting have not been studied in apple but, clearly, important physiological and biochemical changes are occurring that can have important effects on the potential for adventitious shoot regeneration from excised discs.

The effects of explant source on transformation efficiency are rarely mentioned in the literature, although attention has been drawn to the need to use host tissues that are dividing and therefore undergoing DNA synthesis (Chyi and Phillips, 1987). If leaf tissue is being used where cell division has ceased the vast majority of cells will be at G_0 and presumably unlikely to be transformed. Future work should pay more attention to this and efforts should be made to standardize explant source.

The virulence of the bacterial strain used and the type of vector it contains are important factors in transformation. We have used only binary vectors in our work and only one strain of *Agrobacterium*, LBA4404. Fillati *et al.* (1987) have found that binary vectors were ineffective for transforming poplar tissue but that co-integrate vectors could be successfully used. The chromosomal background as well as the virulence functions on the individual plasmid can both contribute towards virulence but only by screening plasmid and bacterium against host can the best combination be found.

It is clear that prolonged selection on kanamycin is not necessary for the regeneration of transgenic apples or strawberries and that limited exposure times can also be used. It seems that selection at kanamycin 50 µg/ml or more for at least a week for apple is necessary (James *et al.*, 1989) and that 2–3 weeks for strawberry is optimal (James *et al.*, unpublished observations). Unfortunately limited selection only marginally reduces the time taken for regeneration and usually many weeks pass before shoot organogenesis occurs and this invariably takes place from callus. The possibility of somaclonal variation (Larkin and Scowcroft, 1981) in plants regenerated in this way must therefore always be considered. First reports from recent work at East Malling suggest that, although phenotypic variation can occur, it may not be a significant problem in strawberry (Beech, 1987) or apple (James, 1989).

We have now dismissed the use of delayed selection procedures for apple and strawberry on the grounds that too many escapes are regenerated presumably because shoot primordia are laid down very quickly (4–5 days) on regeneration media in the absence of the selection agent.

We were guided to our choice of using complex explants rather than protoplasts for transformation because of the availability, simplicity and apparent effectiveness of the leaf disc transformation procedure (Horsch *et al.*, 1985). In large part these assumptions have proved to be correct for apple and strawberry, but it is apparent that the technique, particularly for apple, still leaves a lot to be desired in terms of efficiency. Consequently other methods of transformation need to be considered.

Now that several fruit trees (Ochatt, Cocking and Power, 1987; Patat-Ochatt, Ochatt and Power, 1988) and strawberries (Nyman and Wallin, 1989) can be regenerated from protoplasts, this route, combined with electroporation (Ochatt *et al.*, 1988), could be used for producing transgenic plants. However, regeneration from protoplast culture is still a time-consuming and expensive strategy to employ and it is debatable whether the process would eventually be more efficient in human or biological terms, particularly since the likelihood of somaclonal variation is not lessened by this technology (Larkin and Scowcroft, 1981).

Two other methods are also available; 'the particle gun' used to fire genes into cells at high velocity using microprojectiles (Klein *et al.*, 1987), and shoot tip transformation (Ulian *et al.*, 1988). The former may have application to fruit crops where micropropagation procedures are not available, e.g. olive, but where they are, the latter is likely to be more applicable since its successful implementation would have far reaching consequences for species that cannot be regenerated from cells or complex explants *in vitro*. Any cultivar that can be micropropagated (which includes almost all commercially useful temperate top and soft fruit species) should be amenable to the technique which involves infecting shoot apices directly with disarmed *Agrobacteria* and selecting for transformants on antibiotic-containing media in micropropagating cultures. Since no callus phase is involved somaclonal variation should be avoided and clonal fidelity retained. The obvious potential drawback of chimaera formation was not observed in the first documented report of this technique in *Petunia* where marker genes were shown to be transmitted to F_1 progeny in a Mendelian fashion (Ulian *et al.*, 1988).

In conclusion, explant transformations may be of use in inserting foreign genes into apple and strawberry if the technique becomes more reproducible, more efficient, more widely applicable to other cultivars and if somaclonal variation is not a sizeable problem. Whether or not they are, it is almost certain that other methods of transformation such as electroporation, microprojectiles and shoot apices will eventually be used for the genetic improvement of fruit cultivars by non-conventional means.

References

Beech, M. G. (1987) *Strawberry Research Report 1987, Somacloning*, Institute of Horticultural Research, East Malling, Maidstone, Kent, p. 25

Bevan, M. W. (1984) Binary *Agrobacterium* vectors for plant transformation. *Nucleic Acids Research*, **12**, 8711–8721

Bevan, M. W., Flavell, R. B. and Chilton, M. D. (1983) A chimaeric antibiotic resistance gene as a selectable marker for plant cell transformation. *Nature*, **303**, 184–187

Boxus, P. (1974) The production of strawberry plants by *in vitro* micropropagation. *Journal of Horticultural Science*, **49**, 209–210

Chyi, Y.-S. and Phillips, G. C. (1987) High efficiency *Agrobactrium*-mediated transformation of *Lycopersicon* based on conditions favourable for regeneration. *Plant Cell Reports*, **6**, 105–108

Fasolo, F., Zimmerman, R. H. and Fordham, I. (1989) Adventitious shoot formation on excised leaves of *in vitro* grown shoots of apple cultivars. *Plant Cell Tissue Organ Culture*, **16**, 75–87

Fillatti, J. J., Sellmer, J., McCown, B., Haissig, B. and Comai, L. (1987) *Agrobacterium*-mediated transformation and regeneration of *Populus*. *Molecular and General Genetics*, **206**, 192–199

Fraley, R., Rogers, S. and Horsch, R. (1986) Genetic transformation in higher plants. In *Critical Reviews in Plant Sciences*, **4**, 1–46

Fraley, R., Rogers, S., Horsch, R., Sanders, P., Flick, J., Adams, S. *et al.* (1983) Expression of bacterial genes in plant cells. *Proceedings of the National Academy of Sciences, USA*, **80**, 4803–4807

Hepburn, A. G., Clarke, L. E., Pearson, L. and White, J. (1983) The role of cytosine methylation in the control of nopaline synthase gene expression in a plant tumour. *Journal of Molecular and Applied Genetics*, **2**, 315–329

Horsch, R. B., Fry, J. E., Hoffman, N. L., Eichholtz, D., Rogers, S. G. and Fraley, R. T. (1985) A simple and general method for transferring genes into plants. *Science*, **227**, 1229–1231

James, D. J. (1987) Cell and tissue culture technology for the genetic manipulation of temperate fruit trees. In *Biotechnology and Genetic Engineering Reviews*, vol. 5, (ed. G. E. Russell). Intercept, Newcastle-upon-Tyne, pp. 33–79

James, D. J. (1989) Manipulation of propagation potential in temperature fruit plants using tissue culture technology. *International Congress on Genetic Manipulation in Plant Breeding – Biotechnology for the Breeder*. EUCARPIA, Helsingor, Denmark, September 11–16, (in press)

James, D. J., Passey, A. J. and Rugini, E. (1988) Factors affecting high frequency plant regeneration from apple leaf tissues cultured *in vitro. Journal of Plant Physiology,* **132**, 148–154

James, D. J., Passey, A. J., Barbara, D. J. and Bevan, M. W. (1989) Genetic transformation of apple (*Malus pumilla* Mill.) using a disarmed Ti-binary vector. *Plant Cell Reports,* **7**, 658–661

Jones, O. P. (1976) Effect of phloridzin and phloroglucinol on apple shoots. *Nature,* **262**, 392–393

Klee, H., Horsch, R. and Rogers, S. (1987) *Agrobacterium*-mediated plant transformation and its further applications to plant biology. *Annual Review of Plant Physiology,* **38**, 467–486

Klein, T. M., Wolf, E. D., Wu, R. and Sanford, J. C. (1987) High velocity microprojectiles for delivering nucleic acids into living cells. *Nature,* **327**, 70–73

Larkin, P. J. and Scowcroft, W. R. (1981) Somaclonal variation: a novel source of variability from cell cultures for plant improvement. *Theoretical and Applied Genetics,* **60**, 197–214

Maniatis, T., Fritsch, E. F. and Sambrook, J. (1982) *Molecular Cloning: a laboratory manual.* Cold Spring Harbour, New York

Mathias, R. J. and Boyd, L. A. (1986) Cefotaxime stimulates callus growth, embryogenesis and regeneration in hexaploid bread wheat (*Triticum aestivum* L.EM.Thell). *Plant Science,* **46**, 217–223

Nyman, M. and Wallin, A. (1989) Plant regeneration from strawberry (*Fragaria* × *ananassa*) mesophyll protoplasts. *International Congress on Genetic Manipulation in Plant Breeding – Biotechnology for the Breeder,* EUCARPIA, Helsingor, Denmark, September 11–16. Programme and Abstracts p. 97, (in press)

Ochatt, S. J., Cocking, E. C. and Power, J. B. (1987) Isolation, culture and plant regeneration of Colt cherry (*Prunus avium* × *pseudocerasus*) protoplasts. *Plant Science,* **50**, 139–143

Ochatt, S. J., Shand, P. K., Rech, E. L. and Power, J. B. (1988) Electroporation-mediated improvement of plant regeneration from Colt cherry (*Prunus avium* × *pseudocerasus*) protoplasts. *Plant Science,* **54**, 165–169

Otten, L. A. B. M. and Schilperoort, R. A. (1978) A rapid microscale method for the detection of lysopine and nopaline dehydrogenase activities. *Biochemica et Biophysica Acta,* **527**, 494–500

Patat-Ochatt, E. M., Ochatt, S. J. and Power, J. B. (1988) Plant regeneration from protoplasts of apple rootstocks and scion varieties (*Malus* × *domestica* Borkh.). *Journal of Plant Physiology,* **133**, 460–465

Predieri, S., Malavasi, F. F., Passey, A. J., Ridout, M. S. and James, D. J. (1989) Regeneration from *in vitro* leaves of 'Conference' and other pear cultivars (*Pyrus communis* L.). *Journal of Horticultural Science,* **64**, (in press)

Reed, K. C. and Mann, D. A. (1985) Rapid transfer of DNA from agarose gels to nylon membranes. *Nucleic Acids Research,* **13**, 7207–7221

Stachel, S. E., Messens, E., Van Montagu, M. and Zambryski, P. (1985) Identification of the signal molecules produced by wounded plant cells that activate T-DNA transfer in *Agrobacterium tumefaciens. Nature,* **318**, 624–629

Ulian, E. C., Smith, R. H., Gould, J. H. and McKnight, T. D. (1988) Transformation of plants via the shoot apex. *In Vitro Cellular and Developmental Biology,* **24**, 951–954

Welander, M. (1988) Plant regeneration from leaf and stem segments of shoots raised *in vitro* from mature apple trees. *Journal of Plant Physiology,* **132**, 738–744

22

THE COMPLEXITY OF THE REGULATION OF EXPRESSION OF NODULATION GENES OF RHIZOBIUM

A. W. B. JOHNSTON, E. O. DAVIS, A. ECONOMOU, J. E. BURN and
J. A. DOWNIE
AFRC Institute of Plant Science Research, John Innes Institute, Norwich, UK

Introduction

It makes good adaptive sense that a microorganism that interacts with a plant, whether for good or for ill (i.e. symbiotic or pathogenic), expresses the genes that are required for the interaction only when it is close to or in its host organism. In the field of plant–microbe interactions there are three documented examples in which such differential gene expression occurs when certain bacteria are exposed to their host plants.

Several promoters for genes in the bacterial phytopathogen *Xanthomonas campestris* appear to be activated when the bacteria are in their hosts but are not expressed by bacteria growing in the free-living state (Osbourn, Barber and Daniels, 1986). The environmental signal(s) that activate(s) these promoters has not been identified nor have the precise roles (if any) in pathogenesis of the genes that they activate (A. E. Osbourn, personal communication).

Another example comes from studies on another plant pathogen, *Agrobacterium tumefaciens*, the bacterium that is responsible for the induction of crown gall tumours. Several virulence (*vir*) genes, that are located on the Ti plasmid are required for the processing and transfer of the T-DNA to the host plants. Normally, most of these *vir* genes are not expressed. The infection sites for these bacteria are wounds in the plants and it is a nice adaptation that certain phenolics, such as acetosyringone, which are present at high levels at wound sites, can activate transcription of the *vir* genes. This activation is mediated by the regulatory genes *virG* and *virA*; the product of the former 'senses' the presence of the phenolic inducers in the medium and transduces the signal to the product of *virA* which, in turn activates transcription of the other *vir* genes (*see* Stachel and Zambryski, 1986 for review).

Perhaps the most detailed studies, though, concern the bacterial symbionts that comprise the 'rhizobia'. Thanks to studies in several laboratories, working with different strains, species or genera or rhizobia (the types of rhizobia that form nodules on legumes are, in taxonomic terms, a rather diverse group and already comprise four different genera; *see* Young and Johnston, 1989), we have a considerable amount of comparative data concerning the mechanisms whereby the bacterial nodulation (*nod*) genes are regulated. It is clear that, although there are similarities in the modes of *nod* gene regulation in different rhizobia, there are also some real differences and these can have significant biological consequences. The topic of *nod* gene regulation has been reviewed (Kondorosi and Kondorosi, 1986; Rossen, Davis and Johnston, 1987) so only certain elements of this regulatory system will be described here.

It was found some years ago that most of the genes that are required for the bacteria to nodulate are not transcribed in bacteria that are grown in normal media. However, when they are exposed to the root exudates of their hosts, these *nod* genes are transcribed at high levels (Innes *et al.*, 1985; Mulligan and Long, 1985; Spaink *et al.*, 1988; Kosslak *et al.*, 1987; Rossen *et al.*, 1985). In all cases that have been examined to date, this activation requires the regulatory gene *nodD* (Mulligan and Long, 1985; Rossen *et al.*, 1985; Spaink *et al.*, 1988). The deduced polypeptide products derived from the sequences of the rhizobial *nodD* genes are similar to those of several other prokaryotic regulatory, DNA-binding proteins, the conservation being most pronounced at the NH$_2$-terminal region of the proteins, which have been implicated in binding to regulatory DNA sequences (Shearman *et al.*, 1986; Henikoff *et al.*, 1988).

Preceding the *nod* operons that are activated by *nodD* is a conserved DNA sequence, known as the *nod*-box that is important in the *nodD*-mediated activation (Rostas *et al.*, 1986). It has in fact been shown that the products of *nodD* of *R. meliloti* (which nodulates alfalfa) and *R. leguminosarum* bv. *viciae* (whose hosts are peas, lentils, vetches and field beans) bind to these regulatory DNA sequences (Hong, Burn and Johnston, 1987; Fisher *et al.*, 1988). The inducing molecules in root exudates which are responsible for the activation of *nod* genes have been identified for several rhizobia; they are flavonoid or flavonoid-like compounds and can be extremely potent (Firmin *et al.*, 1986; Peters, Frost and Long, 1986; Redmond *et al.*, 1986; Spaink *et al.*, 1988; Kosslak *et al.*, 1987). In *R. leguminosarum* bv. *vidiae*, the flavanone hesperitin is active at concentrations as low as 50 nM. It has not been shown directly (to our knowledge) that flavonoids bind directly to the product of *nodD*; this is a possibility, but it has not been ruled out that there is another step whereby the presence of the inducer is sensed by another system and the signal is transduced to the NodD polypeptide. All that has been stated above concerning the activation of *nod* genes has, to date, been true for all the rhizobia that have been examined, stressing that there are real similarities in the regulatory systems needed for *nod* gene transcription. In what follows, though, we will describe the differences in the organization and behaviour of the *nodD*-mediated activation of *nod* genes in different rhizobial strains.

First, different rhizobia do not have the same number of *nodD* genes; in *R. leguminosarum* bv. *viciae* and *R. leguminosarum* bv. *trifolii* (which forms nodules on clovers) there is a single copy of *nodD* (Figure 22.1), but in *R. meliloti* and *R. leguminosarum* bv. *phaseoli* (which nodulates *Phaseolus* beans) there are at least three *nodD*s (Honma and Ausubel, 1987; E. O. Davis, unpublished observations and *see below*). This presumably explains why mutations in the single *nodD* of *R. leguminosarum* bv. *viciae* and *R. leguminosarum* bv. *trifolii* completely abolish nodulation (Schofield *et al.*, 1983; Downie *et al.*, 1985), whereas mutations in individual *nodD*s of *R. meliloti* have only a slight effect on nodulation. Only when all three *nodD*s are inactivated by mutation do the bacteria lose their ability to nodulate alfalfa and other hosts of this species (Honma and Ausubel, 1987). In *Sinorhizobium fredii*, there are two adjacent copies of *nodD*. Mutations in one of these genes abolish nodulation of soyabeans whereas mutations in the other do not reduce nodulation but do block symbiotic nitrogen fixation and, for reasons that

R. leguminosarum nodulation genes

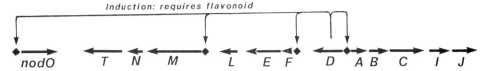

Figure 22.1 Representation of the location of the *nod* genes on the *R.leguminosarum* bv. *viciae* Sym plasmid pRL1JI. The locations and direction of transcription of 13 *nod* genes are shown. The *nod* box sequences preceding the *nodABCIJ*, *nodFEL*, *nodMNT* and *nodO* genes are indicated as diamonds. *nodD* activates these operons in the presence of flavonoid inducers

are not clear, have an effect on the production of the acidic exopolysaccharide by these bacteria (Applebaum *et al.*, 1988).

A second difference in the *nodD*s in different rhizobia is that in *R. leguminosarum* bvs. *viciae* and *trifolii*, *nodD* is autoregulatory; both in the presence and the absence of flavonoid inducers, it represses its own transcription (Rossen *et al.*, 1985; Spaink *et al.*, 1988). In contrast, in *R. meliloti*, *nodD* is not autoregulatory (Mulligan and Long, 1985). Also, in certain strains of *R. meliloti*, it has been shown that the levels of *nodD*-mediated activation of *nod* gene transcription are reduced by the presence of a repressor gene whose product binds to a region close to the *nod* box (Kondorosi *et al.*, 1988).

Finally, it has been shown that, although there are similarities in the molecules that activate *nod* gene transcription, there are differences in the NodD proteins in different rhizobial strains in their ability to recognize specific flavonoid molecules. In initial studies, it appeared that *nodD* genes of different rhizobial strains were equivalent; transfer of cloned *nodD* of *R. meliloti* to *nodD* mutants of *R. leguminosarum* bv. *trifolii* restored their ability to nodulate clovers and transfer of *nodD* of *R. leguminosarum* bv. *phaseoli* to *R. leguminosarum* bv. *viciae nodD* mutants allowed the transconjugants to nodulate peas (Fisher, Tu and Long, 1985; Lamb, Downie and Johnston, 1985). Moreover, some of the flavonoid molecules were shown to be inducers for different species of rhizobia; for example the flavone luteolin, the most potent inducer for *R. meliloti* (Peters, Frost and Long, 1986), is also a potent *nod* gene activator for *R. leguminosarum* bv. *viciae* (Firmin *et al.*, 1986). In contrast, the flavanone hesperitin, which is a very potent inducer for *R. leguminosarum* bv. *viciae* is relatively poor with *R. meliloti* (Firmin *et al.*, 1986; Peters, Frost and Long, 1986). More striking is the fact that certain iso-flavones, such as genistein and daidzein act as inducers for *S. fredii* (Kosslak *et al.*, 1987) but for *R. leguminosarum* bv. *trifolii* and *R. leguminosarum* bv. *viciae*, such molecules, far from acting as inducers, antagonize *nod* gene induction in cells that are grown in the presence of the *bona fide* inducer molecules (Firmin *et al.*, 1986; Djordjevic *et al.*, 1987). Such differences in the responses to individual flavonoids in different rhizobia can have important effects on the determination of host-range specificity. For example, there are certain rhizobial strains that have very wide host-ranges (that can even extend to the non-legume *Parasponium*); the transfer of the *nodD* of such a strain to *R. meliloti* conferred on the recipient the ability to nodulate siratro, a host of the wide host-range strain, indicating that these legumes exude molecules that are sensed by the *nodD* product of the wide host-range strain but not by those of the *nodD*s of *R. meliloti* (Horvath *et al.*, 1987). It has, in fact been found that wide host-range strains respond to a greater variety of plant-specified molecules (such as coumarins and simpler phenolics) than do the rhizobia with more restricted host-range specificity (Lestrange *et al.*, 1988). Similarly Spaink *et al.* (1988) found that the introduction of the *nodD1* gene of *R. meliloti* into a *nodD* mutant of *R. leguminosarum* bv. *trifolii* restored the ability to nodulate white clover but not red clover; presumably, the former host, but not the latter, exudes molecules that activate the *nodD1* regulatory system of *R. meliloti* and Surin and Downie (1988) showed that the transfer of *nodD* of *R. leguminosarum* bv. *trifolii* to *R. leguminosarum* bv. *viciae* was sufficient to confer on the recipient the ability to nodulate red clover.

Here, we present further observations on *nod* gene regulation in *R. leguminosarum* bv. *viciae* and *R. leguminosarum* bv. *phaseoli*. First, we describe evidence that the carboxy-terminal region of NodD interacts with flavonoid inducers and that this regimen may interact with the amino-terminal region of the protein which is likely to interact with DNA. Second, we describe the presence of three copies of *nodD* in *R. leguminosarum* bv. *phaseoli*; these have novel regulatory properties and are linked to other, newly defined *nod* genes. Third, we present evidence that in *R. leguminosarum* bv. *viciae*, the presence of inducer flavonoids that activate transcription of *nod* genes causes the reduction of transcription of other genes that are associated with nodulation ability in *R. leguminosarum* bv. *viciae*.

Isolation and characterization of doubly mutant forms of *nodD* in *R. leguminosarum* bv. *viciae*

Burn, Rossen and Johnston (1987) mutagenized *nodD* of *R. leguminosarum* bv. *viciae* and isolated several derivatives of the gene that were altered in their regulatory properties. Those of class I were defective in both *nodD* autoregulation and flavonoid-dependent activation of the other *nod* genes; class II mutants were specifically deficient in autoregulation and class III mutants conferred the 'reciprocal' phenotype, i.e. such strains were unable to activate *nod* genes, but the autoregulation of *nodD* was not affected. Finally, one mutation (of class IV) was isolated which allowed *nodD* to activate other *nod* genes even in the absence of inducers and to 'hyper-respond' when inducer flavonoids were present. Strikingly, with this class IV mutant strain, molecules such as the iso-flavones genistein and daidzein (which with wild-type *R. leguminosarum* bv. *viciae* are anti-inducers), actually enhance *nod* transcription in this mutant (Burn, Rossen and Johnston, 1987).

The class IV mutation is caused by an Asn–Asp substitution in the carboxy-terminal region of NodD (Burn, Rossen and Johnston, 1987). We have now sequenced the other mutant forms of *nodD* and found that mutations of classes I and II are at the 5′ end of the gene, close to a region that encodes the proposed DNA-binding domain of the protein and that two class III mutations map towards the 3′ end of the gene, upstream of the class IV mutation. These results suggest, but do not prove, that the carboxy-terminal region of the *nodD* product is involved in sensing the presence of inducer flavonoids and that the amino-terminal region interacts with the *nod* box sequences. A similar conclusion was drawn by Horvath *et al.* (1987) who made hybrid *nodD* genes comprising parts of *nodD*s from strains of rhizobia whose *nodD*s responded to different flavonoid inducers; the type of flavonoid that was recognized by the hybrid *nodD*s depended on the source of the 3' end of the hybrid *nodD*.

Taken together, these observations indicate that the carboxy-terminal region of NodD interacts, directly or indirectly with flavonoid inducers and 'anti-inducers' (or derivatives of them) and that this may cause an allosteric shift at its DNA-binding region such that the protein opens up the *nod* promoters and makes them available for transcription. We have obtained circumstantial evidence that there is in fact an interaction between the amino- and carboxy-terminal regions of NodD of *R. leguminosarum* bv. *viciae*.

In the *nodD* gene of this biovar, there is a *Sac*I site between the location of the class IV mutation (*see above*) and those of the class I, II and III mutations. It was possible therefore to construct double mutant forms of *nodD* which had a class IV mutation plus a mutation of one of the other types. Evidence for an interaction between the two terminal domains of the NodD polypeptide came from the observation that two double mutant forms of *nodD* containing the class IV plus one of two different class I mutations, conferred an essentially wild-type phenotype; i.e. activation of the other *nod* genes required the presence of inducer flavonoids and autoregulation was to the same level as with wild-type *nodD*. Thus, in these cases, the 'null' phenotype of these two class I mutation can be corrected by the presence of the class IV mutation and the 'constitutive' phenotype due to the class IV mutation is abolished by the coupling to these class I mutations. This suggests that there is a direct interaction between the two corresponding regions of the *nodD* gene product.

Identification of three *nodD* genes in *R. leguminosarum* bv. *phaseoli*

Lamb, Downie and Johnston (1985) identified two recombinant plasmids containing overlapping cloned DNA from *R. leguminosarum* bv. *phaseoli* which corrected the Nod⁻ defect of a *nodD* mutant of *R. leguminosarum* bv. *viciae*. This *nodD* gene (termed *nodD1*), in the cloned region of *R. leguminosarum* bv. *phaseoli*, was sequenced and was shown to have some novel features. First, *nodD1* activated transcription of *nodC–lacZ* fusions of both *R. leguminosarum* bv. *phaseoli* and *R. leguminosarum* bv. *viciae* in the absence of any inducer molecules, but there was an enhancement of activation when *Phaseolus* exudate was added. This was true with such fusions containing the *nodC*s of either *R. leguminosarum* bv. *phaseoli* or *R. leguminosarum* bv. *viciae*.

Also, in contrast to *nodD* of *R. leguminosarum* bv. *viciae*, which represses its own transcription, *nodD1* of *R. leguminosarum* bv. *phaseoli* actually activated its own expression in the presence or absence of root exudates of *Phaseolus* beans, although induction was more pronounced when inducer was added to the cells. In this connection, it was noted that upstream of *nodD1* was a sequence that was similar to a *nod* box. Further, between this *nod* box and the start of *nodD1* was a short ORF, corresponding to a gene (*nolE*) that, from codon usage analysis, was predicted to encode a polypeptide. By constructing *nolE–lacZ* fusions it was shown that transcription of *nolE* was activated by *nodD1* and that its expression was enhanced by exposure to *Phaseolus* exudate. The deduced amino-terminal region of the polypeptide product of *nolE* resembled a transit leader sequence suggesting that it might be a periplasmic protein. Evidence for this was obtained by using the transposon Tn5*phoA* (Manoil and Beckwith, 1985). This derivative of Tn5 contains, close to the end of the transposon, a truncated form of *Escherichia coli phoA* (a gene which specifies alkaline phosphatase); the deleted form of *phoA* in this construct lacks the 5′ terminus of the gene which encodes the signal peptide which confers the ability of the enzyme to be processed and hence exported into the periplasm. Thus alkaline phosphatase activity is only observed if the transposon inserts (in frame) into an actively transcribed gene whose product is transported into the periplasm or one of the bacterial membranes such that the fusion protein is 'presented' to the alkaline periplasmic space. Following mutagenesis of a recombinant plasmid containing the cloned *nolE* gene with Tn5*phoA*, it was found that two *nolE*::Tn5*phoA* insertions expressed high levels of alkaline phosphatase showing that *nolE* specified a protein that was transported into the periplasm. Thus, it is apparent that *nodD1* of *R. leguminosarum* bv. *phaseoli* is preceded by a gene whose product is exported from the bacterial cytoplasm and whose synthesis is dependent on the activation of the *nolE*/*nodD1* operon. In other rhizobia whose *nodD* genes have been analysed, *nodD* transcriptional units have comprised only the *nodD* gene itself. The finding, therefore, that *nodD1* of *R. leguminosarum* bv. *phaseoli* is preceded by a gene in the same operon is novel, and the fact that the upstream gene encodes a protein that is secreted from the cytoplasm raises some interesting questions concerning the role of *nolE*. The effects of mutations in *nolE* on symbiotic nitrogen fixation are currently under investigation.

In addition to *nodD1* of *R. leguminosarum* bv. *phaseoli*, a strain of this biovar contains at least two other copies of this regulatory *nod* gene. A region of DNA on the Sym plasmid, approximately 42 kb from *nodD1*, hybridized to *nodD1* DNA. This hybridizing region was sequenced and was shown to contain two additional copies of *nodD*, termed *nodD2* and *nodD3*. An unusual feature of *nodD3* was that, like *nodD1* of *R. leguminosarum* bv. *phaseoli*, it activated *nodC–lacZ* of *R. leguminosarum* bv. *viciae* and *R. leguminosarum* bv. *phaseoli* in the absence of flavonoid inducers and when the bacteria were exposed to *Phaseolus* exudate, there was further enhancement of the *nodC–lacZ* fusions.

The flavonoid-independent activation of *nod* genes by the *nodD1* and *nodD3* genes of *R. leguminosarum* bv. *phaseoli* and the hyperactivation that is observed when the bacteria are exposed to root exudates of the host legume, or to defined inducer flavonoids, is reminiscent of the behaviour of the constitutive class IV mutant derivatives of the *nodD* gene of *R. leguminosarum* bv. *viciae*. However, when recombinant plasmids containing the class IV *nodD* mutations of *R. leguminosarum* bv. *viciae* were transferred to wild-type strains of *R. leguminosarum*, the transconjugants induced only a few nodules and those that were formed failed to fix nitrogen (Burn, Rossen and Johnston, 1987). In contrast, introduction of the cloned *nodD1* or *nodD3* genes of *R. leguminosarum* bv. *phaseoli* into wild-type strains of *R. leguminosarum* bv. *viciae* or *R. leguminosarum* bv. *phaseoli* had no apparent effect on nodulation or nitrogen fixation when these derivatives were used to inoculate peas and *Phaseolus* beans respectively. The reason for this difference in the effects of the two types of 'constitutive' *nodD*s is not known.

Reduction of transcription of genes on the Sym plasmid of *R. leguminosarum* bv. *viciae* following exposure to *nod* gene inducers

It was found that strains of *R. leguminosarum* bv. *viciae*, but not those of the closely related *R. leguminosarum* bv. *trifolii* or *R. leguminosarum* bv. *phaseoli*, make an abundant protein of M_r 24 kDa (Dibb, Downie and Brewin, 1984). In several strains of this biovar, it was shown that production of this protein was specified by genes (*rhi*) on the Sym plasmid and in two strains it was demonstrated that these genes were closely linked to *nod* genes (Dibb, Downie and Brewin, 1984; Hombrecher *et al.*, 1984). Despite the tight genetic association between the ability of strains of *R. leguminosarum* bv. *viciae* to make the 24 kDa protein and the particular host range of these bacteria, mutations in the *rhi* gene had no observable effect on the nodulation of peas or vetches (Dibb, Downie and Brewin, 1984).

We have found a further relationship between *rhi* and *nod* genes, following mutagenesis of a recombinant plasmid, pIJ1089 (Downie *et al.*, 1983), which contains 30 kb of cloned DNA containing *rhi* and *nod* genes from a Sym plasmid of *R. leguminosarum* bv. *viciae*, with the *lacZ*-fusion transposon Tn3HoHo-1. Two insertions had reduced β-galactosidase activities in cells that were grown in the presence of the flavanone inducer hesperitin (Economou *et al.*, 1989). These insertions were in *rhiA*, the structural gene for the 24 kDa polypeptide, in the same orientation as *rhiA* (Economou *et al.*, 1989). For reasons that are not understood, the reduction in transcription of *rhiA* in cells exposed to *nod* gene inducers, as monitored by the expression of the *rhiA–lacZ* fusion, was not reflected in a reduction of the *rhiA* protein seen on protein gels. This reduction in transcription requires the presence of the regulatory gene *nodD*; it is not known if *nodD* acts directly on the promoter of the *rhi* genes or whether it has an indirect effect (e.g. by activating transcription of a gene that represses expression of the *rhi* genes). The finding that the expression of the *rhi* gene(s) is affected by the flavonoids that activate *nod* gene transcription, further implicates the *rhi* system as an important factor in the nodulation of peas and/or the other hosts of *R. leguminosarum* bv. *viciae*.

Further evidence that associates *rhiA* with nodulation comes from the following observation. In *nodO* mutant strains, the amount of 24 kDa protein seen on gels appears to be normal if the bacteria are grown in regular media; however, if the *nod* gene inducer hesperitin is added to the medium, the amount of 24 kDa protein is dramatically reduced. Although *nodO* is preceded by a *nod* box and its transcription is activated in cells that are exposed to *nod* gene inducers, mutations, in *nodO* have little or no effects on nodulation or nitrogen fixation (Economou *et al.*, 1989). Nevertheless, it is now clear that *nodO* has a major role to play in the nodulation process. It is apparent that strains of *R. leguminosarum* bv. *viciae* may have at least two mechanisms to make the necessary signals that provoke the host legume to manufacture root nodules. One of these pathways may be mediated by the *nodO* gene and the other by the products of the *nodFEL* operon. This is deduced by the fact that, whereas mutations in *nodO* have virtually no effect on nodulation and those in the *nodFEL* operon only delay and reduce nodulation (Downie *et al.*, 1985; Surin and Downie, 1988), strains with mutations in *nodO* and in the *nodFEL* operon fail to nodulate; i.e. if only one of these *nod* gene systems is eliminated by mutation, the bacteria can still form nodules by using the other 'back-up' system. The observations that there are intimate interactions between *rhi* genes and other genes that have a key involvement in the efficient induction of nodules, together with the finding that the *rhi* genes are repressed by exposure to flavonoid *nod* gene inducer molecules, lends further support to the suggestion that the *rhi* genes are important in the nodulation of the hosts of *R. leguminosarum* bv. *viciae* even though direct evidence for such a role is still lacking.

Conclusions

The comparative studies on the regulation of *nodD*-mediated transcription of *nod* genes in rhizobia are relatively new but, already, studies on this regulatory system have shown that it represents a complex example of gene regulation in prokaryotes. The new observations presented here lead to the following conclusions and questions.

The isolation of doubly mutant forms of *nodD* provides evidence that the carboxy-terminal region of the gene product interacts with flavonoid inducer molecules, consistent with other observations concerning the role of individual domains of the NodD protein in its ability to sense the presence of flavonoid inducers. Secondly, we have found that in *R. leguminosarum* bv. *phaseoli*, the regulation of *nod* gene expression is complicated by the fact that there are at least three *nodD* genes in this strain and that each of these genes has different regulatory properties.

Lastly, the finding that in *R. leguminosarum* bv. *viciae* there are genes whose transcription is reduced by exposure of cells of *R. leguminosarum* bv. *viciae* to flavonoids that activate transcription of *nod* genes adds another level of complexity to the *nodD*-mediated regulatory system in rhizobia. The various studies that have been conducted on this regulon have shown that it represents an inherently interesting example of gene control in prokaryotes and that much remains to be done to elucidate the precise mechanisms that underpin it.

Acknowledgements

This work was funded by the AFRC, the EEC and the Agricultural Genetics Company.

References

Applebaum, E. R., Thompson, D. V., Idler, K. and Chartrain, N. (1988) *Rhizobium japonicum* USDA 191 has two *nodD* genes that differ in primary structure and function. *Journal of Bacteriology*, **170**, 12–20

Burn, J. E., Rossen, L. and Johnston, A. W. B. (1987) Four classes of mutations in the *nodD* gene of *R. leguminosarum* bv. *viciae* that affect its ability to autoregulate and/or to activate other *nod* genes in the presence of flavonoid inducers. *Genes and Development*, **1**, 456–464

Dibb, N. J., Downie, J. A. and Brewin, N. J. (1984) Identification of a rhizosphere protein encoded by the symbiotic plasmid of *Rhizobium leguminosarum*. *Journal of Bacteriology*, **158**, 621–627

Djordjevic, M. A., Redmond, J. W., Batley, M. and Rolfe, B. G. (1987) Clovers secrete specific phenolic compounds which either stimulate or repress *nod* gene expression. *EMBO Journal*, **6**, 1173–1179

Downie, J. A., Knight, C. D., Johnston, A. W. B. and Rossen, L. (1985) Identification of genes and gene products involved in the nodulation of peas by *Rhizobium leguminosarum*. *Molecular and General Genetics*, **198**, 225–262

Downie, J. A., Ma, Q. S., Knight, C. D., Hombrecher, G. and Johnston, A. W. B. (1983) Cloning of the symbiotic region of *Rhizobium leguminosarum*: the nodulation genes are between the nitrogenase genes and a *nifA*-like gene. *EMBO Journal,* **2**, 947–952

Economou, A., Hawkins, F. K. L., Downie, J. A. and Johnston, A. W. B. (1989) Transcription of *rhiA*, a gene on a *Rhizobium leguminosarum* bv. *viciae* Sym plasmid, requires *rhiR* and is repressed by flavonoids that induce *nod* genes. *Molecular Microbiology,* **3**, 87–93

Firmin, J. L., Wilson, K. E., Rossen, L. and Johnston, A. W. B. (1986) Flavonoid activation of nodulation genes in *Rhizobium* reversed by other compounds present in plants. *Nature,* **324**, 90–92

Fisher, R. F., Egelhoff, T. T., Mulligan, J. T. and Long, S. R. (1988) Specific binding of proteins from *Rhizobium meliloti* cell-free extracts containing NodD to DNA sequences upstream of inducible nodulation genes. *Genes and Development,* **2**, 282–293

Fisher, R. F., Tu, J. K. and Long, S. R. (1985) Conserved nodulation genes in *Rhizobium meliloti* and *Rhizobium trifolii. Applied and Environmental Microbiology,* **49**, 1439–1435

Henikoff, S., Haughn, G. W., Calvo, J. M. and Wallace, J. C. (1988) A large family of bacterial activator proteins. *Proceedings of the National Academy of Sciences, USA,* **85**, 6602–6606

Hombrecher, G., Gotz, R., Dibb, N. J., Downie, J. A., Johnston, A. W. B. and Brewin, N. J. (1984) Cloning and mutagenesis of nodulation genes from *Rhizobium leguminosarum* TOM, a strain with extended host range. *Molecular and General Genetics,* **194**, 293–298

Hong, G. F., Burn, J. E. and Johnston, A. W. B. (1987) Evidence that DNA involved in the expression of nodulation (*nod*) genes in *Rhizobium* binds to the product of the regulatory gene *nodD. Nucleic Acids Research,* **15**, 9677–9691

Honma, M. A. and Ausubel, F. M. (1987) *Rhizobium meliloti* has three functional copies of the *nodD* symbiotic regulatory gene. *Proceedings of the National Academy of Sciences, USA,* **84**, 8558–8562

Horvath, B., Bachem, C. W. B., Schell, J. and Kondorosi, A. (1987) Host-specific regulation of nodulation genes in *Rhizobium* is mediated by a plant signal, interacting with the *nodD* gene product. *EMBO Journal,* **8**, 6841–6845

Innes, R. W., Kuempl, P., Plazinski, J., Canter-Cramers, H., Rolfe, B. G. and Djordjevic, M. A. (1985) Plant factors induce expression of nodulation and host range genes in *Rhizobium trifolii. Molecular and General Genetics,* **201**, 426–432

Kondorosi, E., Gyuris, J., Schmidt, J., John, M., Duda, E., Schell, J. *et al.* (1988) Positive and negative control of nodulation genes in *Rhizobium meliloti* strain 41. In *Molecular Genetics of Plant–Microbe Interactions 1988*, (eds R. Palacios and D. P. S. Verma), APS Press, St Paul, Minnesota, pp. 73–81

Kondorosi, E. and Kondorosi, A. (1986) Nodule induction on plant roots by *Rhizobium. Trends in Biochemical Science,* **11**, 296–299

Kosslak, R. M., Bookland, R., Barkei, J., Paaren, H. E. and Applebaum, E. R. (1987) Induction of *Bradyrhizobium japonicum* common *nod* genes by isoflavones isolated from *Glycine max. Proceedings of the National Academy of Sciences, USA,* **84**, 7428–7432

Lamb, J. W., Downie, J. A. and Johnston, A. W. B. (1985) Cloning of the nodulation (*nod*) genes of *Rhizobium phaseoli* and their homology to *R. leguminosarum nod* DNA. *Gene,* **34**, 367–370

Lestrange, K., Batley, M., Redmond, J., Bender, G., Lewis, W., Rolfe, B. *et al.* (1988) Plant signals for nodulation by *Rhizobium*: broad and narrow host-range strategies. In *Molecular Genetics of Plant–Microbe Interactions 1988*, (eds R. Palacios and D. P. S. Verma), APS press, St Paul, Minnesota, pp. 107–108

Manoil, C. and Beckwith, J. (1985) Tn*phoA*: a transposon probe for protein export signals. *Proceedings of the National Academy of Sciences, USA*, **82**, 8129–8133

Mulligan, J. T. and Long, S. R. (1985) Induction of *Rhizobium meliloti nodC* expression requires *nodD*. *Proceedings of the National Academy of Sciences*, **82**, 6609–6613

Osbourn, A. E., Barber, C. E. and Daniels, M. J. (1986) Identification of plant-induced genes of the phytopathogen *Xanthomonas campestris* pathovar *campestris* using a promoter-probe plasmid. *EMBO Journal*, **6**, 23–28

Peters, N. K., Frost, J. W. and Long, S. R. (1986) A plant flavone, luteolin, induces expression of *Rhizobium meliloti* nodulation genes. *Science*, **233**, 377–380

Redmond, J. W., Batley, M., Djordjevic, M. A., Innes, R. W., Kuempl, P. and Rolfe, B. G. (1986) Flavones induce expression of nodulation genes in *Rhizobium*. *Nature*, **323**, 632–634

Rossen, L., Davis, E. O. and Johnston, A. W. B. (1987) Plant-induced expression of *Rhizobium* genes involved in host specificity and early stages of nodulation. *Trends in Biochemical Science*, **12**, 430–433

Rossen, L., Shearman, C. A., Johnston, A. W. B. and Downie, J. A. (1985) The *nodD* gene of *Rhizobium leguminosarum* is autoregulatory and in the presence of plant exudate induces the *nodA,B,C* genes. *EMBO Journal*, **4**, 3369–3373

Rostas, K., Kondorosi, E., Horvath, B., Simoncsits, A. and Kondorosi, A. (1986) Conservation of extended promoter regions of nodulation genes in *Rhizobium*. *Proceedings of the National Academy of Sciences, USA*, **83**, 1747–1751

Schofield, P. R., Djordjevic, M., Rolfe, B. G., Shine, J. and Watson, J. M. (1983) A molecular linkage map of nitrogenase and nodulation genes in *Rhizobium trifolii*. *Molecular General Genetics*, **192**, 459–465

Shearman, C. A., Rossen, L., Johnston, A. W. B. and Downie, J. A. (1986) The *Rhizobium leguminosarum* nodulation gene *nodF* encodes a polypeptide similar to acyl-carrier protein and is regulated by *nodD* plus a factor in pea root exudate. *EMBO Journal*, **5**, 647–652

Spaink, H. P., Wijfellman, C. A., Pees, E., Okker, R. J. H. and Lugtenburg, B. J. J. (1988) *Rhizobium* nodulation gene *nodD* as a determinant of host specificity. *Nature*, **328**, 337–339

Stachel, S. E. and Zambryski, P. (1986) *Agrobacterium tumefaciens* and the susceptible plant cell: a novel adaptation of extracellular recognition and DNA conjugation. *Cell*, **47**, 155–157

Surin, B. P. and Downie, J. A. (1988) Characterization of the *Rhizobium leguminosarum* genes *nodLMN* involved in efficient host specific nodulation. *Molecular Microbiology*, **2**, 173–183

Young, J. P. W. and Johnston, A. W. B. (1989) *Trends in Evolution and Ecological Science*, (in press)

23

FUNCTION AND REGULATION OF THE EARLY NODULIN GENE ENOD2

FRANCINE GOVERS, HENK J. FRANSSEN, CORNÉ PIETERSE,
JEROEN WILMER and TON BISSELING
Department of Molecular Biology, Agricultural University, Wageningen, The Netherlands

Root nodule formation and nodulin gene expression

The symbiosis between (brady)rhizobia and legumes is known to proceed through a series of characteristic stages. The pre-infection stage includes recognition of two symbiotic partners, attachment of bacteria to the plant roots and deformation of root hairs. In the next stage, deformed root hairs are invaded by bacteria which grow and penetrate within a unique plant-derived infection thread. At the same time cell division is initiated in root cortical cells and this elaborates the formation of meristems. Infection threads grow towards these meristems and a nodule structure is formed wherein differentiation of tissue types occurs. From the growing infection thread bacteria are released into some of the newly formed nodule cells. Finally, the cooperation of nodule cells and differentiated bacteroids leads to a nitrogen fixing plant organ (Vincent, 1980).

The establishment and functioning of such an organ requires the specific expression of a number of genes from both partners. The plant genes, called nodulin genes, are developmentally regulated. Some of these are expressed in stages preceding actual nitrogen fixation when nodule formation is in progress. ENOD2 falls into this category of early nodulin genes. Late nodulin genes, such as the leghaemoglobin genes, are first expressed when the nodule structure is complete and when nitrogen fixation begins.

ENOD2 is an early nodulin

The first ENOD2 cDNA clone to be isolated, pGmENOD2, was selected from a soyabean (*Glycine max*) nodule cDNA library (Franssen *et al.*, 1987). This library was differentially screened with probes made from 10-day-old nodule RNA and root RNA. Most of the clones that hybridized specifically to the nodule probe, appeared to represent the same gene which was named ENOD2 (Early NODulin 2). Northern blot analysis showed that ENOD2 mRNA is first detectable in soyabean roots of 8-day-old plants that were inoculated at day 0 with *Bradyrhizobium japonicum*. At day 10 small nodule bumps are visible on the roots and by then the accumulation of ENOD2 mRNA has reached its maximum. In comparison with the other known soyabean nodulin genes ENOD2 is the earliest activated nodulin gene. Expression starts at least 4 days prior to the expression of the late nodulin genes.

ENOD2 is extremely proline rich

DNA sequence analyses of the cDNA clone pGmENOD2 and genomic GmENOD2 clones revealed that the encoded protein has a very high proline content (Franssen *et al.*, 1987; 1989). The protein has an *N*-terminal part with the typical features of a signal peptide. Forty-four per cent of the rest of the protein consists of proline residues which are organized in two repeating pentapeptides (Table 23.1).

Table 23.1 COMPARISON OF CHARACTERISTIC AMINO ACID REPEAT UNITS[a] BETWEEN ENOD2 FROM SEVERAL LEGUMES AND TWO OTHER PROLINE-RICH PROTEINS

Soyabean ENOD2[b]	PRO-PRO-GLU-TYR-GLN	PRO-PRO-HIS-GLU-LYS
Pea ENOD2[c]	PRO-PRO-GLU-TYR-GLN	PRO-PRO-HIS-GLU-LYS
Alfalfa ENOD2[d]	PRO-PRO-GLU-TYR-GLN	PRO-PRO-HIS-GLU-LYS
Sesbania rostrata ENOD2[e]	PRO-PRO-TYR-GLU-LYS	PRO-PRO-HIS-GLU-LYS
Soyabean SbPRP1, 1A10[f]	PRO-PRO-VAL-TYR-LYS	PRO-PRO-ILE-TYR-LYS
Carrot p33-PRP[g]	PRO-PRO-VAL-TYR-THR	PRO-PRO-VAL-HIS-LYS

[a] Predicted from DNA sequences.
[b] Franssen *et al.*, 1987.
[c] Van de Wiel *et al.*, 1990.
[d] Dickstein *et al.*, 1988.
[e] Strittmatter *et al.*, 1989.
[f] Hong, Nagao and Key, 1987; Averyhart-Fuller *et al.*, 1988.
[g] Chen and Varner, 1985b.

This high proline content has a severe impact on the migration behaviour of ENOD2 on SDS-polyacrylamide gels. In hybrid released translation experiments (Franssen *et al.*, 1987), RNA selected by pGmENOD2 and translated in a cell free system with [^{35}S]-methionine as radioactive precursor gives rise to a protein with an apparent molecular weight of 75 000 and an isoeletric point of 6.5. With [^{3}H]-leucine in the translation mixture two proteins are visualized, one of which is identical to the ^{35}S-labelled protein whereas the other, more prominent, polypeptide is more basic but has the same molecular weight. The two ^{3}H-labelled proteins were previously, in independent experiments, identified as early nodulins on two-dimensional gels of *in vitro* translation products of soyabean nodule RNA. They were named Ngm-75 (Gloudemans *et al.*, 1987). Calculation of the molecular weight of ENOD2 from the deduced amino acid sequence results in a protein of 45 000 molecular weight and this is in agreement with the coding capacity of the mRNA which is 1200 nucleotides in length.

In all proline-rich plant proteins that have been analysed to date, one-half or more of the proline residues are hydroxylated and the overall protein is usually glycosylated. Moreover, all known hydroxyproline-rich glycoproteins (HRGPs) are cell wall components. As yet no conclusive evidence has been presented demonstrating that ENOD2 is also such a HRGP. However, there are at least three reasons to assume that ENOD2 is a cell wall protein which is hydroxylated and glycosylated *in vivo*.

First, the pentapeptide repeats in ENOD2 have a striking similarity with pentapeptide repeats found in a proline-rich cell wall protein designated 1A10 that was isolated from soyabean by Averyhart-Fuller, Datta and Marcus (1988) (*see* Table 23.1). The deduced amino acid sequence from a partial cDNA clone p1A10 was compared with the amino acid distribution of the isolated cell wall protein and it was shown that one-half of the prolines in 1A10 are hydroxylated and that 1A10 is a glycoprotein. A soyabean gene designated SbPRP1 and encoding a protein with pentapeptide repeats homologous to those found in 1A10 has been analysed by Hong, Nagao and Key (1987). As with ENOD2, SbPRP1 has an *N*-terminal hydrophobic region which is a putative signal peptide, it has no serines but a relatively high content of basic amino acids and the region with the repeats is very hydrophilic.

Second, the hydrophobic signal peptide and the hydrophilic regions with hydroxyprolines organized in pentapeptide repeats are also found in the cell wall proteins extensin and the carrot p33-PRP (Chen and Varner, 1985a,b). A signal peptide is necessary for transport of the newly formed proteins through the cell membrane to reach the cell wall. For extensin, which is the best studied cell wall protein, the post-translational processing pathways resulting in a secreted HRGP have been elucidated (reviewed by Cooper, 1988). Subsequently Varner and coworkers (Cooper, 1988) have shown that soluble extensins are slowly rendered insoluble following excretion into the cell wall. Intra- and intermolecular crosslinks are formed due to the presence of tyrosine derivatives and carbohydrate sidechains on the hydroxyproline residues. This results in a three-dimensional glycoprotein network wherein cell wall polysaccharides are entangled. ENOD2, SbPRP1 and carrot p33-PRP have the same basic components as extensin. Therefore these proteins may be post-translationally modified and translocated in the same way as extensin.

The third plea in support of a hydroxyproline-rich ENOD2 is the report of Cassab (1986) who showed that the cortex of soyabean root nodules has a very high hydroxyproline content which is mainly localized in the cell wall. Recently it was shown by hybridization *in situ* that the ENOD2 mRNA is specifically localized in the inner cortex of the root nodule (Van de Wiel *et al.*, 1990) and this supports the idea that at least a part of the cell wall hydroxyproline is derived from ENOD2.

ENOD2 is highly conserved in legumes

The soyabean cDNA clone pGmENOD2 has been used as a probe to screen nodule RNA from other legumes. All legumes studied to date express a nodulin gene that is highly homologous to the soyabean ENOD2 gene. In nodule RNA isolated from pea (*Pisum sativum*, Govers *et al.*, 1986), white clover (*Trifolium repens*), bird's foot trefoil (*Lotus corniculatus*), vetch (*Vicia sativa*, Moerman *et al.*, 1987), alfalfa (*Medicago sativa*, Dickstein *et al.*, 1988), *Medicago truncatula* (D. Barker, personal communication), common bean (*Phaseolus vulgaris*, Sanchez *et al.*, 1988) and *Sesbania rostrata* (Strittmatter *et al.*, 1989) ENOD2 mRNAs have been detected. The striking difference between ENOD2 mRNAs in different plant species is their length. The smallest ENOD2 mRNA, ± 1200 bases, is found in soyabean, whereas the largest ENOD2 mRNA, ± 2500 bases, is observed in white clover and bird's foot trefoil. If this increase in length is only due to an increase in the number of repeats or to larger non-coding regions then this will probably not affect the functioning of ENOD2.

ENOD2 cDNA clones have been isolated and sequenced from pea (Van de Wiel *et al.*, 1990), alfalfa (Dickstein *et al.*, 1988) and *Sesbania rostrata* (Strittmatter *et al.*, 1989). As shown in Table 23.1 their predicted amino acid sequences have characteristic pentapeptide repeats which always begin with two proline residues and strongly resemble the pentapeptide repeats from soyabean ENOD2. The strong conservation of ENOD2 genes in legumes is restricted to the region where the pentapeptide repeats are encoded. Sequence comparison has shown that there is no significant homology between the 3' non-coding region of soyabean and pea ENOD2 (Van de Wiel *et al.*, 1990), and soyabean and *Sesbania rostrata* ENOD2 (Strittmatter *et al.*, 1989). On the contrary, two soyabean ENOD2 genes are exactly identical in the 3' non-coding region (Franssen, 1989) and the degree of homology between two *Sesbania rostrata* ENOD2 cDNA clones is even higher in the 3' non-coding than in the coding region (Strittmatter *et al.*, 1989).

The function of ENOD2

To obtain further insight into the function of ENOD2, ENOD2 gene expression has been analysed in a variety of aberrant nodules or nodule-like structures. In general, rhizobia mutated in *nif* or *fix* genes induce the formation of nodules which are morphologically indistinguishable from wild-type nodules. *Rhizobium meliloti exo⁻* mutants and some *Agrobacterium* transconjugants carrying the *sym* plasmid or the *nod* region of the *sym* plasmid induce the formation of so-called empty nodules (Finan *et al.*, 1985; Hirsch *et al.*, 1985; Govers *et al.*, 1986). Empty nodules have most of the morphological characteristics of wild-type nodules, i.e. vascular bundles in the periphery of the nodule, cortical cell layers and a central region. However, in the central region the plant cells lack bacteria and mostly no infection threads are found. Nodule-like structures do not have infected cells nor peripherally located vascular bundles. These structures, which are induced on soyabean upon inoculation with *R. fredii* USDA257, arise from a combination of cell swellings and randomly orientated cortical cell divisions (Franssen *et al.*, 1987). In all *fix⁻* nodules, empty nodules and nodule-like structures induced on a variety of plants, ENOD2 genes are expressed (pea, Govers *et al.*, 1986, 1987; alfalfa, Dickstein *et al.*, 1988; *Vicia sativa*, Moerman *et al.*, 1987; soyabean, Franssen *et al.*, 1987 and common bean, Sanchez *et al.*, 1988). All these different nodule types have one factor in common, they all passed through one of the first developmental steps in root nodule organogenesis, that is cell division in root cortical cells and development in certain tissue types of the nodule. Because ENOD2 mRNA is specifically localized in the inner cortical cells, the function of ENOD2 might be related to the morphogenesis of the inner cortex. The observation that ENOD2 mRNA is first detectable in cells that differentiate from meristematic cells to inner cortex cells supports this suggestion (Van de Wiel *et al.*, 1990). Moreover, ENOD2 has the characteristics of a cell wall protein and it is likely that it represents a specific cell wall protein in the inner cortex.

What then is the function of ENOD2 in the inner cortex of root nodules? A root nodule is a plant organ wherein a strict regulation of the O_2 concentration is a prerequisite for proper functioning of the enzyme nitrogenase. The plant manages to regulate the O_2 concentration in nodules by raising a barrier against O_2 diffusion from the atmosphere to the bacteroids. Within this barrier the O_2 concentration is low and the oxygen binding protein leghaemoglobin is thought to facilitate the transport of O_2 at low O_2 concentration. Measurements with O_2-specific microelectrodes have shown that the inner cortex is the major site of resistance to gaseous diffusion (Witty *et al.*, 1986). Inner cortical cells are more densely packed in comparison with the loosely packed cells of the outer cortex. Dense cell packing in the inner cortex might require extra or other cell wall material of which, possibly, ENOD2 is a specific component. The function of ENOD2 is thus related to the function of the inner cortex.

Regulation of ENOD2 gene expression

A fascinating aspect of the *Rhizobium*-legume symbiosis is the coordination of gene regulation of two interacting symbionts. This is undoubtedly accompanied by an exchange of signals between the two organisms. Since nodule formation is a multistep process, it is conceivable that there is a continuous flow of different signals in successive stages of the symbiosis. In the elucidation of the regulation of nodulin gene expression by signals from *Rhizobium*, bacterial mutants are essential. In all empty nodules induced by *Rhizobium exo*⁻ mutants and *Agrobacterium* transconjugants ENOD2 genes are expressed (*see* previous section) and this led to the conclusion that genes located on the *nod* region of the *sym* plasmid are responsible for the induction of ENOD2 genes (Nap, Van Kammen and Bisseling, 1987). It is well established that the *nod* genes are also required for the earliest stages of nodule development, i.e. root hair curling and cortical cell divisions. It is, however, unknown how the *nod* gene products induce these processes. Although the functions of the *nod* gene products have yet to be determined it seems plausible that the *nod* genes either produce one or more signals or are involved in the production of signals that trigger nodule development in the plant. With regard to the induction and regulation of ENOD2 genes, the question arises whether substances produced by *Rhizobium* directly activate ENOD2 genes or whether ENOD2 gene expression is one of the targets of a signal transduction pathway that in the earliest stages of nodule development is initiated by *Rhizobium* signals. Another question is whether the same activation or the same initiation of a pathway can be achieved by certain compounds in the absence of rhizobia.

Recently Hirsch *et al.* (1989) presented evidence that the latter is indeed possible. They repeated experiments described four decades ago in which substituted benzoic acids were used to induce the formation of nodule-like structures on legume roots. *N*-1-naphthylphthalamic acid (NPA) and 2,3,5-tri-idobenzoic acid (TIBA) are compounds that presumably modify hormone levels in the plant by inhibiting the polar transport of auxins. Hirsch *et al.* (1989) found that nodules which are formed upon adding either TIBA or NPA to alfalfa roots are morphologically similar to empty nodules that are induced by *R. meliloti exo*⁻ mutants. Moreover, expression of ENOD2 genes is observed in the TIBA- or NPA-elicited nodules and the ENOD2 mRNA concentration is more or less the same as the concentration present in *exo*⁻ nodules. The expression of ENOD2 genes in the absence of *Rhizobium* suggests that ENOD2 genes are developmentally regulated and that *Rhizobium* is able to control the plant's developmental programme. The resemblance between *Rhizobium exo⁻* nodules and TIBA- or NPA-induced nodules leads to the speculation that *Rhizobium* initiates nodule development by changing the auxin/cytokinin ratio in root cortical cells which then causes cell division, followed by differentiation of cells and ENOD2 gene expression in inner cortical cells. Whether endogenous auxins and cytokinins are indeed present in a specific ratio in those cells where ENOD2 genes are activated, and whether this by itself is sufficient for the induction of ENOD2 gene expression are still open questions.

ENOD2 gene expression in transgenic plants

Another approach which has been successfully used in elucidating the regulation of various plant genes is making use of transgenic plants. For analysing ENOD2 gene expression we have transformed leguminous and non-leguminous plants with a soyabean ENOD2 gene. In our laboratory, two soyabean ENOD2 genes have been isolated from a genomic library and the genes, designated GmENOD2A and GmENOD2B, have been analysed (Franssen, *et al.*, 1989). The 5′ non-coding region up to 600 bp upstream of the start codon, the coding region and 400 bp 3′ non-coding region are exactly identical in both genes. Moreover, the nucleotide sequence of the coding region is 100% homologous to the cDNA clone pGmENOD2. On Southern blots of genomic soyabean DNA two fragments of the same length as those found in the two genomic clones hybridize to pGmENOD2. It was concluded that the soyabean genome contains two ENOD2 genes and that both genes have been cloned. A 4.5 kb soyabean genomic DNA fragment with the 1 kb coding region of GmENOD2A in the middle was introduced into a binary Ti-plasmid and conjugated to *Agrobacterium tumefaciens* LBA4404 and *Agrobacterium rhizogenes* LBA9402. The disarmed *A. tumefaciens* transconjugants were used to transform tomato, whereas the *A. rhizogenes* transconjugants, which have both a wild-type Ri-plasmid and the disarmed binary Ti-plasmid, were used to obtain hairy root cultures from white clover (*Trifolium repens*) and bird's foot trefoil (*Lotus corniculatus*).

Transgenic tomato plants, genotype MsK93 and MsK9, were regenerated from transformed leaf discs (Koornneef *et al.*, 1986). From five transgenic plants for which it was shown that one or more copies of the GmENOD2A gene are stably integrated in the genome, RNA was isolated from shoots, roots and callus derived from transgenic leaf or root segments. Northern blot analyses showed that the GmENOD2A gene is expressed in all tissues analysed (Figure 23.1). The length of the ENOD2 mRNA is the same in soyabean nodules and transgenic tomato plants, and primer extension analysis has shown that the startpoint of transcription is also identical. The amount of ENOD2 mRNA accumulating in transgenic roots, shoots and callus appears to be correlated to the number of integrated GmENOD2A genes, indicating that all introduced genes are constitutively expressed. Moreover, the overall ENOD2 mRNA concentration is higher in roots compared with shoots. If this indeed mirrors the transcription level of the GmENOD2A gene in the two different organs, then it is tempting to speculate that the expression is influenced by the endogenous auxin/cytokinin ratio which is different in roots and shoots.

Why the GmENOD2A gene is constitutively expressed in plants which do not interact with *Rhizobium* is not clear. One possible explanation is that in soyabean, expression of ENOD2 genes is normally repressed and that the repressor is

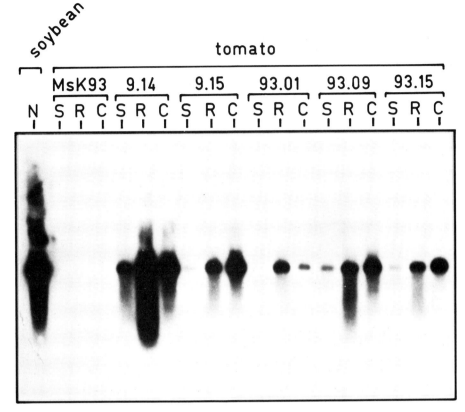

soybean

tomato

MsK93 9.14 9.15 93.01 93.09 93.15

N S R C S R C S R C S R C S R C S R C

Figure 23.1 Expression of the GmENOD2A gene in transgenic tomato plants. Autoradiograph of a Northern blot containing RNA from 14-day-old soyabean nodules (N) and from shoots (S), roots (R) and callus (C) from wild-type tomato (genotype MsK93) and transgenic tomato plants (9.14, 9.15, 93.01, 93.09, 93.15). The blot was hybridized with the cDNA clone pGmENOD2. The length of the ENOD2 mRNA is 1200 bases

inactivated during nodule formation. In non-leguminous plants such as tomato this repressor might be absent. A second explanation is that the GmENOD2A gene is not the ENOD2 gene which is expressed in soyabean nodules. In soyabean it might be a silent gene because we have never observed ENOD2 gene expression in shoots, roots or callus. Although both genes, GmENOD2A and GmENOD2B are identical over a large region extending beyond the coding region, we have no evidence that both genes are indeed activated in soyabean nodules. Regulatory sequences further upstream from the conserved region might be required for nodule specific expression. The only experimental tool to determine this is analysis of nodule RNA from transgenic leguminous plants which contain the GmENOD2 genes. Therefore we regenerated transgenic *Lotus corniculatus* plants from hairy root cultures that were obtained upon infection of seedlings with *A. rhizogenes* LBA9402. By co-transformation of a binary Ti-plasmid, the GmENOD2A gene and the β-glucuronidase (GUS) gene fused to the cauliflower mosaic virus (CaMV) 35S promoter, were introduced. From GUS positive hairy root cultures transgenic plants were regenerated according to Petit *et al.* (1987). Nodule, root and shoot RNA isolated from the transgenic plants was hybridized to a subclone of the 3' non-coding region of pGmENOD2. This clone specifically recognizes soyabean ENOD2 RNA and does not cross-hybridize with ENOD2 RNA of *L. corniculatus*. On Northern blots no expression of GmENOD2A can be detected (Figure 23.2).

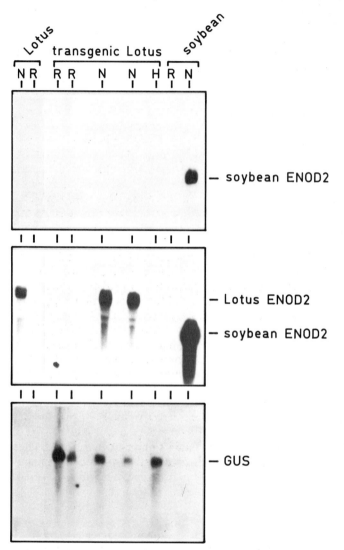

Figure 23.2 Expression analyses of the GmENOD2A gene in transgenic *Lotus corniculatus* plants. Autoradiographs of a Northern blot containing RNA from roots (R) and nodules (N) from wild-type *Lotus corniculatus* plants, transgenic *Lotus corniculatus* plants and wild-type soyabean plants. The lane marked with H contains RNA isolated from the hairy root culture from which the transgenic *Lotus* plants were regenerated. The same blot was hybridized subsequently with a subclone from the 3' non-coding region of the soyabean cDNA clone pGmENOD2 which specifically recognizes soyabean ENOD2 mRNA (upper panel), with pGmENOD2 which hybridizes with soyabean ENOD2 mRNA (1200 b) and cross-hybridizes with *Lotus* ENOD2 mRNA (2500 b) (middle panel) and with a GUS probe (lower panel)

When the same Northern blot is hybridized to the complete cDNA clone pGmENOD2, nodule specific expression of the endogenous *Lotus* ENOD2 gene(s) is observed and hybridization with a GUS probe shows that the chimaeric 35S CaMV–GUS gene, which is located on the same T-DNA next to GmENOD2A, is expressed in all tissues of the transgenic plants (Figure 23.2). Apparently the signal that induces expression of *L. corniculatus* ENOD2 gene(s) is not able to induce expression of the soyabean ENOD2A gene. This might be due to the fact that either GmENOD2A, located on a 4.5 kb genomic soyabean DNA fragment, does not contain all the required regulatory sequences for nodule specific expression, or GmENOD2A is also not activated in nodules in soyabean. To investigate these possibilities transgenic *Lotus* plants transformed with the GmENOD2B gene have to be analysed. These experiments are still in progress (in collaboration with K. Marcker and co-workers in Åarhus and F. de Bruyn and co-workers in Cologne).

Because we have used the *A. rhizogenes* co-transformation system to obtain transgenic leguminous plants there is, besides the T-DNA of the binary Ti-plasmid, at least one other T-DNA integrated in the plant genome. That is the T_L-DNA from the *A. rhizogenes* Ri-plasmid. The *rol A, B, C* and *D* loci, which are involved in hairy root formation, are located on the T_L-DNA. Due to the integration of these loci, hairy root cultures can be cultivated on hormone-free medium. The presence of a third T-DNA, the T_R-DNA of the Ri-plasmid, is not essential for the growth of hairy root cultures nor for regeneration. However, this T-DNA, which harbours among others two genes involved in auxin synthesis, may be transferred as well (Birot *et al.*, 1987). It has been shown by Shen *et al.* (1988) that hairy roots of *Lotus corniculatus* are 100–1000 times more sensitive to auxin than normal roots. The most plausible explanation for this is that hairy roots have either an increased number of auxin receptors or an increased efficiency of the transduction system for the auxin signal. For the increased auxin sensitivity only the T_L-DNA-borne genes are responsible and not the T_R-DNA-borne genes. The integration of T_R-DNA and/or T_L-DNA in transgenic plants appears to be a major drawback for studying the regulatory mechanism controlling early nodulin genes. We have found that in several hairy root cultures of white clover and *Lotus corniculatus* endogenous clover or *Lotus* ENOD2 genes are expressed. So far we have only observed this phenomenon in hairy root cultures transformed with wild-type *A. rhizogenes* and co-transformed with micro-Ti-vectors. It will be interesting to see if expression of a transformed ENOD2 gene is coupled to expression of endogenous ENOD2 genes. The observation that ENOD2 genes can be activated in hairy root cultures supports the hypothesis that expression of ENOD2 genes is regulated by phytohormones or changes in the phytohormone balance.

Concluding remarks

We have summarized the current knowledge on the early nodulin gene ENOD2 which is expressed during the formation of legume root nodules. ENOD2 genes are highly conserved in legumes and encode extremely proline rich proteins which most likely belong to the group of hydroxyproline-rich glycoproteins localized in the cell wall. Hybridization *in situ* in wild-type nodules and expression analyses in aberrant nodule types indicate that the function of ENOD2 is related to the function of the inner cortex. Expression of ENOD2 genes seems to be regulated by changes in the phytohormone balance.

Acknowledgements

We thank G. Heitkönig for typing the manuscript, J. Hille and R. Weide for advice and support with the transformation experiments, R. Jefferson for providing binary Ti-plasmids with the GUS gene and our colleagues for support and helpful discussions. This work was supported by the Netherlands Foundation for Biological Research (BION), with financial aid from the Netherlands Organization for Scientific Research (NWO).

References

Averyhart-Fullard, V., Datta, K. and Marcus, A. (1988) A hydroxyproline-rich protein in the soybean cell wall. *Proceedings of the National Academy of Sciences, USA,* **85**, 1083–1085

Birot, A.-M., Bouchez, D., Casse-Delbart, F., Durand-Tardif, M., Jouanin, L., Pautot, V. *et al.* (1987) Studies and uses of the Ri plasmids of *Agrobacterium rhizogenes. Plant Physiology and Biochemistry,* **25**, 323–335

Cassab, G. I. (1986) Arabinogalactan proteins during the development of soybean root nodules. *Planta,* **168**, 441–446

Chen, J. and Varner, J. E. (1985a) An extracellular matrix protein in plants: characterization of a genomic clone for carrot extensin. *EMBO Journal,* **4**, 2145–2151

Chen, J. and Varner, J. E. (1985b) Isolation and characterization of cDNA clones for carrot extensin and a proline-rich 33 kDa protein. *Proceedings of the National Academy of Sciences, USA,* **82**, 4399–4403

Cooper, J. B. (1988) Cell wall extension genes. In *Plant Gene Research, V,* (eds D. P. S. Verma and R. Goldberg). Springer-Verlag, New York, pp. 235–251

Dickstein, R., Bisseling, T., Reinhold, V. N. and Ausubel, F. M. (1988) Expression of nodule-specific genes in alfalfa root nodules blocked at an early stage of nodule development. *Genes and Development,* **2**, 677–687

Finan, T. M., Hirsch, A. M., Leigh, J. A., Johansen, E., Kuldau, G. A., Deegan, S. *et al.* (1985) Symbiotic mutants of *Rhizobium meliloti* that uncouple plant from bacterial differentiation. *Cell,* **40**, 869–877

Franssen, H. J., Nap, J. P., Gloudemans, T., Stiekema, W., Van Dam, H., Govers, F. *et al.* (1987) Characterization of cDNA for nodulin-75 of soybean: a gene product involved in early stages of root nodule development. *Proceedings of the National Academy of Sciences, USA,* **84**, 4495–4499

Franssen, H. J., Thompson, D. V., Idler, K., Kormelink, R., Van Kammen, A. and Bisseling, T. (1989) Nucleotide sequence of two soybean ENOD2 early nodulin genes encoding Ngm-75. *Plant Molecular Biology,* (in press)

Gloudemans, T., De Vries, S. C., Bussink, H.-J., Malik, N. S. A., Franssen, H. J., Louwerse, J. *et al.* (1987) Nodulin gene expression during soybean (*Glycine max*) nodule development. *Plant Molecular Biology,* **8**, 395–403

Govers, F., Moerman, M., Downie, J. A., Hooykaas, P., Franssen, H. J., Louwerse, J. *et al.* (1986) *Rhizobium nod* genes are involved in inducing an early nodulin gene. *Nature,* **323**, 564–566

Govers, F., Nap, J. P., Moerman, M., Franssen, H. J., Van Kammen, A. and Bisseling, T. (1987) cDNA cloning and developmental expression of pea nodulin genes. *Plant Molecular Biology,* **8**, 425–435

Hirsch, A. M., Drake, D., Jacobs, T. W. and Long, S. R. (1985) Nodules are induced on alfalfa roots by *Agrobacterium tumefaciens* and *Rhizobium trifolii* containing small segments of the *Rhizobium meliloti* nodulation region. *Journal of Bacteriology*, **161**, 223–230

Hirsch, A. M., Bhuvaneswari, T. V., Torrey, J. G. and Bisseling, T. (1989) Early nodulin genes are induced in alfalfa root outgrowths elicited by auxin transport inhibitors. *Proceedings of the National Academy of Sciences, USA*, **86**, 1244–1248

Hong, J. C., Nagao, R. T. and Key, J. L. (1987) Characterization and sequence analysis of a developmentally regulated putative cell wall protein gene isolated from soybean. *Journal of Biological Chemistry*, **262**, 8367–8376

Koornneef, M., Hanhart, C., Jongsma, M., Toma, I., Weide, R., Zabel, P. *et al.* (1986) Breeding of a tomato genotype readily accessible to genetic manipulation. *Plant Science*, **45**, 201–208

Moerman, M., Nap, J. P., Govers, F., Schilperoort, R., Van Kammen, A. and Bisseling, T. (1987) *Rhizobium nod* genes are involved in the induction of two early nodulin genes in *Vicia sativa* root nodules. *Plant Molecular Biology*, **9**, 171–179

Nap, J. P., Van Kammen, A. and Bisseling, T. (1987) Towards nodulin function and nodulin gene regulation. In *Plant Molecular Biology*, (eds D. Von Wettstein and N.-H. Chua). Plenum Press, New York

Petit, A., Stougaard, J., Kühle, A., Marcker, K. A. and Tempé, J. (1987) Transformation and regeneration of the legume *Lotus corniculatus*: a system for molecular studies of symbiotic nitrogen fixation. *Molecular and General Genetics*, **207**, 245–250

Sanchez, F., Quinto, C., Vázquez, H., Spaink, H., Wijffelman, C. A., Cevallos, M. A. *et al.* (1988) The symbiotic association of *Phaseolus vulgaris* and *Rhizobium Leguminosarum* bv. *phaseoli*. In *Molecular Genetics of Plant–Microbe Interactions*, (eds R. Palacios and D. P. S. Verma). APS Press, St Paul, Minnesota, pp. 370–375

Shen, W. H., Petit, A., Guern, J. and Tempé, J. (1988) Hairy roots are more sensitive to auxin than normal roots. *Proceedings of the National Academy of Sciences, USA*, **85**, 3417–3421

Strittmatter, G., Chia, T.-F., Trinh, T. H., Katagiri, F., Kuhlemeier, C. and Chua, N.-H. (1989) Characterization of nodule-specific cDNA clones from *Sesbania rostrata* and expression of the corresponding genes during the initial stages of stem nodules and root nodule formation. *Molecular Plant–Microbe Interactions*, **2**, 122–127

Van de Wiel, C., Scheres, B., Franssen, H., Van Lierop, M. T., Van Lammeren, A., Van Kammen, A. and Bisseling, T. (1990) The early nodulin ENOD2 transcript is located in the nodule-specific parenchyma (inner cortex) of pea and soybean root nodules. *The EMBO Journal*, **9**, (in press)

Vincent, J. M. (1980) Factors controlling the legume-*Rhizobium* symbiosis. In *Nitrogen Fixation, II*, (eds N. E. Newton and W. H. Orme-Johnson). University Park Press, Baltimore, pp. 103–129

Witty, J. F., Minchin, F. R., Skøt, L. and Sheehy, J. E. (1986) Nitrogen fixation and oxygen in legume root nodules. *Oxford Surveys of Plant Molecular and Cell Biology*, **3**, 275–315

LIST OF PARTICIPANTS

Al-Atabee, J.	University of Nottingham, Department of Botany, University Park, Nottingham, NG7 2RD, UK
Al-Mallah, M. K.	University of Nottingham, Department of Botany, University Park, Nottingham, NG7 2RD, UK
Albrechtsen, M.	Statens Plantevaerncentret, Lottenborgvej, 2, DK-2800 Lyngby, Denmark
Alderson, P. G.	University of Nottingham, Department of Agriculture and Horticulture, Faculty of Agricultural and Food Science, Sutton Bonington, Loughborough, LE12 5RD, UK
Asma, K.	University of Nottingham, Department of Botany, University Park, Nottingham, NGY 2RD, UK
Atherton, J.	University of Nottingham, Department of Agriculture and Horticulture, Faculty of Agricultural and Food Science, Sutton Bonington, Loughborough, LE12 5RD, UK
Austin, J. E.	Ministry of Agriculture, Fisheries and Food, Whitehouse Lane, Huntington Road, Cambridge, CB3 0LF, UK
Bailey, H. M.	University of Nottingham, Department of Botany, University Park, Nottingham, NG7 2RD, UK
Balcells, L.	AFRC Institute of Plant Science Research, Maris Lane, Trumpington, Cambridge, CB2 2JB, UK
Bang, J.	Chungnam National University, Korea; c/o Rothamsted Experimental Station, Harpenden, Hertfordshire, AL5 2JQ, UK
Baset, A.	University of Nottingham, Department of Botany, University Park, Nottingham, NG7 2RD, UK
Bass, P.	University of Nottingham, Department of Physiology and Environmental Science, Faculty of Agricultural and Food Science, Sutton Bonington, Loughborough, LE12 5RD, UK
Batty, N.	AFRC Institute of Plant Science Research, Maris Lane, Trumpington, Cambridge, CB2 2JB, UK
Baulcombe, D.	The Sainsbury Laboratory, John Innes Institute, Colney Lane, Norwich, NR4 7UH, UK
Bayliss, M. W.	ICI Seeds, Jealotts Hill Research Station, Bracknell, RG12 6EY, UK

Beachy, R. N.	Washington University, Department of Biology, St Louis, MO 63130, USA
Bell, P. J.	ICI Seeds, Jealott's Hill Research Station, Bracknell, RG12 6EY, UK
Bennett, A.	BP Nutrition Plant Science Research Laboratory, Unit B, Silwood Park, Buckhurst Road, Ascot, SL5 7TB, UK
Beri, R.	ICI Joint Laboratory, University of Leicester, University Road, Leicester, UK
Bevan, M.	AFRC Institute of Plant Science Research, Maris Lane, Trumpington, Cambridge, CB2 2LQ, UK
Blackhall, N. W.	University of Nottingham, Department of Botany, University Park, Nottingham, NG7 2RD, UK
Bond, S.	University of Nottingham, Department Agriculture and Horticulture, Faculty of Agricultural and Food Science, Sutton Bonington, Loughborough, LE12 5ED, UK
Boulter, D.	University of Durham, Department of Biological Sciences, South Road, Durham, DH1 3LE, UK
Boyd, P.	Strathclyde University, Department of Bioscience and Biotechnology, Todd Centre, 31 Taylor Street, Glasgow, G4 0NR, UK
Brace, J.	Manchester Polytechnic, Department of Biological Sciences, Chester Street, Manchester, M1 5GD, UK
Bradshaw, T.	Shell Research Ltd., Sittingbourne Research Centre, Sittingbourne, Kent, ME9 8AG, UK
Brears, T.	Plant Breeding International, Maris Lane, Trumpington, Cambridge, CB5 8SR, UK
Brooks, F. J.	AFRC Institute of Plant Science Research, Maris Lane, Trumpington, Cambridge, CB2 2JB, UK
Brown, C.	BP Nutrition Ltd., Unit B, Silwood Park, Buckhurst Road, Ascot, SL5 7TB, UK
Brown, S.	Leicester Polytechnic, Applied Biology and Biotechnology Department, Leicester, UK
Carter, D.	Advanced Technologies (Cambridge), 210 Cambridge Science Park, Milton Road, Cambridge, CB4 4WA, UK
Chakravarty, A.	University of Nottingham, Department of Botany, University Park, Nottingham, NG7 2RD, UK
Chartier-Hollis, J. M.	Derbyshire College of Higher Education, Kedlestone Road, Derby, DE3 1GB, UK
Chatha, M. R.	University of Nottingham, Department of Agriculture and Horticulture, School of Agriculture, Sutton Bonington, Loughborough, LE12 5RD, UK
Chen, D. F.	AFRC Institute of Plant Science Research, Maris Lane, Trumpington, Cambridge, CB2 2JB, UK
Cheng, L.	University of Nottingham, Department of Agriculture and Horticulture, School of Agriculture, Sutton Bonington, Loughborough, LE12 5RD, UK
Chisholm, D.	Floranova Ltd., Norwich Road, Foxley, Dereham, Norfolk, UK
Clark, J.	Wye College, University of London, Wye, Nr. Ashford, Kent, TN25 5AH, UK

Clerk, S. P. P.	Shell Research Limited, PBB/2 Sittingbourne Research Centre, Broad Oak Road, Sittingbourne, Kent ME9 8AG, UK
Cocking, E. C.	University of Nottingham, Department of Botany, University Park, Nottingham, NG7 2RB, UK
Cooper-Bland, S.	Scottish Crop Research Institute, Invergowrie, Dundee, DD2 5DA, UK
Cotton, C.	AFRC Institute of Plant Science Research, Maris Lane, Trumpington, Cambridge, CB2 2JB, UK
Covey, S.	AFRC Institute of Plant Science Research, Colney Lane, Norwich, NR4 7UH, UK
Cuong, V. V.	ISAB Laboratoire de Biologie Végétale, 19 Rue P. Waguet, 60026 Beauvais Cedex, France
D'Utra Vaz, F. B.	University of Nottingham, Department of Botany, University Park, Nottingham, NG7 2RD, UK
Darby, P.	IHR Department of Hop Research, Wye College, Wye, Ashford, Kent, TN25 5AH, UK
Daveran-Mingot, M.	CNRS-INRA, Laboratoire de Biologie Moléculaire, BP 27, 31326 Castenet-Tolosan Cedex, France
Davey, M. R.	University of Nottingham, Department of Botany, University Park, Nottingham, NG7 2RD, UK
Davies, J. W. W.	AFRC Institute of Plant Science Research, Colney Lane, Norwich, NR5 0NF, UK
Davis, A.	University of Nottingham, Department of Botany, University Park, Nottingham, NG7 2RD, UK
De Both	BIOSEM, Campus Universitaire des Cezeaux, 24, Avenue des Landais, 63170 Aubiere, France
De March, G.	Institut Supérieure Agricole, ISAB BP 313, 60026 Beauvais Cedex, France
De Miranda, J. R.	University of Liverpool, Department of Genetics, PO Box 147, Liverpool, L69 3BX, UK
Dick, E.	Department of Biology, Queen's University, Belfast, BT7 1NN, UK
Dowson Day, M. J.	IPSR Nitrogen Fixation Laboratory, University of Sussex, Brighton, E. Sussex, UK
Du, J.	AFRC Institute of Plant Science Research, Cambridge Laboratory, Maris Lane, Trumpington, Cambridge, CB2 2JB, UK
Dumar, D.	University of Nottingham, Department of Agriculture and Horticulture, School of Agriculture, Sutton Bonington, Loughborough, LE12 5RD, UK
Escandon, A.	Institute de Biologie Moléculaire des Plantes, 12, rue du General Zimmer, 6700 Strasbourg, France
Evans, I. J.	ICI Seeds, Plant Biotechnology, ICI Seeds, Jealott's Hill Research Station, Bracknell, RG12 6ET, UK
Eyles, P. S.	University of Nottingham, Department of Botany, University Park, Nottingham, NG7 2RD, UK
Falcetti, M.	Istituto Agrario Provinciale, Via E. Mach, 1,38010 San. Michele All'Adige, Trento, Italy

Faluyi, M. A.	Ondo State University, c/o Department of Agriculture, University College of Wales, Penglais, Aberystwyth, Dyfed, SY23 3DD, UK
Feix, G.	University of Freiburg, Institute for Biology III, D-78 Freiburg, W. Germany
Fenning, T.	University of Nottingham, Department of Botany, University Park, Nottingham, NG7 2RD, UK
Finch, R. P.	University of Nottingham, Department of Botany, University Park, Nottingham, NG7 2RD, UK
Forde, B. G.	Biochemistry Department, Rothamsted Experimental Station, Harpenden, AL5 2JQ, UK
Fowler, M. R.	Leicester Polytechnic, Applied Biology and Biotechnology Department, Leicester, UK
Fray, R. G.	University of Nottingham, Department of Plant Science, PES, Faculty of Agricultural and Food Science, Sutton Bonington, Loughborough, LE12 5RD
Freeman, J.	AFRC Institute of Plant Science Research, Maris Lane, Trumpington, Cambridge, CB2 2JB, UK
Gartland, J.	Leicester Polytechnic, Applied Biology and Biotechnology Department, Leicester, UK
Gartland, K.	Leicester Polytechnic, Applied Biology and Biotechnology Department, Leicester, UK
Gasser, G. S.	Monsanto Company, 700, Chesterfield Parkway, St Louis, MO 67017, USA
Gibbs, M.	University of Durham, Department of Biological Sciences, South Road, Durham, DH1 3LE, UK
Gielen, J.	Zaadunie, B.V., Plant Biotechnology Division, PO Box 26, 1600 AA Enkuizen, The Netherlands
Gleadle, A.	University of Nottingham, Department of Botany, University Park, Nottingham, NG7 2RD, UK
Godwin, I.	School of Biological Sciences, University of Birmingham, PO Box 363, Birmingham, B15 2TT, UK
Goebel, E.	Plant Genetic Systems NV, Plateaustraat, 22,B-9000 Gent, Belgium
Gonzalez, M. E.	University of Nottingham, Department of Agriculture and Horticulture, Faculty of Agricultural and Food Science, Sutton Bonington, Loughborough, LE12 5RD
Gough, M.	Shell Research Ltd., Sittingbourne Research Centre, Sittingbourne, Kent, ME9 8AG, UK
Govers, F.	Agricultural University, Department of Molecular Biology, De Dreijen 11, 6703 BC Wageningen, The Netherlands
Gray, J. C.	University of Cambridge, Botany School, Downing Street, Cambridge, CB2 3EA, UK
Greenland, A.	ICI Seeds, Jealott's Hill Research Station, Bracknell, RG12 6EY, UK
Grierson, D.	University of Nottingham, Department of Physiology and Environmental Science, School of Agriculture, Sutton Bonington, Loughborough, LE12 5RD, UK
Grieve, T. M.	Leicester Polytechnic, Applied Biology and Biotechnology Department, Leicester, UK

Grumbles, R. M.	Yale University, Molecular Biophysics/Biochemistry, POB 6666, 260 Whitney Avenue, New Haven, CT 06511, USA
Hamill, J.	Institute of Food Research, Colney Lane, Norwich, NR4 7UA, UK
Hamilton, W. D. O.	Agricultural Genetics Company Ltd., Department of Molecular Genetics, Cambridge Laboratory, Maris Lane, Cambridge, CB2 2JB, UK
Harding, K.	University of Nottingham, Department of Genetics, Queens Medical Centre, Nottingham, UK
Heaney, C.	Department of Biology, Queen's University, Belfast, BT7 1NN, UK
Hebblethwaite, P. D.	University of Nottingham, Department of Agriculture and Horticulture, Faculty of Agricultural and Food Science, Sutton Bonington, Loughborough, LE12 5RD, UK
Heide, M.	Statens Plantevaerncentret, Lottenborgvej, 2, DK-2800 Lyngby, Denmark
Hilder, V.	University of Durham, Department of Biological Sciences, South Road, Durham, DH1 3LE, UK
Hirst, B.	University of Nottingham, Department of Botany, University Park, Nottingham, NG7 2RD
Horwood, E.	University of Nottingham, Department of Plant Science, PES, Faculty of Agricultural and Food Science, Sutton Bonington, Loughborough, LE12 5RD, UK
Hosemans, D.	BIOSEM, Laboratoire de Biologie, Cellulaire et Moléculaire, Campus Universitaire des Cezeaux, 24, Ave des Landais, 63170 Aubiere, France
Howley, P.	University of Reading, Botany Department, Plant Science Laboratories, Whiteknights, PO Box 221, Reading, RG6 2AS, UK
Hunter, C.	Shell Research Ltd., PBB 2, Sittingbourne Research Centre, Sittingbourne, Kent, ME9 8AG, UK
Husnain, T.	University of Nottingham, Department of Botany, University Park, Nottingham, NG7 2RD, UK
Hutchings, C.	UMIST, PO Box 88, Sackville Street, Manchester, M60 1QD, UK
Huttly, A.	AFRC Institute of Plant Science Research, Maris Lane, Trumpington, Cambridge, CB2 2LQ, UK
Jaiswal, S. K.	University of Nottingham, Department of Botany, University Park, Nottingham, NG7 2RD, UK
James, D. C.	Department of Biochemistry, Wye College, University of London, Ashford, Kent, TN25 5AH, UK
James, D. J.	AFRC Institute of Horticultural Research, Breeding and Genetics Department, East Malling, Maidstone, Kent, ME19 6BJ, UK
Jepson, I.	ICI Seeds, Jealott's Hill Research Station, Bracknell, RG12 6EY, UK
Johnston, A.	AFRC Institute of Plant Science Research, Colney Lane, Norwich, NR4 7UH, UK
Jones, B.	University of Nottingham, Department of Botany, University Park, Nottingham, NG7 2RD, UK

Jones, M. G. K.	Institute of Arable Crops Research, Biochemistry Department, Rothamsted Experimental Station, Harpenden, AL5 2JQ, UK
Jotham, J.	University of Nottingham, Department of Botany, University Park, Nottingham, NG7 2RD, UK
Kajiwara, H.	National Institute of Agrobiological Resource, Kannondai, Tsukuba, Ibaraki, 305, Japan
Khan, M. F.	University of Nottingham, Department of Physiology and Environmental Science, School of Agriculture, Sutton Bonington, Loughborough, LE12 5RD, UK
Korlyuk, S.	University of Nottingham, Department of Botany, University Park, Nottingham, NG7 2RD, UK
Kumar, I. K.	Scottish Crop Research Institute, Department of Virology, Invergowrie, Dundee, DD2 5DA, UK
Layfield, P.	University of Sussex, Cinder Rough, Cornwells Bank, North Chailey, Lewes, BN8 4RH, UK
Lazzeri, P. A.	Max-Planck-Institut für Züchtungsforschung, Egelspfad, D-5000 Cologne 30, W. Germany
Leemans, J.	Plant Genetic Systems NV, Jozef Plateaustraat, 22,B-9000 Gent, Belgium
Legg, T.	Plant Cell and Molecular Sciences Group, King's College (London), Campden Hill, London W8 7AH, UK
Lin, Q.	AFRC Institute of Arable Crops Research, Department of Plant Pathology, Rothamsted Experimental Station, Harpenden, AL5 2JQ, UK
Linn, F.	Max-Planck-Institut für Züchtungsforschung, Egelspfad, D-5000 Cologne 30, W. Germany
Liu, X.	Institut für Genbiologische Forschung, Ihnestrasse 63, D-1000 Berlin 33, W. Germany
Llewellyn, D.	CSIRO Division of Plant Industry, GPO Box 2600, ACT 2601, Canberra, Australia
Lowe, J.	University of Nottingham, Department of Botany, University Park, Nottingham, NG7 2RD, UK
Lycett, G. W.	University of Nottingham, Department of Physiology and Environmental Science, School of Agriculture, Sutton Bonington, Loughborough, LE12 5RD, UK
Lynch, P. T.	University of Nottingham, Department of Botany, University Park, Nottingham, NG7 2RD, UK
McInnes, E.	University of Nottingham, Department of Botany, University Park, Nottingham, NG7 2RD, UK
McLauchlan, W. R.	AFRC Institute of Food Research, Department of Genetics and Microbiology, Colney Lane, Norwich, NR4 7UA, UK
McLellan, M.	University of Nottingham, Department of Botany, University Park, Nottingham, NG7 2RD, UK
McPartlan, H. C.	AFRC Institute of Plant Science Research, Department of Biotechnology and Physiology, Maris Lane, Trumpington, Cambridge, CB2 2JB, UK
Malaure, R. S.	University of Nottingham, Department of Botany, University Park, Nottingham, NG7 2RD, UK

Manders, G.	University of Nottingham, Department of Botany, University Park, Nottingham, NG7 2RD, UK
Marks, M.	AFRC Institute of Plant Science Research, Maris Lane, Trumpington, Cambridge, CB2 2JB, UK
Martinelli, L.	Istituto Agrario Provinciale, Via E. Mach, 1,38010 S. Michele All'Adige, Trento, Italy
Mathias, R.	AFRC Institute of Plant Science Research, Maris Lane, Trumpington, Cambridge, CB2 2JQ, UK
Meadows, J. W.	Plant Biology, University of Warwick, Coventry, UK
Meakin, P.	Biological Sciences Department, University of Durham, South Road, Durham, DL1 3LE, UK
Mendin, M. H.	University of Nottingham, Department of Botany, University Park, Nottingham, NG7 2RD, UK
Merodio, C.	Instituto del Frio, Ciudad Universitaria, 28040 Madrid, Spain
Millam, S.	Scottish Crop Research Institute, Tissue Culture and Cytology, SCRI, Invergowrie, Dundee, DD2 5DA, UK
Morris, E. J.	AECI Limited, Private Bag ×2, Modderfontein 1645, South Africa
Morris, P.	Welsh Plant Breeding Station, Institute for Grassland and Animal Production, Plas Gogerddan, Aberystwyth, SY23 3EB, UK
Mulligan, B.	University of Nottingham, Department of Botany, University Park, Nottingham, NG7 2RD, UK
Nahar, M. A.	University of Nottingham, Department of Botany, University Park, Nottingham, NG7 2RD, UK
Nall, C. S.	AFRC Institute of Plant Science Research, Cambridge Laboratory, Maris Lane, Trumpington, Cambridge, CB2 2LQ, UK
Nedkovska, M.	University of Nottingham, Department of Botany, University Park, Nottingham, NG7 2RD, UK
Newbury, H. J.	School of Biological Sciences, University of Birmingham, PO Box 363, Birmingham, B15 2TT, UK
Nichols, B.	British Sugar, British Sugar Research Laboratories, Colney Lane, Colney, Norwich, UK
Norton, G.	University of Nottingham, Department of Applied Biochemistry and Food Science, School of Agriculture, Sutton Bonington, Loughborough, LE12 5RD, UK
Ochatt, S. J.	University of Nottingham, Department of Botany, University Park, Nottingham, NG7 2RD, UK
Ockendon, D. J.	AFRC Institute of Horticultural Research, Wellesbourne, Warwick, CV35 9EF, UK
Okawara, R.	University of Nottingham, Department of Botany, University Park, Nottingham, NG7 2RD, UK
Patat-Ochatt, E. M.	University of Nottingham, Department of Botany, University Park, Nottingham, NG7 2RD, UK
Paul, W.	IPSR Nitrogen Fixation Laboratory, University of Sussex, Brighton, E. Sussex, BN1 9RH, UK
Pearson, D.	AFRC Institute of Plant Science Research, Maris Lane, Trumpington, Cambridge, CB2 2JB, UK

Pedersen, H. C. R.	Maribo Seed, Cell and Tissue Culture Laboratory, 14 Hojbygardvej, PO Box 29, DJ-4960 Holeby, Denmark
Pethe, V. V.	National Chemical Laboratory, Division of Biochemical Sciences, National Chemical Laboratory, Pune 411 008, India
Phelpstead, J.	University of Nottingham, Department of Botany, University Park, Nottingham, NG7 2RD, UK
Phillips, J.	Leicester Polytechnic, Applied Biology and Biotechnology Department, Leicester, UK
Pillai, V.	University of Nottingham, Department of Botany, University Park, Nottingham, NG7 2RD, UK
Power, J. B.	University of Nottingham, Department of Botany, University Park, Nottingham, NG7 2RD, UK
Puddephat, I.	University of Nottingham, Department of Agriculture and Horticulture, Faculty of Agricultural and Food Science, Sutton Bonington, Loughborough, LE12 5RD, UK
Ramsay, G.	Scottish Crop Research Institute, Invergowrie, Dundee, DD2 5DA, UK
Ramsay, M.	Scottish Crop Research Institute, Invergowrie, Dundee, DD2 5DA, UK
Rasheed, J. H.	University of Nottingham, Department of Botany, University Park, Nottingham, NG7 2RD, UK
Raynaerts, A.	Plant Genetic Systems NV, Plateaustraat, 22,B-9000 Gent, Belgium
Rech, E. L.	University of Nottingham, Department of Botany, University Park, Nottingham, NG7 2RD, UK
Riggs, T. J.	AFRC Institute of Horticultural Research, Wellesbourne, Warwick, CV35 9EF, UK
Robbins, M. P.	Welsh Plant Breeding Station, Cell Biology Department, Plas Gogerddan, Aberystwyth, Dyfed, SY23 13J, UK
Roberts, J. A. A.	University of Nottingham, Department of Physiology and Environmental Science, Faculty of Agricultural and Food Science, Sutton Bonington, Loughborough, LE12 5RD, UK
Rommens, C.	Free University, Department of Genetics, De Boelelaan 1087, 1081 HV Amsterdam, The Netherlands
Saalbach, G.	Zentralinstitut für Genetik und Kulturpflanzenforschung, Akademie der Wissenschaften der DDR, Correnstrasse 3, 4325 Gatersleben, E. Germany
Safford, R.	Unilever Research, (Bioscience Dept.), Colworth House, Sharnbrook, Bedford, MK44 1LQ, UK
Saleh, N. M.	University of Nottingham, Department of Botany, University Park, Nottingham, NG7 2RD, UK
Sallis, P. A.	University of Lancaster, Division of Biological Sciences, Bailrigg, Lancaster, LA1 4YQ, UK
Sanders, G. E.	University of Nottingham, Department of Environmental Science, Faculty of Agricultural and Food Science, Sutton Bonington, Loughborough, LE12 5RD, UK
Satyajit Kumar, B.	University of Nottingham, Department of Botany, University Park, Nottingham, NG7 2RD, UK

Schöffl. F.	University of Bielefeld, Biologie VI, Genetics, D-4800 Bielefeld 1, W. Germany
Schreier, P.	Bayer AG, Pflanzenschutzzentrum Monheim, PF-A/BF, D-5090 Leverkusen-Bayerwerk, W. Germany
Schuch, W.	ICI Seeds, Jealott's Hill Research Station, Jealott's Hill, Bracknell, RG12 6EY, UK
Severin, K.	University of Bielefeld, Lehrstuhl für Genetik, Universitätstrasse, D-4800 Bielefeld, W. Germany
Seymour, G.	University of Nottingham, Applied Biochemistry and Food Science, Faculty of Agricultural and Food Science, Sutton Bonington, Loughborough, LE12 5RD, UK
Shen, W.	AFRC Institute of Arable Crops Research, Biochemistry Department, Rothamsted Experimental Station, Harpenden, AL5 2JQ, UK
Shirsat, A. H.	University of Durham, Department of Biological Sciences, South Road, Durham, DH1 3LE, UK
Siggens, K. W.	Dalgety plc, Leicester Biocentre, University of Leicester, University Road, Leicester, LE1 7RH, UK
Slamet, I.	University of Nottingham, Department of Botany, University Park, Nottingham, NG7 2RD, UK
Slater, A.	Leicester Polytechnic, Applied Biology and Biotechnology Department, Leicester, UK
Smith, C. J.	University of Nottingham, Department of Physiology and Environmental Science, School of Agriculture, Sutton Bonington, Loughborough, LE12 5RD, UK
Spencer, A.	AFRC Institute of Food Research, Colney Lane, Norwich, NR4 7UA, UK
Stacey, J.	Nickerson International Seed Co. Ltd., Cambridge Science Park, Milton Road, Cambridge, CB4 4GZ, UK
Stanford, A.	AFRC Institute of Plant Science Research, Department of Molecular Genetics, Maris Lane, Trumpington, Cambridge, CB2 2JB, UK
Stevenson, L.	University of Nottingham, Department of Agriculture and Horticulture, Faculty of Agricultural and Food Science, Sutton Bonington, Loughborough, LE12 5RD, UK
Sy, M. H.	Instituto Agrario Provinciale, Via E. March, 01 38010 San Michele All Adige, Trento, Italia
Tassie, A.	AFRC Institute of Plant Science Research, Maris Lane, Trumpington, Cambridge, CB2 2JB, UK
Taylor, I. B.	University of Nottingham, Plant Science, PES, Faculty of Agricultural and Food Science, Sutton Bonington, Loughborough, LE12 5RD, UK
Taylor, J. E. E.	University of Nottingham, Plant Science, PES, Faculty of Agricultural and Food Science, Sutton Bonington, Loughborough, LE12 5RD, UK
Todd, G.	School of Biological Sciences, University of Birmingham, PO Box 363, Birmingham, B15 2TT, UK
Tomsett, A. B.	University of Liverpool, Department of Genetics and Microbiology, PO Box 147, Liverpool, L69 3BX, UK

Tor, M.	Wye College, University of London, Department of Horticulture, Wye, Ashford, Kent, TN25 5AH, UK
Tucker, G. A.	University of Nottingham, Applied Biochemistry and Food Science, Faculty of Agricultural and Food Science, Sutton Bonington, Loughborough, LE12 5RD, UK
Walker, A.	University of Nottingham, Department of Botany, University Park, Nottingham, NG7 2RD, UK
Walker, N.	Sanofi EIF Biorecherches, BP 137, 31328 Labege, France
Wang, G.	University of Nottingham, Department of Botany, University Park, Nottingham, NG7 2RD, UK
Ward, A.	University of Nottingham, Department of Botany, University Park, Nottingham, NG7 2RD, UK
Ward, K.	AFRC Institute of Food Research, Colney Lane, Norwich, NR4 7UA, UK
Watson, C. F.	University of Nottingham, Plant Science, PES, Faculty of Agricultural and Food Science, Sutton Bonington, Loughborough, LE12 5RD, UK
Webb, C. L.	University of Nottingham, Department of Botany, University Park, Nottingham, NG7 2RD, UK
Whittington, W. J.	University of Nottingham, Plant Science, PES, Faculty of Agricultural and Food Science, Sutton Bonington, Loughborough, LE12 5RD, UK
Wibberley, M.	University of Nottingham, Department of Botany, University Park, Nottingham, NG7 2RD, UK
Wienand, U.	Max-Planck-Institut für Züchtungsforschung, D-5000 Cologne 30, W. Germany
Wilkie, S.	Hatfield Polytechnic, Biology Department, C.P. Snow Building, College Lane, Hatfield, AL10 9AB, UK
Willmitzer, L.	Institut für Genbiologische Forschung GmbH, Ihnestrasse, 63, Berlin 33, W. Germany
Wilson, I.	University of Nottingham, Applied Biochemistry and Food Science, Faculty of Agricultural and Food Science, Sutton Bonington, Loughborough, LE12 5RD, UK
Wilson, Z.	University of Nottingham, Department of Botany, University Park, Nottingham, NG7 2RD, UK
Wiltshire, J. J.	University of Nottingham, Physiology and Environmental Science, Faculty of Agricultural and Food Science, Sutton Bonington, Loughborough, LE12 5RD, UK
Wise, R.	Max-Planck-Institut für Züchtungsforschung, Egelspfad, D-5000 Cologne 30, W. Germany
Wood, R.	University of Nottingham, Department of Botany, University Park, Nottingham, NG7 2RD, UK
Woodman, K. J.	University of Nottingham, Plant Science, PES, Faculty of Agricultural and Food Science, Sutton Bonington, Loughborough, LE12 5RD, UK
Woudt, B.	Zaadunie B.V., Plant Biotechnology Division, PO Box 26, 1600 AA Enkhuizen, The Netherlands
Wright, C. J.	University of Nottingham, Agriculture and Horticulture, Faculty of Agricultural and Food Science, Sutton Bonington, Loughborough, LE12 5RD, UK

Wright, S.	ICI Seeds, Jealott's Hill Research Station, Bracknell, RG12 6EY, UK
Xu, Y.	AFRC Institute of Arable Crops Research, Biochemistry Department, Rothamsted Experimental Station, Harpenden, AL5 2JQ, UK
Yot, P.	CNRS-INRA, Laboratoire de Biologie Moléculaire, Boite Postale 27, 31326 Castanet-Tolosan Cedex, France
Zalensky, A. O.	Institute of Agricultural Microbiology, Leningrad, USSR, c/o Department of Molecular Biology, Wageningen Agricultural University, Wageningen, The Netherlands
Zhang, H.	University of Nottingham, Department of Botany, University Park, Nottingham, NG7 2RD, UK
Zhang, J.	University of Nottingham, Department of Botany, University Park, Nottingham, NG7 2RD, UK
Zhang, S.	University of Nottingham, Department of Botany, University Park, Nottingham, NG7 2RD, UK
Zhang, X.	University of Durham, Department of Biological Science, South Road, Durham, DH1 3LE, UK

INDEX